ANALYTICAL APPLICATIONS
OF COMPLEX EQUILIBRIA

ELLIS HORWOOD SERIES IN ANALYTICAL CHEMISTRY

General Editor: Dr. R. A. Chalmers
University of Aberdeen

Founded as a library of fundamental books on important and growing subject areas in analytical chemistry, this series will serve chemists in industrial work and research, and in teaching or advanced study

PUBLISHED OR IN ACTIVE PREPARATION

Particle Size Analysis
Z. K. Jelínek, *Organic Research Institute, Pardubice*

Automatic Chemical Analysis
J. K. Foreman
P. B. Stockwell } *Laboratory of the Government Chemist, London*

Theoretical Foundations of Chemical Electroanalysis
Z. Galus, *Warsaw University*

Electroanalytical Chemistry
G. F. Reynolds, *University of Reading*

Analysis of Organic Solvents
V. Šedivec
J. Flek } *Institute of Hygiene and Epidemiology, Prague*

Handbook of Process Stream Analysis
K. J. Clevett, *Crest Engineering (U. K.) Inc.*

Methods of Catalytic Analysis
G. Svehla, *Queen's University of Belfast*
H. Thompson, *University of New York*

Organic Reagents in Inorganic Analysis
Z. Holzbecher et alii, *Institute of Chemical Technology, Prague*

Analysis of Synthetic Polymers
J. Urbanski et alii, *Warsaw Technical University*

Spectrophotometric Determination of Elements
Z. Marczenko, *Warsaw Technical University*

Gradient Liquid Chromatography
C. Liteanu
S. Gocan } *University of Cluj*

Operational Amplifiers in Chemical Instrumentation
Robert Kalvoda, *J. Heyrovský Institute of Polarography, Prague*

Stripping Voltammetric Analysis
F. Vydra et alii, *Charles University, Prague*

ANALYTICAL APPLICATIONS OF COMPLEX EQUILIBRIA

J. INCZÉDY, D. Sc.
Professor of Analytical Chemistry,
University of Chemical Engineering,
Veszprém, Hungary

Translation Editor:
JULIAN TYSON, Ph. D.
Department of Chemistry,
Imperial College of Science and Technology,
University of London

ELLIS HORWOOD LIMITED
Publisher • Chichester

Halsted Press: a division of
JOHN WILEY & SONS INC.
New York • London • Sydney • Toronto

The publisher's colophon is reproduced from James Gillison's drawing of ancient Market Cross, Chichester

First published in English in 1976 by
ELLIS HORWOOD LIMITED, PUBLISHER
Coll House, Westergate, Chichester, Sussex, England
and
AKADÉMIAI KIADÓ, Budapest

Distributors:

Australia, New Zealand, South-east Asia:
JOHN WILEY & SONS AUSTRALASIA PTY LIMITED
110 Alexander Street, Crow's Nest, N. S. W., Australia

Canada:
JOHN WILEY & SONS CANADA LIMITED
22 Worcester Road, Rexdale, Ontario, Canada

Europe, Africa:
JOHN WILEY & SONS LIMITED
Baffins Lane, Chichester, Sussex, England

East European Countries, China, People's Republic of Viet-Nam, Korean People's Republic, Cuba and Mongolia:
AKADÉMIAI KIADÓ, Budapest

U. S. A., South America and the rest of the World:
HALSTED PRESS a division of
JOHN WILEY & SONS INC.
605 Third Avenue, New York, N. Y. 10016, U. S. A.

© Akadémiai Kiadó, Budapest 1976

Library of Congress Cataloging in Publication Data

Inczédy János
Analytical applications of complex equilibria.
Translation of Komplex egyensúlyok analitikai alkalmazása
Bibliography: p.
Includes index.
1. Chemistry, Analytic. 2. Complex compounds. 3. Chemical equilibrium. I. Title.
OD77. L5213 1976 543 75-25687 ISBN 0 470-42713-2 (Halsted Press)
85312-019-6 (Ellis Horwood, Publisher)

This work was originally published as *Komplex Egyensúlyok Analitikai Alkalmazása* by Műszaki Könyvkiadó, Budapest
Translated by Dr Andrea Páll–Hrabéczy

All rights reserved. No part of this publication may be reproduced, stored in a retrieval system or transmitted, in any form or by any means, electronic, mechanical photocopying, recording or otherwise, without prior permission

Printed in Hungary

TO THE MEMORY OF MY PARENTS

CONTENTS

Preface   11

List of symbols   13

Chapter 1. Complexes and their Properties   17

  1.1. Elements of the chemistry of complexes   17
  1.2. Equilibria   22
    1.2.1. Complex-formation equilibria   22
      1.2.1.1. Equilibria of mononuclear complexes   25
      1.2.1.2. Equilibria of polynuclear complexes   31
    1.2.2. Equilibria of acid–base reactions: protonation of the ligand   34
    1.2.3. Equilibria of redox reactions   39
    1.2.4. Practical instructions for equilibrium calculations   43
    1.2.5. Worked examples   46
  1.3. Conditional equilibrium constants   51
    1.3.1. Role of the hydrogen ion concentration in complex formation   59
    1.3.2. Worked examples   63
  1.4. Factors affecting complex formation   67
    1.4.1. The nature of the central ion   69
    1.4.2. The properties of the ligand   70
  1.5. Kinetics of the reactions of complexes   74
  1.6. Energy changes involved in complex formation reactions   77
    1.6.1. Calculation of equilibrium constants from thermodynamic data   78
    1.6.2. Temperature dependence of the equilibrium constant   79
    1.6.3. Worked examples   81
  1.7. Dependence of equilibrium constants on the solvent   86

Chapter 2. Determination of Equilibrium Constants   89

  2.1. Determination of protonation constants   89
    2.1.1. Potentiometric methods   89
      2.1.1.1. Practical instructions   98
    2.1.2. Spectrophotometric methods   101
      2.1.2.1. Practical advice   104

Contents

- 2.1.3. Worked examples — 105
- 2.2. Determination of the stability constants of mononuclear complexes — 111
  - 2.2.1. Potentiometric methods — 112
    - 2.2.1.1. Direct methods — 112
    - 2.2.1.2. Methods based on measurement of pH — 116
    - 2.2.1.3. Worked examples — 127
  - 2.2.2. Spectrophotometric methods — 137
    - 2.2.2.1. Method of continuous variation — 137
    - 2.2.2.2. The mole-ratio method — 140
    - 2.2.2.3. Bjerrum's method — 141
    - 2.2.2.4. Other methods — 143
    - 2.2.2.5. Worked examples — 143
  - 2.2.3. Polarographic methods — 147
    - 2.2.3.1. Practical aspects — 151
    - 2.2.3.2. Worked examples — 153
  - 2.2.4. Extraction methods — 155
    - 2.2.4.1. Worked example — 164
  - 2.2.5. Ion-exchange methods — 165
    - 2.2.5.1. Worked example — 170
  - 2.2.6. Determination of the stability constant of mixed-ligand complexes — 174
    - 2.2.6.1. Worked example — 177
- 2.3. Determination of the stability constants of polynuclear complexes — 181

Chapter 3. Analytical Applications — 182

- 3.1. Gravimetric analysis — 182
  - 3.1.1. Worked examples — 187
- 3.2. Acid–base titrations — 194
  - 3.2.1. Calculation of the hydrogen ion concentration — 194
  - 3.2.2. Determination of acids and bases — 199
  - 3.2.3. Indirect determination of complex-forming ions by acid–base titration — 207
  - 3.2.4. Worked examples — 208
- 3.3. Precipitation titrations — 213
  - 3.3.1. Electrodes of the second kind and membrane electrodes — 216
  - 3.3.2. Worked examples — 218
- 3.4. Complexometric titrations — 219
  - 3.4.1. Theory of complexometric titrations using chelating agent as titrant — 220
  - 3.4.2. End-point detection — 224
  - 3.4.3. Analysis of mixtures of metals — 227

	3.4.4. Worked examples	229
3.5.	Redox titrations	232
	3.5.1. Worked examples	240
3.6.	Polarography	247
	3.6.1. Worked examples	249
3.7.	Spectrophotometry	252
	3.7.1. Worked examples	255
3.8.	Liquid–liquid extraction	258
	3.8.1. Extraction equilibrium for simple or ion-association complexes	259
	3.8.2. Extraction equilibrium for chelates	260
	3.8.3. Selectivity and degree of extraction	264
	3.8.4. Worked examples	266
3.9.	Ion-exchange separations	270
	3.9.1. Cation-exchange equilibria	270
	3.9.1.1. Chromatographic separation of metal ions on cation-exchange columns	273
	3.9.1.2. Chromatographic separation of organic bases on cation-exchange columns	278
	3.9.2. Anion-exchange equilibria	281
	3.9.2.1. Separation of anions and organic acids on anion-exchange resin columns	283
	3.9.2.2. Chromatographic separation of metal ions on anion-exchange resin columns	284
	3.9.3. Separation of more than two similarly charged ions	285
	3.9.3.1. Separation by a stepwise change in eluent	286
	3.9.3.2. Separation by gradient elution	286
	3.9.4. Extraction with liquid ion-exchangers	289
	3.9.4.1. Cation-exchange equilibria	290
	3.9.4.2. Anion-exchange equilibria	291
	3.9.5. Ion-exchange paper chromatography	296
	3.9.6. Worked examples	297
3.10.	Electrophoresis	313

Chapter 4. Tables of Equilibrium Constants   317

4.1.	Protonation and complex formation equilibrium constants	317
4.2.	Precipitate formation constants	369
4.3.	Redox equilibrium constants	375
4.4.	Extraction constants	379
4.5.	Ion-exchange equilibrium constants	384

References   387

Index   408

PREFACE

The development of the pure sciences during the past few years has produced a large number of new results from fundamental research which have altered the methods of the applied sciences, including the engineering sciences, in that empirical relationships are being replaced more and more by fundamental relationships.

In chemical analysis especially, the development of coordination chemistry has brought about considerable changes.

In 1963 Ringbom, in his excellent book entitled "Complexation in Analytical Chemistry", coined the term 'conditional constants' and showed by many examples how α-functions of side reactions have shown many advantages in calculations concerning complex formation equilibria. The solution of various analytical problems can be predicted, and the optimum conditions (solution composition, pH, etc.) can be simply calculated before the experimental work.

The appropriate choice of analytical method and the proper planning of the experimental conditions are of great importance, particularly in connection with the application of instrumental methods of analysis.

Lengthy and sometimes fruitless experimental work may be avoided, enabling both mental effort and instruments to be made more effective.

The reliability of the calculations depends on the reliability of the fundamental relationships, on the accuracy of the data used and also on the knowledge and critical sense of the person doing the calculations, who has to consider various factors according to their importance. Calculations done with appropriate circumspection, however, will yield results which can be verified experimentally.

Although a large number of fundamental relationships have been described mathematically in various scientific publications, they are not as widely used as would be desirable. This reluctance to do calculations can be ascribed partly to the fact that the necessary data and constants were not available earlier in sufficiently great number, and their reliability was not satisfactory. However, the number and reliability of the data increase daily, and the collection given at the end of this book is only a sample of the enormous quantity of data in the literature available to the analytical chemist.

In writing this book, my aim has been to guide and aid the analytical chemist working in industrial and research laboratories in solving everyday problems, and by working through typical problems to show how rapid exploratory calculations may be performed with a slide rule (or other simple means).

Acid–base, complexation, precipitation and redox reactions are all treated in similar terms. By this uniform approach, my aim has been to make the calculations simpler and easier to understand. With only a little effort, experience in doing calculations using protonation, precipitate-formation and redox equilibrium constants may be gained in a short time by those who have not done them before.

Very often, complex stability and protonation constants have to be determined by analytical chemists. This is why the basic methods for their determination and some practical instructions have also been included in this book.

I have made some alterations to the earlier, Hungarian, edition for the purposes of the English edition. In this book intended for English-speaking readers I have cited monographs and textbooks which are widely known. Section 3.10, where electromigration methods are dealt with, is a new section in this edition. Finally, I have updated the collection of equilibrium constants at the end of the book.

I remember with gratitude the late Professor László Erdey, D. Sc. (Budapest, Hungary), in whose Institute, together with my co-workers, I have done the greater part of the experimental work which has served as the basis of the book.

I pay respect and express my gratitude to the late Professor A. Ringbom (Åbo, Finland) for permission to use methods of calculation developed by himself as well as the term 'conditional constant', which I find very appropriate. I greatly acknowledge his advice and remarks in connection with the Hungarian edition.

I wish to thank A. Páll-Hrabéczy, Ph. D. for doing the translation, and R. A. Chalmers, D. Sc., Editor of the Ellis Horwood Series in Analytical Chemistry, for helping in the preparation of the manuscript by his valuable advice.

Finally, I wish to express my thanks to all my co-workers and students for their part in the experimental work.

János Inczédy, D. Sc.

LIST OF SYMBOLS

a	degree of titration [(Eq. (2.7)]; void fraction of the chromatographic column
a_M	activity of the species M
A^-	negatively charged base
$A_{M(L)}$	complex formation function of metal ion M with ligand L [Eq. (2.34)]
b	concentration gradient (in chromatography) (mole ml^{-2}) [Eq. (3.193)]
B	neutral base species
C_L	total (analytical) concentration of ligand L (mole litre^{-1}) [Eq. (1.16)]
C_L^0	initial total concentration of ligand L (mole litre^{-1})
C_M	total (analytical) concentration of metal ion M (mole litre^{-1}) [Eq. (1.14)]
d_A	distribution constant (partition coefficient) of the species A [Eq. (2.94)]
D_A	distribution coefficient of substance A [Eq. (2.95)]
e	electron
E	optical absorbance; potential (V)
E^0	standard electrode or redox potential (V)
f_A	activity coefficient of the species A
F	Faraday constant (23.06 kcal equiv^{-1} V^{-1})
ΔG^0	standard free enthalpy change
h	plate height of the chromatographic column (cm)
ΔH	enthalpy change
i	current intensity; species number
i_d	diffusion current in polarography
I	ionic strength
k	dissociation constant
K	stability constant
K'	conditional stability constant
K_d	separation factor

List of symbols

K_e	extraction constant
K_e'	conditional extraction constant
K_r	oxidation-reduction constant
K_r'	conditional oxidation-reduction constant
K_S	precipitation formation constant
K_S'	conditional precipitation formation constant
K^x	ion-exchange constant
$K^{x'}$	conditional ion-exchange constant
L	ligand
[HL]	concentration of species HL in aqueous solution (mole litre^{-1})
(HL)	concentration of species HL in organic or ion-exchange resin phase (mole litre^{-1})
m	proton number of an acid
m_A	mass of substance A
M	metal ion
n	ligand number
$\bar{n}$	average ligand number or protonation number [Eq. (1.17)]
N	maximum ligand number
N	theoretical plate number of a chromatographic column
p	constant parameter
P	normalized constant parameter
Q	capacity of ion-exchange resin (equiv litre^{-1})
R	gas constant ($1 \cdot 987 \times 10^{-3}$ kcal mole^{-1} deg^{-1})
R_f	retention factor (in paper chromatography)
s	ligand number of the neutral complex
S_0	solubility product
ΔS	entropy change
T	temperature (K)
v_{max}	retention volume (ml)
V	volume (ml)
V_0	void volume of a chromatographic column (ml)
x	independent variable; chromatographic column volume (ml)
X	normalized independent variable; total volume of the chromatographic column (ml)
y	variable
Y	chelate-forming ligand; normalized variable y
z	ionic charge number; variable
$\alpha_{L(H)}$	side-reaction function of species L in reactions with hydrogen ion
$\alpha_{M(L)}$	side-reaction function of species M in reactions with ligand L

List of symbols

β_n	overall complex formation constant of a complex of composition ML_n
β'_n	conditional overall complex formation constant of complex ML_n
$\beta_{m,n}$	overall complex formation constant of complex M_mL_n [Eq. (1.25)]
γ_n	nth overall complex formation constant
$\Delta\%$	percent error of the determination
θ	separation efficiency ratio [Eq. (3.165)]
σ	column density of the chromatographic column (g ml^{-1})
Φ_{ML}	mole fraction of species ML

Chapter 1

COMPLEXES AND THEIR PROPERTIES

1.1. Elements of the Chemistry of Complexes

The term 'complexes' is generally taken as meaning compounds of more or less involved composition, formed by the combination of simple ions or molecules which can also exist by themselves.

However, it should be noted that simple ions occur very rarely as such — and then only in the gaseous phase — since in solution they are invariably solvated. The molecules of the solvent are bound more or less firmly to the ions. The innermost molecules in the solvation sheath are arranged regularly. Thus ions in solution form solvocomplexes (in aqueous solution aquocomplexes) with solvent molecules. The number and orientation of the solvent molecules around an ion are determined by the volume and charge density of the ion, and the spatial requirement of the solvent molecules.

Hence, complexation reactions in solution can be handled as exchange of solvent molecules for ligands.

$$M(H_2O)_n + L \rightarrow M(H_2O)_{n-1}L + H_2O$$

where M is the central metal ion and L the ligand. In metal complexes the central ion is a metal ion or possibly a proton, the ligand an organic or inorganic anion or neutral molecule. In the course of a complex forming reaction the solvent molecules surrounding the central ion may be successively exchanged for other ligand ions or molecules, leading finally to the complex ML_n, where n indicates the number of ligands in the complex. This is equal to the coordination number if the ligand is bound to the central ion at one site (see later). The charge of the complex formed may differ from that of the original solvated ion. The sign and magnitude of the charge of the complex formed are given by the algebraic sum of the charges of the ions constituting the complex. For example

$[Cu(NH_3)_4]^{2+}$ $\qquad$ $[Fe(CN)_6]^{3-}$ $\qquad$ $[ZnCl_2(H_2O)_2]^0$

$+2 + 4 \times 0 = +2$ $\qquad$ $+3 + 6(-1) = -3$ $\qquad$ $+2 + 2(-1) + 2 \times 0 = 0.$

In the chemical formulae of complexes the numbers of metal ions and ligands composing the complex, and their charges, are expressed. In the case of some metal ions the coordination number in the complex formed is characteristic of the particular metal ion. Thus, the coordination number of the complexes of cobalt(III), platinum(IV) and chromium(III) is always six. The coordination number depends on the oxidation state of the metal ion, but very often also on the nature of the ligand. For example, palladium(IV) forms complexes of coordination number (C. N.) six, whereas palladium(II) generally has a C. N. of four. The C. N. of nickel(II) in its cyanide complex is four, whereas in its ammine complex it is six.

For simplicity, when electrolyte solutions are dealt with, solvated metal ions will be referred to as free metal ions, and denoted by M^{z+}. However, it should be noted that very few metal salts are present in aqueous solution completely in the form of hydrated ions. Most anions have complexing ability and form complexes with the majority of metal ions. The formation of complexes in dilute aqueous solutions of simple ions is indicated for example by the difference in the colours of dilute nickel(II) sulphate and nickel(II) perchlorate solutions. Of the cations the alkali metal ions, and of the anions the perchlorate ion, show the least tendency to form complexes. For this reason perchlorates are of great importance in the investigation of complexes. The properties of solutions which contain complexes are often compared with those of perchlorate solutions. However, it must be noted that perchlorate does form complexes with some metal ions [mercury(II) and iron(III) for example].

It is usual to treat complex formation reactions in terms of the *Lewis acid–base theory*. According to the theory, acids are ions or compounds which accept a pair of electrons, i.e. they are electron acceptors, and bases the ions or molecules which can donate a pair of electrons, i.e. are electron donors. Metal ions can be considered as polybasic acids with a deficiency of electron pairs corresponding to the maximum coordination number, which therefore can bind bases. The acid strength increases with the valency. The number of bases (ligands) to be bound depends on the coordination number and the space available. The electron donor bases are neutral molecules or ions containing an electronegative group (or groups).

Protons in aqueous solution are always bound to water molecule, i.e. are present as hydronium (H_3O^+) ions (a more correct, but inconvenient formulation is $H_9O_4^+$). They can be considered as the strongest acids in aqueous solution, according to the *Brønsted theory*. When the proton is bound to other molecules or ions, a weaker acid is formed. Similarly, if the behaviour of a metal ion is studied in aqueous solutions containing different ions, it can be stated that the hydrated metal ion is the strongest acid, and the acid strength gradually decreases as the hydrate-water molecules are exchanged for ions or molecules of bases stronger than water. Perchlorate is a very weak base, weaker than water. Consequently, perchlorate complexes are not usually formed in aqueous solution. For characterizing the Lewis-acidity of solutions, pM can equally well be used, as pH is for characterizing the Brønsted acidity.

Ligands which are bound to the central ion by one pair of electrons are called unidentate ligands, e.g. H_2O, NH_3, Cl^-, CN^-. Some molecules or ions contain more than one group which may displace hydrate-water molecules or other ligands. This type of ligand is called a multidentate ligand. In the complexes of multidentate ligands rings may be formed. This type of complex is called a chelate. Chelate-forming multidentate ligands may contain neutral and negatively charged groups within one molecule. The charge of the complex formed can be calculated again as the algebraic sum of those of the components.

The neutral, uncharged, so-called inner complexes are usually slightly soluble in water, e.g. the [bis(8-hydroxyquinolinato)copper(II)] complex.

Complexation with multidentate ligands may lead to complexes which also contain protons in addition to the metal ion and ligand. These are called protonated complexes. For example iron(III) forms not only the complex with composition FeL but also FeHL with ethylenediamine-tetra-acetate.

Complexes which contain more than one central ion are called polynuclear complexes. Complexes of this type are mainly formed when ligands are present which may serve as bridges, or if the number of lone-pair electrons on a multidentate ligand is greater than the coordination number of the metal ion. An example of the first case is [octoammine-μ-amido-μ-nitrodicobalt(III)]:

and of the second [dizinc triethylenetetraminehexa-acetate]. Polynuclear complexes are often formed by multivalent metal ions and hydroxide ions. For example:

$$\left[\begin{array}{c} \text{HO} \diagdown \quad \overset{\text{H}}{\underset{\text{O}}{\mid}} \quad \diagup \text{OH} \\ \quad \text{Zn} \quad \quad \text{Zn} \\ \text{H}_2\text{O} \diagup \quad \text{OH}_2 \quad \text{HO} \diagup \quad \diagdown \text{OH}_2 \end{array}\right]^0$$

By 'mixed-ligand' complexes is meant complexes which contain different ligands. Water and solvent molecules are usually not considered as foreign ligands. A mixed-ligand complex is [trichlorotriamminecobalt(III)]:

$$\left[\begin{array}{c} \quad \quad \text{NH}_3 \\ \text{NH}_3 \diagdown \mid \diagup \text{NH}_3 \\ \quad \quad \text{Co} \\ \text{Cl} \diagup \mid \diagdown \text{Cl} \\ \quad \quad \text{Cl} \end{array}\right]^0$$

Mixed-ligand complexes include, of course, those which contain hydroxide ions as ligands beside other ligands (basic complexes); e.g. ethylene diamine-tetraacetate forms with mercury(II) complexes of composition HgL^{2-} as well as of $Hg(OH)L^{3-}$.

In solid crystalline complex compounds, the ligands are situated in a certain geometrical form around the central ion, and the characteristic geometry of the complexes is usually retained in solution. When the coordination number is four, the configuration may be tetrahedral or square-planar, and for C. N. six an octahedral orientation is usual. With mixed-ligand complexes geometric isomerism may occur. The formation of complexes with different geometry can be accounted for by the crystal-field and ligand-field theories.

Square-planar   Tetrahedral

[tetrachloroplatinate (II)]   [tetra-amminezinc (II)]

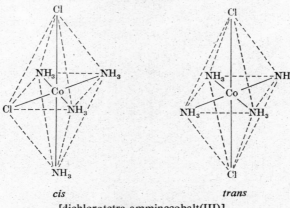

[dichlorotetra-amminecobalt(III)]

For the nomenclature of complexes the following general rules are to be taken into consideration.

(i) In the name of a complex the order is as follows: ligand, metal ion and the oxidation number of the latter in Roman numerals. Before the name of the ligand is given the number of ligands indicated by di, tri, tetra, penta, hexa, etc., prefixes of Greek origin. Negatively charged ligands are indicated by the suffix 'o' (chloro, bromo, cyano, etc.).

$[Ni(NH_3)_6]^{2+}$ [hexa-amminenickel(II)]

$[ZnCl_4]^{2-}$ [tetrachlorozincate(II)].

(ii) In the case of mixed-ligand complexes the order is as follows: the names of negatively charged ligands are followed by those of neutral ones, multidentate ligands precede unidentate ones.

$[CoCl_2(NH_3)_4]^+$ [dichlorotetra-amminecobalt(III)]

$[Cuen(NH_3)_4]^{2+}$ [ethylenediaminetetra-amminecopper(II)]

(iii) In polynuclear complexes the bridging ligand is indicated by the Greek letter μ (see above).

(iv) In the case of geometrical isomers the *cis* form is the one in which similar ligands are next to each other, whereas in the *trans* form they are opposite each other (see above).

(v) If the ligand forms a negatively charged (anionic) complex with the metal ion, the name of the metal is given the suffix 'ate', the

Latin name of the metal being used when necessary, e.g. [hexacyanoferrate(III)], [tetrachlorozincate(II)].

Organic molecule-complexes (or adducts) which are formed between organic molecules by donation and acceptance of a lone pair of electrons, mainly in non-polar solvents, are usually treated separately from metal complexes; e.g. [naphthalene–trinitrobenzene], [pyridine–tetracyanoethylene], [naphthalene–iodine] etc. As the importance of such adducts in analytical chemistry is at present less than that of metal complexes, they will not be dealt with here. For more details see [1, 2].

1.2. Equilibria

1.2.1. COMPLEX-FORMATION EQUILIBRIA

The formation of metal complexes can be represented by the following general reaction:

$$mM + nL \rightleftarrows M_mL_n \qquad (1.1)$$

The thermodynamic equilibrium constant of the reaction is

$$\beta_t = \frac{a_{M_mL_n}}{a_M^m a_L^n} \quad (p, T = \text{const.}) \qquad (1.2)$$

where a stands for the activities of the species indicated by subscript.

Expressing the activities as the product of concentrations and activity coefficients we obtain:

$$\beta_t = \frac{[M_mL_n]}{[M]^m[L]^n} \cdot \frac{f_{M_mL_n}}{f_M^m \cdot f_L^n} \qquad (1.3)$$

where the square brackets denote concentration in mole/l, and f the activity coefficient. For simplicity, charges are omitted.

According to the *Debye–Hückel theory*, to a first approximation, the activity coefficients in dilute solutions depend only on the ionic strength. In the solutions of medium concentration used in analytical determinations the activity coefficients of the individual ions can be calculated with fair accuracy by using the Davies equation [3] as follows:

$$-\log f_{z\pm} = Az^2 \left[\frac{\sqrt{I}}{1+\sqrt{I}} - 0.2\, I \right] \qquad (1.4)$$

where I is the ionic strength:

$$I = 1/2(c_1 z_1^2 + c_2 z_2^2 + \ldots) = 1/2 \Sigma c_i z_i^2 \tag{1.5}$$

(where c is the concentration of the individual ions in the solution in mole/l and z the charge); the constant A is equal to 0·509 for aqueous solutions at room temperature (see Example 1).

The activity coefficient of non-ionic, uncharged compounds can be approximated by the equation

$$\log f_0 = bI$$

where b is an experimental constant depending on the nature of the substance. Its value generally varies between 0·01 and 0·1. Therefore, if $I < 1$, the activity coefficient need not be taken into consideration in the case of non-ionic substances.

At finite but constant ionic strengths the first concentration term in Eq. (1.3) (which is called the concentration constant) differs from β_t for ionic strength $I = 0$ by a constant factor. At $I = 0$ every activity coefficient is equal to unity, and the concentration term is equal to the thermodynamic constant. The concentration constant at given ionic strength will be called the equilibrium constant of complex formation (or overall stability constant, see later)

$$\beta_{M_m L_n} = \frac{[M_m L_n]}{[M]^m [L]^n} \quad (p, T, I = \text{const.}) \tag{1.6}$$

In analytical practice we deal with solutions of reagents or samples (metal ions, complexes, etc.) of finite concentration. In determining complex formation equilibrium constants finite analytical concentrations or physico-chemical quantities proportional to them are measured by means of analytical methods. It is obvious therefore that the procedure involves the least error if the equilibrium constants determined by analytical methods, and corresponding to finite ionic strength, are directly used in equilibrium calculations. A further advantage of the use of equilibrium constants at finite ionic strength, defined by Eq. (1.6), is due to the fact that the average activity coefficients do not change much in the range of ionic strengths from 0·1 to 0·5 and can be considered as practically constant in the concentration range most often used in analytical practice. Literature equilibrium constants are mostly referred to

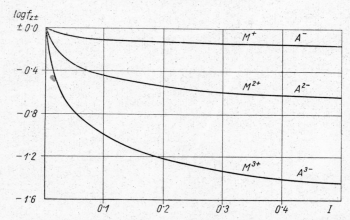

Fig. 1.1. Logarithm of the mean activity coefficients of ions with different charge numbers (calculated by the Davies equation), *vs.* ionic strength

$I = 0.1$. In most cases the error is not very great if these data are used in calculations without correction.

In Fig. 1.1 the logarithms of the mean activity coefficients calculated by using the Davies equation for ions with different numbers of charges are plotted versus the ionic strength. As can be seen from the figure, the change in the activity coefficient is smaller on changing the ionic strength from 0.1 to 0.5 than from 0 to 0.1. The change is especially small in the case of univalent ions.

Since in practice it is the hydrogen ion activity that is determined by an electrometric method, using a pH-meter, whereas concentrations of ligands and complexes are usually measured by spectrophotometry, etc., on the basis of Ringbom's suggestion [15] it is reasonable to use mixed constants, i.e. activities are used instead of concentrations in the case of hydrogen and hydroxyl ions.

The equilibrium constant at zero ionic strength (the thermodynamic constant β_t) can generally be obtained by determining the equilibrium constant in solutions of different ionic strengths and extrapolating to zero ionic strength, or by calculation.

If the mean activity coefficients are known [Eq. (1.3)] (see Examples 3 and 4), the reverse calculation is also possible.

To facilitate calculations, the logarithmic form of the equation is usually used:

$$\log \beta_t = \log \beta + \log f_{M_mL_n} - m \log f_M - n \log f_L \qquad (1.7)$$

Sec. 1.2]   **Equilibria**

and
$$\log \beta = \log \beta_t - \log f_{M_mL_n} + m \log f_M + n \log f_L \tag{1.8}$$

The calculation of the activity coefficients of ions with different numbers of charges becomes easier if it is remembered that

$$-\log f_z = -z^2 \log f_1 \qquad (I = \text{const.})$$

[See Eq. (1.4).]

1.2.1.1. *Equilibria of mononuclear complexes*

The equilibria of mononuclear complexes will be treated in more detail than those of polynuclear ones. Although polynuclear complexes also occur frequently in practice, in the majority of calculations they can be neglected, especially at low concentrations.

The complexes are usually formed in a stepwise manner, according to the following reaction equations (it is understood that water completes the coordination sphere)

$$M + L \rightleftharpoons ML$$
$$ML_2 + L \rightleftharpoons ML_2$$
$$\vdots \qquad \vdots \qquad \vdots$$
$$ML_{n-1} + L \rightleftharpoons ML_n$$

The successive stability constants characterizing the formation of the different species are as follows:

$$K_1 = \frac{[ML]}{[M][L]} \; ; \quad K_2 = \frac{[ML_2]}{[ML][L]} \; ; \quad K_n = \frac{[ML_n]}{[ML_{n-1}][L]} \tag{1.9}$$

By successive substitution we obtain

$$[ML] = K_1[M][L]$$
$$[ML_2] = K_1K_2[M][L]^2$$
$$\vdots$$
$$[ML_n] = K_1 \ldots K_n[M][L]^n \tag{1.10}$$

Denoting the products of the stepwise stability constants by β and the corresponding subscript, we obtain

$$\beta_1 = \frac{[ML]}{[M][L]} = K_1;$$

$$\beta_2 = \frac{[ML_2]}{[M][L]^2} = K_1 K_2;$$

$$\vdots$$

$$\beta_n = \frac{[ML_n]}{[M][L]^n} = K_1 \ldots K_n = \prod_1^n K_i \qquad (1.11)$$

β_n is called the overall complex stability constant of the nth complex.

$$\log \beta_n = \log K_1 + \log K_2 + \ldots + \log K_n = \sum_{i=1}^{n} \log K_i \qquad (1.12)$$

In general, the value of the first stability constant is the largest and the values of the successive stability constants decrease, i.e. $K_1 > K_2 > K_3 > \ldots$ However, there are exceptions, e.g. for the silver ammine complex $K_1 < K_2$. It also happens that one or more intermediate species practically do not exist, their stability constants being too small to be determined.

In solutions in which various species are simultaneously present in amounts depending on the K values, the amounts of the particular species can be calculated if all the stability constants and the total metal ion, ligand and hydrogen ion concentrations are known, or conversely the stability constants can be calculated if the composition of the system can be determined. For this reason the relationships between the constants (K or β) and quantities measurable by analytical means ([L], n, etc.) are very important.

The total concentration of the metal ion in the solution, whether it is complexed or not, is given by

$$C_M = [M] + [ML] + [ML_2] + \ldots = \sum_{i=0}^{i=N} [ML_i] \qquad (1.13)$$

Using Eqs. (1.10) and (1.11), Eq. (1.13) can be transformed into:

$$C_M = [M] + \beta_1[M][L] + \beta_2[M][L]^2 + \ldots \qquad (1.14)$$

By introducing the stability constant $\beta_0 = 1$, we have

$$[M] = \beta_0[M][L]^0$$

Sec. 1.2] Equilibria

Then Eq. (1.14) can be expressed in the simple form

$$C_M = [M] \sum_{i=0}^{i=N} \beta_i [L]^i \tag{1.15}$$

The total concentration of the ligand can be expressed as

$$C_L = [L] + [ML] + 2[ML_2] + \ldots = [L] + \sum_{i=1}^{i=N} i[ML_i]$$

or, by inserting the overall stability constants:

$$C_L = [L] + \beta_1[M][L] + 2\beta_2[M][L]^2 + \ldots$$

$$= [L] + [M] \sum_{i=1}^{i=N} i\beta_i [L]^i \tag{1.16}$$

where N indicates the maximum number of ligands.

The average ligand number has been suggested by Niels Bjerrum [4] for expressing the degree of complex formation, which, in the case of given concentrations and stability constants is characteristic of the extent of complexation. The average ligand number gives the mean number of ligands bound (in form of various complex species) to one metal ion; i.e.

$$\bar{n} = \frac{C_L - [L]}{C_M} \tag{1.17}$$

For unidentate ligands $\bar{n}$ gives the average coordination number.

The average ligand number can also be expressed by using Eqs. (1.14) and (1.16)

$$\bar{n} = \frac{\beta_1[M][L] + 2\beta_2[M][L]^2 + \ldots}{[M] + \beta_1[M][L] + \beta_2[M][L]^2 + \ldots}$$

Cancelling [M], we obtain

$$\bar{n} = \frac{\beta_1[L] + 2\beta_2[L]^2 + \ldots}{1 + \beta_1[L] + \beta_2[L]^2 + \ldots} = \frac{\sum_1^N i\beta_i[L]^i}{1 + \sum_1^N \beta_i[L]^i} \tag{1.18}$$

Equation (1.18) is the so-called complex formation function. This, at given overall stability constants, presents a relationship between the

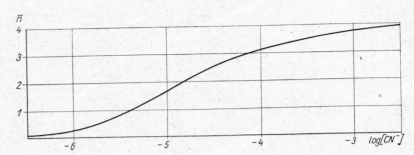

Fig. 1.2. Variation of the average ligand number with the logarithm of the ligand concentration for [cyanocadmium(II)] complexes

free ligand concentration and average ligand number. As shown by the function, $\bar{n}$ depends only on the concentration of the free ligand and is independent of that of the metal ion. This statement, as well as Eq. (1.18), is only true for the formation of mononuclear complexes.

If the concentrations of the ligand and the overall stability constants are known, $\bar{n}$ can be calculated by using the complex formation function. In Fig. 1.2 the average coordination number of [cyanocadmium(II)] complexes is plotted against the logarithm of the free ligand concentration.

When the total concentration of the ligand is much higher than that of the metal ion, i.e. $C_L \gg C_M$ and the ligand is not involved in reactions other than the complex formation, then $[L] \simeq C_L$, i.e. the total ligand concentration can be used instead of the free ligand concentration (see also pp. 59, 114).

If the metal and the ligand are present in comparable amount ($C_L < 10\, C_M$), then the concentration of the complexed ligand cannot be neglected in calculating the free ligand concentration, which can be done by solving the equation (given below) obtained by combining Eqs. (1.17) and (1.18) (see also Example 6):

$$\frac{C_L - [L]}{C_M} = \frac{\sum_1^N i\beta_i [L]^i}{1 + \sum_1^N \beta_i [L]^i}$$

i.e.

$$C_M(\beta_1[L] + 2\beta_2[L]^2 + \ldots) = (C_L - [L])(1 + \beta_1[L] + \beta_2[L]^2 + \ldots)$$
(1.19)

Sec. 1.2] Equilibria

The mole-fraction of the ith complex species in the solution, which gives the fraction of the total metal ion in the complex of composition ML_i, is given by:

$$\Phi_i = \frac{[ML_i]}{C_M} = \frac{\beta_i[M][L]^i}{[M] + \beta_1[M][L] + \beta_2[M][L]^2 + \ldots}$$

$$= \frac{\beta_i[L]^i}{1 + \beta_1[L] + \beta_2[L]^2 + \ldots} = \frac{\beta_i[L]^i}{1 + \sum_1^N \beta_i[L]^i} \quad (1.20)$$

Obviously, the mole-fraction of the free metal ion is given by

$$\Phi_0 = \frac{[M]}{C_M} = \frac{1}{1 + \beta_1[L] + \beta_2[L]^2 + \ldots} \quad (1.21)$$

It follows from the definition of mole-fractions that

$$\Phi_0 + \Phi_1 + \Phi_2 + \ldots = \sum_0^N \Phi_i = 1, \quad (1.22)$$

and from the combination of Eqs. (1.18) and (1.20)

$$\bar{n} = \sum_0^N i\Phi_i \quad (1.23)$$

It appears from Eqs. (1.20) and (1.21) that the mole-fractions of the free metal ion and of the different complex species depend only on the free ligand concentration for given overall stability constants. The equations enable the percentage distribution of the different species (metal ion and complexes) in the solution to be calculated for any free ligand concentration if the constants are known. The mole-fraction multiplied by 100 gives the amount of the species in question as a percentage of the total metal ion concentration (see Example 5).

The equations can also be used when the mole-fraction (or percentage) of a particular species is studied as function of the free ligand concentration. In Fig. 1.3 the mole-fractions of the cyanocadmium(II) complex species, as calculated by using Eqs. (1.20) and (1.21), are plotted against the logarithm of the free ligand cyanide concentration. At ligand concentrations where one of the species is present in maximum amount (at the peaks of the curves), the value of $\bar{n}$ is nearly equal to the number of ligands bound to the species in question. (Compare Figs. 1.2 and 1.3.) The abscissa values of the intersections of the mole-fraction curves,

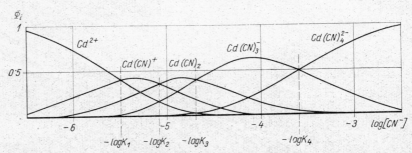

Fig. 1.3. Mole-fraction distribution of [cyanocadmium(II)] complexes *vs.* logarithm of the ligand concentration

where the concentrations of two consecutive complex species are the same, give the negative logarithms of the stability constants.

If
$$[ML_{n-1}] = [ML_n],$$
then, according to Eq. (1.9)
$$K_n = \frac{1}{[L]},$$
or
$$-\log K_n = \log [L]$$

The stability constants of the [cyanocadmium(II)] complex species are as follows [5]:

$$\log K_1 = 5\cdot 5, \ \log K_2 = 5\cdot 1, \ \log K_3 = 4\cdot 7 \text{ and } \log K_4 = 3\cdot 6$$

In another form of representation the mole fractions of the complexes at a given free ligand concentration are drawn one above another. As the sum of the mole-fractions of the different species must total 1 [see Eq. (1.22)], the top of the diagram is a straight line parallel to the abscissa. The relative amounts of the species are given by the vertical distances between the curves, and their range of existence can be estimated on the basis of the areas and relative positions of the fields between the curves (see Fig. 1.4).

If the metal ion forms complexes with more than one complexant present in the solution, and the stability constants are known, the amounts of the species can be calculated similarly. Equations (1.20) and (1.21) are to be modified by introduction of terms corresponding to

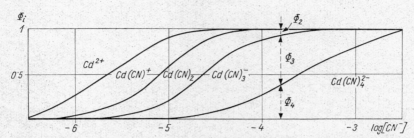

Fig. 1.4. Mole-fraction distribution of [cyanocadmium(II)] complexes plotted against logarithm of the ligand concentration

the complexes formed with the second, third, etc. complexants, into the denominator of the fraction. If L and Y indicate the two ligands, and β and γ the corresponding overall stability constants, Eq. (1.20) takes the form:

$$\Phi_i = \frac{\beta_i[L]^i}{1 + \beta_1[L] + \beta_2[L]^2 + \ldots \gamma_1[Y] + \gamma_2[Y]^2 + \ldots} \quad (1.24)$$

(see Example 8).

1.2.1.2. *Equilibria of polynuclear complexes*

The formation of polynuclear complexes can be described by the following general reaction equation:

$$m\mathrm{M} + n\mathrm{L} \rightleftharpoons \mathrm{M}_m\mathrm{L}_n$$

The formation of these complexes, like that of mononuclear complexes, proceeds stepwise, and m and n change with the concentration. At very low metal ion concentrations only mononuclear complexes are formed. At higher concentrations polynuclear complexes also appear, according to the successive equilibria. The law of mass action for the general reaction takes the form

$$\beta_{m,n} = \frac{[\mathrm{M}_m\mathrm{L}_n]}{[\mathrm{M}]^m[\mathrm{L}]^n} \quad (1.25)$$

$\beta_{m,n}$ is the overall formation or stability constant of the m-nuclear complex of composition $\mathrm{M}_m\mathrm{L}_n$. The overall stability constant is equal to the product of the successive stability constants.

The total metal ion and total ligand concentration in the solution of polynuclear complexes can be given as follows:

$$C_M = [M] + \sum_{j=1}^{j=m}\sum_{i=1}^{i=n} j[M_jL_i] = [M] + \sum_{j=1}^{j=m}\sum_{i=1}^{i=n} j\beta_{j,i}[M]^j[L]^i \quad (1.26)$$

$$C_L = [L] + \sum_{j=1}^{j=m}\sum_{i=1}^{i=n} i[M_jL_i] = [L] + \sum_{j=1}^{j=m}\sum_{i=1}^{i=n} i\beta_{j,i}[M]^j[L]^i \quad (1.27)$$

In the course of the summation j and i can take any value from 1 to m, and 1 to n, respectively. The value of $\bar{n}$ as defined by Eq. (1.17) in the solution of polynuclear complexes gives the average number of ligands bound to one metal ion.

In the case of the formation of mononuclear complexes the average ligand number is independent of the total concentration of the metal ion, but in the case of polynuclear complexes the average ligand number depends on the total metal ion concentration. It can be deduced [6] that

$$\bar{n} = \frac{[M]\left(\dfrac{\partial \theta}{\partial [M]}\right)_{[L]}}{[L]\left[1 + \left(\dfrac{\partial \theta}{\partial [L]}\right)_{[M]}\right]} = \frac{\left(\dfrac{\partial(\theta + [L])}{\partial \ln [M]}\right)_{[L]}}{\left(\dfrac{\partial(\theta + [L])}{\partial \ln [L]}\right)_{[M]}} \quad (1.28)$$

where

$$\theta = \sum_{j=1}^{j=m}\sum_{i=1}^{i=n} \beta_{j,i}[M]^j[L]^i \quad (1.29)$$

In plots of the average ligand number *vs.* the logarithm of the free ligand concentration at different but constant total metal ion concentrations, the position of the formation curves will depend on C_M. The curves are shifted; the greater the concentration of metal ion, the greater being the number of nuclei in the complex.

Formation of polynuclear complexes is found in the course of formation of hydroxo-complexes. In Fig. 1.5 the formation curves of the hydroxo-complexes of thorium(IV) are shown at different total metal ion concentrations.

According to the investigations by Sillén [6] hydroxo-complexes are often of the 'core plus links' type, with the composition $M(ML_t)_s$, where t is an integer giving the number of ligands in a link, and s is an integer giving the number of links in the complex. The negative shift of the $\bar{n}$ *vs.* log [OH] curves (complex formation curves) is usually proportional to the total metal ion concentration within a certain concentration range.

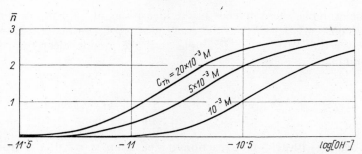

Fig. 1.5. Formation curves of [hydroxothorium(IV)] complexes for different total metal ion concentrations [7]

It is important from the analytical point of view that the formation curves of the complexes converge to one curve, the so-called 'mononuclear wall', with increasing dilution, i.e. decreasing total metal ion concentration. For concentrations below a certain limit, the position of the formation curve will not change with further dilution and only mononuclear complexes are formed. In analytical hydrolytic reactions the free metal ion concentration can be very low, owing to precipitate formation and the fact that the hydroxo-complexes are formed in the presence of other complexing agents. Thus the formation of polynuclear complexes can often be ignored in solving such analytical problems.

The limiting concentration below which only mononuclear complexes are to be considered, depends on the overall stability constant of the polynuclear complex first formed, namely the binuclear one, and that of the mononuclear one last formed, which has the same number of ligands as the binuclear species.

For the mononuclear species $M(OH)_n$ to be of higher concentration than the binuclear one, the following condition must be fulfilled:

$$2[M_2(OH)_n] < [M(OH)_n]$$

Substituting the corresponding overall stability constants,

$$2[M]^2[OH]^n \beta_{2,n} < [M][OH]^n \beta_n \qquad (1.30)$$

From this

$$[M] < \frac{\beta_n}{2\beta_{2,n}}$$

or in logarithmic form

$$\log [M] < \log \beta_n - \log \beta_{2,n} - 0 \cdot 3 \qquad (1.31)$$

3 Inczédy

If, for example, the difference in the logarithms of the two overall stability constants is -5, then the concentration of the metal ion should be lower than $10^{-5.3}$ (see Example 9) if the mononuclear form is to predominate.

1.2.2. Equilibria of acid–base reactions: protonation of the ligand

In dealing with acid–base reactions it is easiest to start with the Brønsted theory, according to which acids are compounds capable of releasing protons, and bases are compounds capable of accepting protons. In the reaction

$$L + H^+ \rightleftharpoons HL$$

L is a base and HL the corresponding acid, irrespective of whether they are charged ions or uncharged molecules. Acid–base reactions involve proton transfer from an acid to a base, by which a new acid and new base are formed. In aqueous solutions protons never occur free; although they are denoted as H^+, they are always present as the solvated ion $[H(H_2O)_x]^+$.

The complexing ligands used in analytical chemistry are bases according to the Brønsted theory, and most of them are capable of accepting protons in aqueous solution. If the ligand is an anion, then a protonated anion or neutral molecule is formed; if it is neutral, a positively charged protonated cation is formed.

Chemical calculations including complexes are greatly simplified by using protonation constants (which are similar to stability constants) instead of the usual acidic or basic dissociation constants. Protonation constants are simply the inverse of the acid dissociation constants. In the following treatment, protonation constants will be denoted by K, and dissociation constants by k and the appropriate subscript.

The acetate ion is transformed into acetic acid by protonation:

$$CH_3COO^- + H^+ \rightleftharpoons CH_3COOH$$

The equilibrium constant of the reaction is the protonation constant of acetate ion:

$$K = \frac{[CH_3COOH]}{[CH_3COO^-][H^+]} \tag{1.32}$$

which is equal to the reciprocal value of the dissociation constant of acetic acid.

$$K = \frac{1}{k_a}$$

In logarithmic form:

$$\log K = -\log k_a \tag{1.33}$$

In the protonation of an uncharged univalent base the conjugate acid is formed, e.g.,

$$NH_3 + H^+ \rightleftharpoons NH_4^+$$

The corresponding protonation constant is:

$$K = \frac{[NH_4^+]}{[NH_3][H^+]} \tag{1.34}$$

Equation (1.33) is also valid in this case, i.e. the protonation constant of the base is equal to the reciprocal of the dissociation constant of the conjugate acid. The protonation constant is also related to the basic dissociation constant, k_b. By multiplying both the numerator and denominator of the fraction by $[OH^-]$ we get

$$K = \frac{[NH_4^+][OH^-]}{[NH_3][H^+][OH^-]}$$

and since

$$[H^+][OH^-] = k_w,$$

which is a constant at constant temperature, and

$$\frac{[NH_4^+][OH^-]}{[NH_3]} = k_b \tag{1.35}$$

then

$$K = \frac{k_b}{k_w} \tag{1.36}$$

Since at room temperature $\log k_w = -14$, the logarithmic form of Eq. (1.36) is

$$\log K = 14 + \log k_b \tag{1.37}$$

The protonation constant of a base can be calculated from the dissociation constant of the conjugate acid by using Eq. (1.33) or from that of the base by using Eq. (1.37).

The calculations are simplified if it is noted that for a base and its conjugate acid in aqueous solution,

$$\log k_b + \log k_a = -14 \tag{1.38}$$

Most bases can take up more than one proton successively. The first proton is bound most strongly, subsequent bond strengths gradually decrease. The free ligand is the strongest base, i.e. has the greatest proton affinity.

In the case of phosphate ions, the protonation reactions and the corresponding protonation constants are as follows:

$$PO_4^{3-} + H^+ \rightleftharpoons HPO_4^{2-}$$

$$HPO_4^{2-} + H^+ \rightleftharpoons H_2PO_4^-$$

$$H_2PO_4^- + H^+ \rightleftharpoons H_3PO_4$$

$$K_1 = \frac{[HPO_4^{2-}]}{[PO_4^{3-}][H^+]} = \frac{1}{k_3} = 10^{11 \cdot 7}$$

$$K_2 = \frac{[H_2PO_4^-]}{[HPO_4^{2-}][H^+]} = \frac{1}{k_2} = 10^{6 \cdot 9} \tag{1.39}$$

$$K_3 = \frac{[H_3PO_4]}{[H_2PO_4^-][H^+]} = \frac{1}{k_1} = 10^2$$

The reciprocal of the last dissociation constant of phosphoric acid is equal to the first protonation constant.

The protonation constant products can be derived similarly to the stability constant products as follows:

$$K_1 = \frac{1}{k_3} = \frac{[HPO_4^{2-}]}{[PO_4^{3-}][H^+]} = 10^{11 \cdot 7}$$

$$K_1 K_2 = \frac{1}{k_3 k_2} = \frac{[H_2PO_4^-]}{[PO_4^{3-}][H^+]^2} = 10^{18 \cdot 6}$$

$$K_1 K_2 K_3 = \frac{1}{k_3 k_2 k_1} = \frac{[H_3PO_4]}{[PO_4^{3-}][H^+]^3} = 10^{20 \cdot 6} \tag{1.40}$$

In dealing with the equilibria of multivalent bases and complexants which contain both acidic and basic groups, the use of the successive protonation constants is expedient and unambiguous, since the variously defined acidic and basic dissociation constants become unnecessary. As protons are always bound to one ligand, the relationships deduced for mononuclear complexes can be used in calculations in treating the protonation of ligands as well as the dissociation of acids and bases.

The complex formation function defined by Eq. (1.18) can be used for describing the protonation of ligands if the hydrogen ion concentrations are written in the place of the ligand concentration, and the protonation constant products in the place of the overall stability constants: $\bar{n}$ gives information on the degree of protonation of the ligand.

$$\bar{n} = \frac{[H^+]K_1 + 2[H^+]^2 K_1 K_2 + \ldots + n[H^+]^n K_1 \ldots K_n}{1 + [H^+]K_1 + [H^+]^2 K_1 K_2 + \ldots + [H^+]^n K_1 \ldots K_n} \quad (1.41)$$

Similarly, Eqs. (1.20) and (1.21) can be used after suitable modification to express the mole-fraction or to calculate the percentage of the dissociated species in solutions of acids and bases.

The mole-fraction of the non-protonated species is given by

$$\Phi_0 = \frac{1}{1 + [H^+]K_1 + [H^+]^2 K_1 K_2 + \ldots [H^+]^n K_1 \ldots K_n} \quad (1.42)$$

The mole-fractions of the protonated species are given by

$$\Phi_i = \frac{[H^+]^i K_1 \ldots K_i}{1 + [H^+]K_1 + [H^+]^2 K_1 K_2 + \ldots} \quad (1.43)$$

As shown by Eq. (1.43), the percentage distribution of the protonated species in the solution of an acid–base system depends only on the hydrogen ion concentration, and not on the total amount of acid or base present, at constant ionic strength. (See Example 7.)

In practice, pH is used to express the acidity of solutions, which is, by definition

$$\text{pH} = -\log [H^+] \quad (1.44)$$

The degree of dissociation α_D expresses that fraction of an acid $H_n L$ which is in the dissociated form, and can be calculated as follows:

$$\alpha_D = \frac{C_{H_n L} - [H_n L]}{C_{H_n L}} = 1 - \Phi_n \quad (1.45)$$

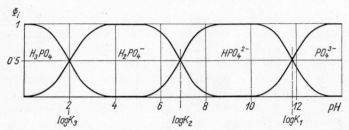

Fig. 1.6. Mole-fraction distribution of the dissociation products of phosphoric acid as a function of pH

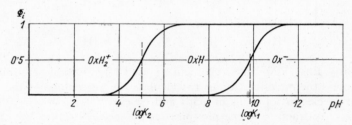

Fig. 1.7. Mole-fraction distribution of the dissociation products of 8-hydroxyquinoline as a function of pH

where C_{H_nL} is the total concentration of the acid. For a monobasic acid

$$\alpha_D = 1 - \Phi_1 = \Phi_0 \tag{1.46}$$

[see Eq. (1.22)]. If the protonation constants are known, the distribution curves of the products of dissociation can be plotted against pH, just as distribution curves of mononuclear complexes are plotted as a function of the logarithm of the free ligand concentration. Figures 1.6 and 1.7 show the distribution curves of the dissociation products of phosphoric acid and 8-hydroxyquinoline.

In Fig. 1.6 the abscissa values of the intersections of the curves represent the logarithms of the protonation constants or the negative logarithms of the dissociation constants (cf. Fig. 1.3).

8-Hydroxyquinoline contains both an acidic and a basic group; it gives a negative ion on releasing a proton and a positive ion on accepting one:

$$\text{[quinoline-O}^-\text{]} \xrightleftharpoons{-H^+} \text{[quinoline-OH]} \xrightleftharpoons{+H^+} \text{[quinoline-OH, NH}^+\text{]} \tag{1.47}$$

Sec. 1.2] Equilibria

In protonation of an anionic complexing ligand, the order of proton uptake will be determined by the decreasing order of the protonation constants. In this reaction the order from left to right is: $\log K_1 = 9\cdot9$, $\log K_2 = 5\cdot0$ [8].

The method of presentation used in Fig. 1.7 is especially advantageous when the range of existence of the dissociation products of a polybasic acid is to be estimated. If the protonation or dissociation constants are known, the diagram can be constructed very easily by remembering that at $\text{pH} = \log K_n$ the mole-fractions of both $H_n L$ and $H_{n-1} L$ are 0·5, at pH one unit lower 0·91 and 0·09, and at pH one unit higher 0·09 and 0·91 respectively, if the difference in the logarithm of the successive protonation constants is at least 3.

1.2.3. Equilibria of redox reactions

For uniformity, redox equilibria in electrolyte solutions will be dealt with similarly to complexation or acid–base reactions. Redox reactions, unlike the reactions mentioned so far, involve the transfer of electrons. In general

$$\text{Ox} + ze \rightleftharpoons \text{Red} \qquad (1.48)$$

where e stands for the electron, z for the number of electrons involved in the reaction, 'Ox' for the oxidized form and 'Red' for the reduced form of the substance. Acceptance of electrons constitutes reduction, and oxidation is removal of electrons. The process may be compared to acid–base reactions in aqueous solution, where an acid is formed by acceptance and a base by release of protons. Just as free protons are not present in solutions of acids and bases, free electrons are not present in solutions of reducing or oxidizing substances.

If redox reactions are regarded as complexation reactions, the oxidized form corresponds to the central ion, which forms the reduced species with the electrons as ligands.

The equilibrium constant of the reaction described by Eq. (1.48) is

$$K_r = \frac{[\text{Red}]}{[\text{Ox}][e]^z} \quad (p, T, I = \text{const.}) \qquad (1.49)$$

where [e] indicates the 'electron activity' [9]. The higher the value of K_r the greater is the oxidizing power of the redox system.

Taking the logarithm of Eq. (1.49) we obtain

$$-\log [e] = \frac{1}{z} \log K_r + \frac{1}{z} \log \frac{[Ox]}{[Red]} \qquad (1.50)$$

The negative logarithm of the electron activity will be denoted by **pe**. The pe scale can be used for characterizing the oxidizing power of solutions just as the pH and pM scales are used for the Brønsted acidity and Lewis acidity, respectively.

Since

$$-\log [e] = \text{pe} \qquad (1.51)$$

then as pe increases, the electron activity decreases, i.e. the oxidizing power of the solution increases.

It can be seen from Eq. (1.50) that in a solution in which the concentrations of the oxidized and reduced forms of a redox system are equal to one another (more precisely the activity of both forms is 1) the pe value is characteristic of the redox system, and can be calculated if K_r and z are known from

$$\text{pe}^0 = \frac{1}{z} \log K_r \qquad (1.52)$$

This value is the standard pe of the redox system in question, pe^0.

Thus

$$\text{pe} = \text{pe}^0 + \frac{1}{z} \log \frac{[Ox]}{[Red]} \qquad (1.53)$$

The Nernst equation for the electrochemical equilibrium redox potential is

$$E = E^0 + \frac{2 \cdot 3RT}{zF} \log \frac{[Ox]}{[Red]} \qquad (1.54)$$

Comparing Eq. (1.53) with Eq. (1.54) we can equate terms to obtain:

$$E = \frac{2 \cdot 3RT}{F} \text{pe};$$

$$E^0 = \frac{2 \cdot 3RT}{F} \text{pe}^0 = \frac{2 \cdot 3RT}{zF} \log K_r \qquad (1.55)$$

i.e. there is a close relationship between the redox potential E and pe, and the standard redox potential E^0 and pe^0. The value of the term $2 \cdot 3RT/F$ is $0 \cdot 059$ V at 25°C.

It should be noted that in the case of oxy-ion redox systems, hydrogen ions also appear in the reaction equation:

$$Ox + ze + mH^+ \rightleftharpoons Red + \frac{m}{2} H_2O$$

In this case Eqs. (1.49) and (1.50) are modified

$$K_r = \frac{[Red]}{[Ox][e]^z[H^+]^m} \qquad (1.56)$$

$$pe = pe^0 + \frac{1}{z} \log \frac{[Ox]}{[Red]} - \frac{m}{z} pH \qquad (1.57)$$

If pH = 0, then Eq. (1.57) reduces to Eq. (1.50).

The value of the equilibrium constant of the redox process

$$H^+ + e \rightleftharpoons 1/2\, H_2,$$

is by definition equal to unity, hence the corresponding pe^0 value is the zero point of the pe scale.

The mole-fractions of the oxidized and reduced forms of a redox system can be calculated for solutions with different pe values, from considerations similar to those applied in the case of acids, bases and complexes.

The total concentration of an oxidizing or reducing substance can be written as

$$C_r = [Ox] + [Red] = [Ox] + [Ox][e]^z K_r \qquad (1.58)$$

from Eq. (1.49). The mole-fraction of the oxidized form is given by

$$\Phi_{Ox} = \frac{[Ox]}{C_r} = \frac{1}{1 + [e]^z K_r} \qquad (1.59)$$

Equation (1.59) is suitable for the calculation of the relative amounts of the oxidized (or reduced) form of a redox system in solutions with different pe values. (See Example 10.)

In Fig. 1.8 the mole-fractions of the oxidized and reduced forms of the iron(III)–iron(II) and of the dichromate–chromium(III) redox systems are presented as a function of pe. The range of existence of the different species can be estimated well from the figure. The abscissa corresponding to $\Phi = 0.5$ is $pe^0 = \frac{1}{z} \log K_r$. The equilibrium constants

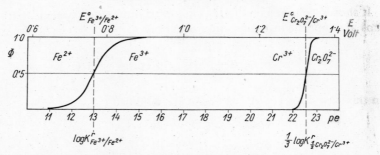

Fig. 1.8. Mole-fraction distribution of the oxidized and reduced forms for the iron(III)/iron(II) couple and dichromate/chromium(III) couple as functions of pe

used are

$$\log K_{r\,Fe(III)/Fe(II)} = 13\cdot 0 \quad [10]$$

$$K_{Cr(VI)/Cr(III)} = 67\cdot 6 \quad [11]$$

For comparison the redox potential is also shown.

In a redox reaction two redox systems must be involved, i.e. the electrons donated by one system are accepted by the other. The equilibrium constant of a redox reaction can be calculated from the individual K_r values. For example

Multiplier

$$\begin{array}{lr} 1/2\,Cr_2O_7^{2-} + 7\,H^+ + 3\,e \rightleftharpoons Cr^{3+} + 7/2\,H_2O & 1 \\ Fe^{3+} + e \rightleftharpoons Fe^{2+} & -3 \\ \hline 1/2\,Cr_2O_7^{2-} + 7\,H^+ + 3\,Fe^{2+} \rightleftharpoons Cr^{3+} + 3\,Fe^{3+} + 7/2\,H_2O & \end{array}$$

The equilibrium constant of the last redox reaction is given by

$$K = \frac{[Cr^{3+}][Fe^{3+}]^3}{[Cr_2O_7^{2-}]^{1/2}[H^+]^7[Fe^{2+}]^3}$$

$$= K_{rCr(VI)/Cr(III)}\,(K_{rFe(III)/Fe(II)})^{-3} = 10^{67\cdot 6}(10^{13})^{-3} = 10^{67\cdot 6 - 39} = 10^{28\cdot 6}$$

In logarithmic form:

$$\log K = \log K_{r\,Cr(VI)/Cr(III)} - 3\log K_{rFe(III)/Fe(II)} = 67\cdot 6 - 3\times 13 = 28\cdot 6$$

Equilibria

This method can also be used for calculating the overall equilibrium constants when other reactions take place in addition to the redox reaction, e.g. complex or precipitate formation.

For example

	Multiplier	log K
$1/2\ Cr_2O_7^{2-} + 7H^+ + 3e \rightleftharpoons Cr^{3+} + 7/2\ H_2O$	1	67·6
$Fe^{3+} + e \rightleftharpoons Fe^{2+}$	-3	13
$Fe^{3+} + HPO_4^{2-} \rightleftharpoons FeHPO_4^+$	3	9·3

$$1/2\ Cr_2O_7^{2-} + 7H^+ + 3Fe^{2+} + 3HPO_4^{2-} \rightleftharpoons Cr^{3+} + 3FeHPO_4^+ + 7/2\ H_2O$$

The equilibrium constant of the overall reaction is given by

$$\log K = 67·6 - 3 \times 13 + 3 \times 9·3 = 56·5$$

1.2.4. Practical instructions for equilibrium calculations

In calculations, concentrations are expressed in mole/l, the mole being Avogadro's number of particles of the species indicated (which may be molecules, atoms or ions). Square brackets round a chemical formula or symbol denote concentration.

As equilibrium constants range over many orders of magnitude, the concentrations of the reaction products may also vary over a wide range, so it is useful to express the constants and concentrations in exponential form, and make the calculations with decimal logarithms. For example $1·6 \times 10^{-3}$ can be written as $10^{0·20} \times 10^{-3} = 10^{-2·8}$. The exponential form can be transformed back as follows:

$$10^{5·35} = 10^{0·35} \times 10^5 = 2·24 \times 10^5$$
$$10^{-6·35} = 10^{0·65} \times 10^{-7} = 4·47 \times 10^{-7}$$

As the error in the data (pH, stability constants, etc.) used in the calculations is about 1%, two- or three-figure logarithms are good enough. For rapid calculations a slide rule or a calculator can be used.

The variable to be determined (ligand concentration, etc.) may occur with an exponent higher than 2 in the equation to be solved. For solving cubic or quartic equations Newton's approximation can be used. The principle of the method is as follows. Let us write the equation in the form:

$$ax^3 + bx^2 + cx + d = 0$$

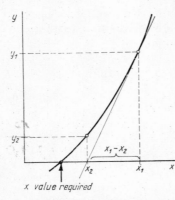

Fig. 1.9. Newton's approximation method

If y is written instead of 0, then on plotting y against x, the function will intersect the x axis at the value of x required.

As a first approximation, we insert an estimated value x_1 into the equation. A value y_1 different from 0 will be obtained. Then we differentiate the function with respect to x

$$\frac{dy}{dx} = 3ax^2 + 2bx + c \quad (1.60)$$

and calculate dy/dx after inserting x_1 into the equation. Then dy/dx gives the slope of the curve at $x_1 y_1$ and since this is given by $y_1/(x_1 - x_2)$ then x_2 is clearly a better approximation than x_1 (Fig. 1.9).

We can then repeat the process with x_2 as the estimate and obtain a better value x_3. We repeat the procedure until the difference between two consecutive x values is smaller than the permissible error. As we do not require accuracy better than 1%, the second or third approximation usually gives the required result.

If the coefficients in the basic equation are negative powers of 10, it is useful to modify the equation by multiplying by a factor which will remove the negative exponents, e.g.

$$x^3 + 2 \times 10^{-4} x^2 + 7 \times 10^{-10} x + 4 \times 10^{-16} = 0$$

becomes

$$z^3 + 200 z^2 + 700 z - 400 = 0$$

where $z = 10^6 x$.

In determination of equilibrium constants many measurements are usually made and the most probable value of the constants is determined by using all the experimental data. In such cases, if the explicit function is known, it is best to use the method of least squares.

We have n corresponding x and y pairs, and look for a function

$$y = a_0 + a_1 x + a_2 x^2 + \ldots \quad (1.61)$$

for which after insertion of the experimental x and y values, the sum

Sec. 1.2] Equilibria

of the squares of the differences, Δ, given by

i. e.
$$\Delta = a_0 + a_1 x + a_2 x^2 + \ldots - y \qquad (1.62)$$

$$\sum_{i=1}^{i=n} [a_0 + a_1 x + a_2 x^2 + \ldots - y]^2 = \Sigma \Delta^2 \qquad (1.63)$$

is the smallest.

The conditions for a minimum are

$$\frac{\partial \Sigma \Delta^2}{\partial a_0} = 0; \quad \frac{\partial \Sigma \Delta^2}{\partial a_1} = 0; \quad \frac{\partial \Sigma \Delta^2}{\partial a_2} = 0, \text{ etc.} \qquad (1.64)$$

In the simplest case, when the function is the equation of a straight line (and only y but not x may be uncertain) the function takes the form:

$$y = a_0 + a_1 x \qquad (1.65)$$

$$\sum^{n} (a_0 + a_1 x - y)^2 = \sum \Delta^2 \qquad (1.66)$$

and the conditions are

$$a_1 \sum_{1}^{n} x + n a_0 - \sum_{1}^{n} y = 0 \qquad (1.67)$$

$$a_1 \sum_{1}^{n} x^2 + a_0 \sum_{1}^{n} x - \sum_{1}^{n} xy = 0 \qquad (1.68)$$

As n, Σx, Σy, Σx^2 and Σxy are known, the most probable values of a_0 and a_1 can be obtained by solving these simultaneous equations (see also Example 6).

It is an essential condition of equilibrium calculations that the equilibrium constants available are reliable. The Committee for Equilibrium Data of the International Union of Pure and Applied Chemistry (IUPAC) deals with the collection, registration and critical evaluation of stability constants, protonation constants, etc. In the collection by Sillén and Martell and its supplement [12] the complex stability, protonation and redox equilibrium constants published up to about 1970 are presented along with the method used for the determination, ionic strength, temperature, reference, etc. but without criticism. The data cited for some systems differ by several orders of magnitude. However, many equilibrium constants may be considered as reliable and can be used in equilibrium calculations.

1.2.5. Worked examples

1. Calculate the ionic strength in a solution which is $0.1M$ in calcium chloride and $0.001M$ in potassium nitrate.

The calculation is done by using Eq. (1.5).

$$I = \frac{1}{2}\{[Ca^{2+}]\,2^2 + [Cl^-]\,1^2 + [K^+]\,1^2 + [NO_3^-]\,1^2\}$$

$$= \frac{1}{2}(0.1 \times 4 + 0.2 + 0.001 + 0.001) = 0.301$$

2. Calculate the activity coefficient of calcium ions in a solution of ionic strength 0.3. The calculation is done by using Eq. (1.4).

$$\log f_2 = -0.509 \times 2^2 \left[\frac{\sqrt{0.3}}{1 + \sqrt{0.3}} - 0.2 \times 0.3\right] = -0.598$$

$$= 0.402 - 1; \quad f_2 = 0.252$$

3. Calculate the overall stability constants of [tetra-amminecopper(II)] and [(tris acetylacetonato)-iron(III)] at zero ionic strength, given the following data:

$$[Cu(NH_3)_4]^{2+}; \quad \log \beta_4 = 12.59 \quad (I = 0.1) \quad [16]$$

$$Fe(Acac)_3; \quad \log \beta_3 = 24.9 \quad (I = 0.1) \quad [17]$$

The calculation is done by using Eq. (1.7); $\log f_1$ is taken from Fig. 1.1.

$$\log \beta_{4,t} = 12.59 - \log f_2 + \log f_2 = 12.59 \quad (I = 0);$$

$$\log \beta_{3,t} = 24.9 - \log f_3 - 3\log f_1$$

$$= 24.9 + 9 \times 0.11 + 3 \times 0.11 = 26.22 \quad (I = 0).$$

4. Calculate the protonation constant of lactate in solutions of ionic strengths 0.1 and 0.5 at $25°C$, if the dissociation constant of lactic acid is $\log k = -3.86$ at $I = 0$ [18].

Sec. 1.2] Equilibria 47

The calculation is done by using Eq. (1.8) and taking Eq. (1.33) into consideration. Since in the calculations $[H^+] = a_H$ (see p. 24) only the activity coefficient of lactate is taken into account; $\log f$ values are taken from Fig. 1.1 for $I = 0{\cdot}1$ and $I = 0{\cdot}5$.

$$\log K = 3{\cdot}86 + \log f = 3{\cdot}86 - 0{\cdot}11 = 3{\cdot}75 \quad (I = 0{\cdot}1);$$

$$\log K = 3{\cdot}86 - 0{\cdot}16 = 3{\cdot}70 \quad (I = 0{\cdot}5).$$

5. Calculate the fraction of indium(III) present as In^{3+}, $InBr^{2+}$, $InBr_2^+$ and $InBr_3$, if the concentration of bromide is $0{\cdot}1M$ and $C_{Br} \gg C_{In}$ (the solution is slightly acidic, so the formation of hydroxo-complexes need not be considered). The overall formation constants are:

$$\log \beta_1 = 1{\cdot}2; \ \log \beta_2 = 1{\cdot}8; \ \log \beta_3 = 2{\cdot}5 \ [19]$$

The calculation is done by using Eqs. (1.20) and (1.21).

$$\Phi_0 = \frac{1}{1 + \beta_1[Br^-] + \beta_2[Br^-]^2 + \beta_3[Br^-]^3}$$

$$= \frac{1}{1 + 10^{1{\cdot}2} \times 10^{-1} + 10^{1{\cdot}8} \times 10^{-2} + 10^{2{\cdot}5} \times 10^{-3}} = \frac{1}{3{\cdot}54} = 0{\cdot}28$$

$$\Phi_1 = \frac{10^{1{\cdot}2} \times 10^{-1}}{3{\cdot}54} = \frac{1{\cdot}59}{3{\cdot}54} = 0{\cdot}45$$

$$\Phi_2 = \frac{10^{1{\cdot}8} \times 10^{-2}}{3{\cdot}54} = \frac{0{\cdot}63}{3{\cdot}54} = 0{\cdot}18$$

$$\Phi_3 = \frac{10^{2{\cdot}5} \times 10^{-3}}{3{\cdot}54} = \frac{0{\cdot}32}{3{\cdot}54} = 0{\cdot}09$$

Thus 28% of the indium(III) is present as the free ion, and 45, 18 and 9% as the 1:1, 1:2 and 1:3 bromo-complexes respectively.

6. Calculate the free ligand concentration in a solution in which both the total bromide and total indium(III) concentrations are $0{\cdot}1M$. For the stability constants, etc. see the preceding problem.

The calculation is done by using Eq. (1.19). To a first approximation, the terms in which β_3 occurs and the concentration exponent is greater than 2 will be omitted, since the species with coordination number 3 is formed in the smallest amount.

$$C_{In}\beta_1[Br^-] + 2C_{In}\beta_2[Br^-]^2 = C_{Br} + C_{Br}\beta_1[Br^-] + C_{Br}\beta_2[Br^-]^2$$
$$- [Br^-] - \beta_1[Br^-]^2;$$

$$10^{-1} \times 10^{1\cdot 2}[Br^-] + 2 \times 10^{-1} \times 10^{1\cdot 8}[Br^-]^2$$
$$= 10^{-1} + 10^{-1} \times 10^{1\cdot 2}[Br^-] + 10^{-1}10^{1\cdot 8}[Br^-]^2 - [Br^-] - 10^{1\cdot 2}[Br^-]^2;$$

$$(10^{0\cdot 8} + 10^{1\cdot 2})[Br^-]^2 + [Br^-] - 10^{-1} = 0,$$

from which

$$Br^- \simeq 0\cdot 048 M$$

A more precise result is obtained if the terms mentioned are not neglected, and the quartic equation is solved by Newton's approximation method (see p. 43).

Let us denote the concentration of free bromide by x.

$$10^{0\cdot 2}x + 2 \times 10^{0\cdot 8}x^2 + 3 \times 10^{1\cdot 5}x^3 = 10^{-1} + 10^{0\cdot 2}x + 10^{0\cdot 8}x^2$$
$$+ 10^{1\cdot 5}x^3 - x - 10^{1\cdot 2}x^2 - 10^{1\cdot 8}x^3 - 10^{2\cdot 5}x^4;$$
$$316x^4 + 126\cdot 3x^3 + 22\cdot 1x^2 + x - 0\cdot 10 = 0.$$

Writing y in the place of 0 on the right-hand side of the equation, and calculating y_1 corresponding to $x = 0\cdot 05$ (i.e. half the total bromide concentration) we obtain

$$0\cdot 001975 + 0\cdot 01578 + 0\cdot 05525 + 0\cdot 050 - 0\cdot 10 = 0\cdot 0230 = y_1$$

Differentiating the original equation with respect to x, then inserting $x_1 = 0\cdot 05$ and calculating $(dy/dx)_1$ gives

$$1264x^3 + 378\cdot 9x^2 + 44\cdot 2x + 1 = 0$$
$$0\cdot 1580 + 0\cdot 9473 + 2\cdot 21 + 1 = 4\cdot 315 = (dy/dx)_1$$

Then

$$x_2 = x_1 - \frac{y_1}{(dy/dx)_1} = 0\cdot 05 - \frac{0\cdot 0230}{4\cdot 315} = 0\cdot 0447.$$

On repetition with x_2 we get

$$0\cdot 00126 + 0\cdot 01128 + 0\cdot 04416 + 0\cdot 0447 - 0\cdot 1 = 0\cdot 0014 = y_2$$
$$0\cdot 1128 + 0\cdot 7578 + 1\cdot 976 + 1 = 3\cdot 846 = (dy/dx)_2$$
$$\therefore x_3 = 0\cdot 0447 - \frac{0\cdot 0014}{3\cdot 846} = 0\cdot 0443$$

The procedure need not be repeated, since x_2 and x_3 differ only in the fourth decimal place, and this number is uncertain anyway. The free bromide concentration is $0.044M$.

7. Calculate the percentage of carbonic acid present as CO_3^{2-} and HCO_3^- in a solution at pH 9.3. The dissociation constants of carbonic acid are $\log k_1 = -6.3$ and $\log k_2 = -10.3$ ($I = 0$).

The calculation is done by using Eqs. (1.42) and (1.43)

$$\Phi_{CO_3} = \frac{1}{1 + 10^{-9.3} \times 10^{10.3} + 10^{-18.6} \times 10^{16.6}} = \frac{1}{11} \approx 0.09;$$

$$\Phi_{HCO_3} = \frac{10^{-9.3} \times 10^{10.3}}{11} = 0.91.$$

Therefore 91% of the carbonic acid is present as HCO_3^- and 9% as CO_3^{2-} at pH 9.3.

8. Calculate the percentage of nickel present as the free ion, and as the thiocyanato and hydroxo-complexes in a solution which is $0.1M$ in potassium thiocyanate, and contains a small concentration of nickel at pH 7 and at pH 9 (adjusted with potassium hydroxide).

The overall stability constants are: for the [thiocyanatonickel(II)] complexes, $\log \beta_1 = 1.2$; $\log \beta_2 = 1.6$ and $\log \beta_3 = 1.8$ [20]; for the [hydroxonickel(II)] complex, $\log \beta_1 = 4.6$ [21].

The calculation is done by using Eq. (1.24). At pH 7 the hydroxide concentration $[OH^-] = 10^{-7}$.

$$\Phi_{Ni} = \frac{1}{1 + 10^{-1} \times 10^{1.2} + 10^{-2} \times 10^{1.6} + 10^{-3} \times 10^{1.8} + 10^{-7} \times 10^{4.6}}$$

$$= \frac{1}{1 + 1.58 + 0.40 + 0.06 + 0.004} = \frac{1}{3.044} = 0.328$$

The first term in the denominator of the fraction is the relative amount of the free metal ion, the second, third and fourth are those of the metal ion in the thiocyanato-complexes and the last the relative amount of nickel ion in the hydroxo-complex. The mole fractions of the thiocyanato and hydroxo-complexes are as follows:

$$\Phi_{Ni(SCN)_x} = \frac{1.58 + 0.40 + 0.06}{3.044} = \frac{2.04}{3.044} = 0.671$$

$$\Phi_{Ni(OH)} = \frac{0.004}{3.044} \simeq 0.001$$

Thus at pH 7, 67·1% of the nickel(II) is in the form of thiocyanato-complexes, 32·8% is present as free metal ion, and only 0·1% as hydroxo-complex. For pH 9 the calculation is done similarly:

$$\Phi_{Ni} = \frac{1}{1 + 10^{-1} \times 10^{1 \cdot 2} + 10^{-2} \times 10^{1 \cdot 6} + 10^{-3} \times 10^{1 \cdot 8} + 10^{-5} \times 10^{4 \cdot 6}}$$

$$= \frac{1}{1 + 1 \cdot 59 + 0 \cdot 4 + 0 \cdot 06 + 0 \cdot 4}$$

$$\Phi_{Ni(SCN)_x} = \frac{2 \cdot 05}{3 \cdot 45} \sim 0 \cdot 59$$

$$\Phi_{Ni(OH)} = \frac{0 \cdot 4}{3 \cdot 45} \sim 0 \cdot 12$$

At pH 9, 59% of the total nickel is in the form of thiocyanato-complexes and 12% in that of the hydroxo-complex.

9. Calculate the limiting concentration below which the hydroxo-complexes of iron(III) can be considered as practically mononuclear. The overall stability constants of the mononuclear [hydroxo-iron(III)] complexes are $\log \beta_1 = 11 \cdot 0$; $\log \beta_2 = 21 \cdot 7$. The overall stability constant of the binuclear hydroxo-complex is $\log \beta_{2,2} = 25 \cdot 1$ [22].

The calculation is done by using the inequality (1.31)

$$\log [Fe] < 21 \cdot 7 - 25 \cdot 1 - 0 \cdot 3 = -3 \cdot 7$$

The formation of binuclear complexes need not be considered at iron(III) concentrations lower than $10^{-3 \cdot 7} = 2 \times 10^{-4}$ M.

10. Calculate the percentage of univalent and tervalent thallium in a solution of pe = 20 at equilibrium. The constant of the redox couple thallium(III) − thallium(I) is $\log K_r = 42 \cdot 6$ [23].

The calculation is done by using Eq. (1.59).

$$\Phi_{Tl(III)} = \frac{1}{1 + (10^{-20})^2 \times 10^{42 \cdot 6}} = \frac{1}{1 + 10^{2 \cdot 6}} = \frac{1}{399} \sim 0 \cdot 0025$$

Thus 0·25% of the thallium is in the tervalent form and 99·75% in the monovalent form.

1.3. Conditional Equilibrium Constants

The dependence of stability constants of complexes on the ionic strength has already been dealt with. Much more important is that very rarely in practice are the metal ions and ligands involved in the complexation the only species present. The constituents of the complex may also take part in equilibrium reactions with other species present, including ions produced by the solvent. The ligand may be protonated or be competed for by other metal ions, and the real concentration of free ligand may differ widely in solutions of the same total ligand concentration.

On the other hand the metal ion may interact with other complexing ligands present, including hydroxide ions, the concentration of free metal ion thereby being reduced.

The analytical chemist, however, is usually not interested in the amount or concentration of all the species present. The degree of completeness of the main reaction is much more important. If, for example, magnesium is titrated with ethylenediaminetetra-acetate (EDTA) it is not important in which form the untitrated magnesium is present (amine, hydroxo, acetato or aquo-complex). Similarly, the extent of protonation of the EDTA ligand is also unimportant. The analytical chemist is only interested in the fraction of the magnesium originally present that reacts with the titrant.

Schwarzenbach [24] introduced the concept of the 'apparent equilibrium constant' in connection with the theory of complexometric titrations. Ringbom [15] developed the theory further, and extended the concept to other fields of analytical chemistry, so that the optimum conditions of analytical operations could be calculated and predicted by using the complex formation functions. The name 'conditional constant' first introduced by Ringbom is recommended, since it well expresses the notion of a term which can be assumed as constant only under given conditions (concentrations, pH). It is now accepted and widely used in the English literature.

The general form of the overall conditional stability constant for a mononuclear complex [cf. Eq. (1.11)] is

$$\beta'_n = \frac{[ML_n]}{[M'][L']^n} \qquad (1.69)$$

where [M'] is the apparent free metal ion concentration, i.e. the concentration of metal ion that has not reacted with complexant L. Similarly, [L′] is the apparent ligand concentration, i.e. the concentration of the ligand not bound to the central ion M, whether L is in protonated or non-protonated form, or in the form of complexes with other metal ions.

To find the relationship between the apparent and real free metal ion concentration, let Y be the interfering complexant which reduces the free metal ion concentration. Therefore

$$\frac{[M']}{[M]} = \frac{[M] + [MY] + [MY_2] + \cdots}{[M]}$$

By use of the overall stability constants, γ_n of the complexes formed in the side-reaction, this equation can be written as

$$\frac{[M']}{[M]} = 1 + [Y]\gamma_1 + [Y]^2\gamma_2 + \cdots = \alpha_{M(Y)} \tag{1.70}$$

The quantity $\alpha_{M(Y)}$ is called the side-reaction coefficient or α-coefficient for M reacting with Y.

If the γ values and the concentration of Y are known, the ratio [M′]/[M] i.e. the value of $\alpha_{M(Y)}$, can be calculated.

The ratio [L′]/[L] can be similarly calculated for the side-reaction of L with B:

$$[L']/[L] = 1 + \gamma_1[B] + \gamma_2[B]_2 + \cdots = \alpha_{L(B)} \tag{1.71}$$

The side-reaction of the ligand is very often protonation. In such cases $\alpha_{L(H)}$ is given by means of the protonation constants:

$$\alpha_{L(H)} = 1 + [H^+]K_1 + [H^+]^2 K_1 K_2 + \cdots \tag{1.72}$$

If [M′] and [L′] from (1.70) and (1.71) are inserted into (1.69), we obtain

$$\beta'_n = \frac{[ML_n]}{[M][L]^n \alpha_{M(Y)} \alpha_{L(B)}^n}$$

By combining this with Eq. (1.11), we get:

$$\beta'_n = \frac{\beta_n}{\alpha_{M(Y)} \alpha_{L(B)}^n} \tag{1.73}$$

or

$$\log \beta'_n = \log \beta_n - \log \alpha_{M(Y)} - n \log \alpha_{L(B)}$$

Sec 1.3] Conditional equilibrium constants

If M or L are involved in more than one side-reaction the α values are summed to give an overall value:

$$\alpha_M = \alpha_{M(Y)} + \alpha_{M(X)} + \ldots - (i - 1) \quad (1.74)$$

and

$$\alpha_L = \alpha_{L(B)} + \alpha_{L(H)} + \ldots - (i - 1) \quad (1.75)$$

where i is the number of side-reactions. The correction $-(i - 1)$ is necessary since all α functions involve 1, representing free M or free L. In practice one or two α terms predominate, and the others, being smaller by some orders of magnitude, can be neglected. Similarly, $-(i - 1)$ can also often be neglected.

The complex itself may undergo side-reactions, e.g. formation of protonated complexes or mixed-ligand complexes, in which case an appropriate α-coefficient is calculated and inserted in the numerator of Eq. (1.73). (See p. 62.)

The α functions are the reciprocal values of the partial mole-fractions of the free metal ion and free ligand, referred to the amount which does not take part in the main reaction [cf. Eqs (1.21) and (1.42)]:

$$\alpha_M = \frac{1}{\Phi'_M} ; \quad \alpha_L = \frac{1}{\Phi'_L} \quad (1.76)$$

If no side-reactions are involved, the value of the α function is 1. In such cases $[M] = [M']$ and $[L] = [L']$. In other cases α may be as large as 10^{20} or more.

The importance of conditional constants is that they can be applied in calculations like true constants, to yield answers that refer to particular conditions.

In Figs. 1.10 and 1.11, $\log \alpha_{M(OH)}$ and $\log \alpha_{M(NH_3)}$ values are plotted vs. the logarithm of the ligand concentration for various metal ions. In Figs. 1.12–1.14 the $\log \alpha_{L(H)}$ values are plotted vs. pH for some complex-forming agents frequently used in analytical chemistry. For constructing the curves the complex products and protonation constants given at the end of this book (see Section 4.1) have been used.

In a similar manner, conditional precipitation and redox constants etc. can be defined.

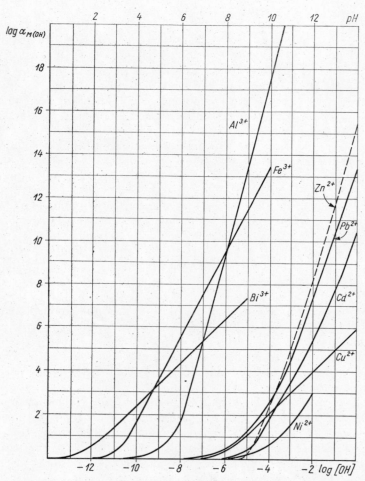

Fig. 1.10. Log $\alpha_{M(OH)}$ values for some metal ions as a function of the logarithm of the ligand concentration for mononuclear complexes

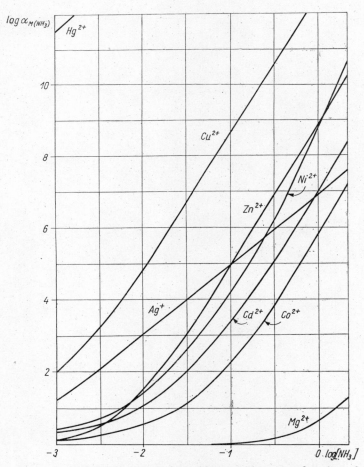

Fig. 1.11. Log $\alpha_{M(NH_3)}$ values for some metal ions as a function of the logarithm of the ammonia concentration [15]

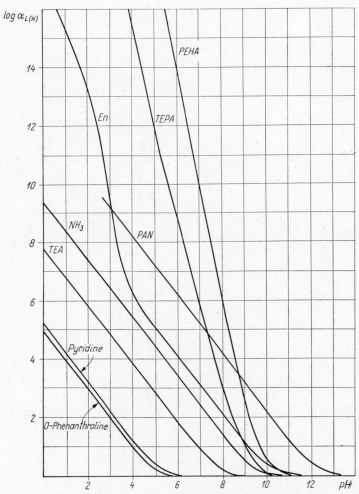

Fig. 1.12. Log $\alpha_{L(H)}$ values for some ligands important in analytical chemistry, as functions of pH. PEHA: pentaethylenehexamine; TEPA: tetraethylenepentamine; En: ethylenediamine; TEA: triethanolamine; PAN: 1-(2-pyridylazo)-2-naphthol

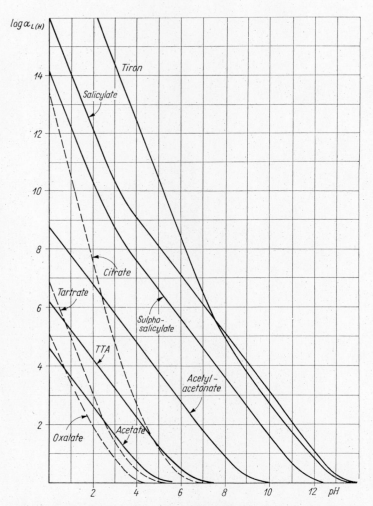

Fig. 1.13. Log $\alpha_{L(H)}$ values of some ligands important in analytical chemistry, as functions of pH. TTA: thenoyl-trifluoroacetonate

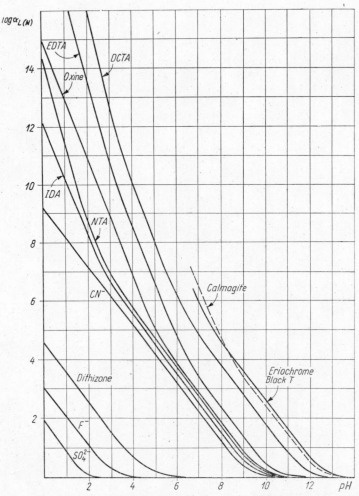

Fig. 1.14. Log $\alpha_{L(H)}$ values of some ligands important in analytical chemistry, as functions of pH. IDA: iminodiacetate; NTA: nitrilotriacetate; EDTA: ethylenediaminetetra-acetate; DCTA: 1,2-diaminocyclohexanetetra-acetate

Sec 1.3] Conditional equilibrium constants 59

1.3.1. Role of the hydrogen ion concentration in complex formation

The hydrogen ion concentration plays an important part in complex formation. Because ligands are usually fairly strong Brønsted bases, and hence easily protonated, the free ligand concentration depends on the pH. In the absence of complexing metal ions, the following relationship holds:

$$[L] = \frac{C_L}{\alpha_{L(H)}} \qquad (1.77)$$

i.e.

$$\log [L] = \log C_L - \log \alpha_{L(H)}$$

The pH above which the ligand is less than 1% protonated depends on the value of the first protonation constant, and is given by

$$\text{pH} \geq 2 + \log K_1 \qquad (1.78)$$

In this case $\alpha_{L(H)} \leq 1\cdot 01$.

In the presence of a metal ion which forms mononuclear complexes, the total ligand concentration is given as follows [see Eq. (1.16)]:

$$C_L = [L] + \beta_1[M][L] + 2\beta_2[M][L]^2 \ldots + [H][L]K_1$$
$$+ [H]^2[L]K_1 K_2' + \ldots$$

i.e.

$$C_L = [M] \sum_{i=1}^{i=N} i\beta_i[L]^i + [L]\alpha_{L(H)} \qquad (1.79)$$

If $C_L \gg C_M$ (practically if $C_L > 20 C_M$), the first term on the right-hand side of Eq. (1.79) may be neglected and the concentration of the free ligand calculated by using Eq. (1.77).

The amount of ligand bound in the complex is given by the first term on the right-hand side of Eq. (1.79). Dividing this by the total metal ion concentrations gives the average number of ligands bound by one metal ion:

$$\bar{n} = \frac{C_L - [L]\alpha_{L(H)}}{C_M} \qquad (1.80)$$

Equation (1.17) is modified in this way if protonation of the ligand is to be taken into account.

Another pH-dependent phenomenon to be considered when dealing with aqueous solutions is the formation of hydroxo-complexes, which may be looked upon as a side-reaction of the metal ion, and has to be taken into account when calculating conditional equilibrium constants. Alkali and alkaline earth metal ions do not form stable hydroxo-complexes. The highest of their stability constants is that of the magnesium complex ($\log K_1 = 2\cdot 6$), but the solubility of magnesium hydroxide is small. For the alkali and alkaline earth metal ions $\alpha_{M(OH)} \simeq 1$ over a wide pH range. Most other metal ions, however, have a greater tendency to form hydroxo-complexes. Multivalent metal ions are often present as hydroxo-complexes even in acid solutions.

The hydroxo-complexes are really formed by the dissociation of protons from the corresponding aquo-complexes. The formation of the [hydroxo-iron(III)] complexes can be described as follows:

$$[Fe(H_2O)_6]^{3+} \underset{}{\overset{k_1}{\rightleftharpoons}} [Fe(H_2O)_5OH]^{2+} + H^+;$$

$$[Fe(H_2O)_5OH]^{2+} \underset{}{\overset{k_2}{\rightleftharpoons}} [Fe(H_2O)_4(OH)_2]^+ + H^+;$$

$$2[Fe(H_2O)_6]^{3+} \underset{}{\overset{k_{2,2}}{\rightleftharpoons}} [Fe_2(H_2O)_8(OH)_2]^{4+} + 2H^+ + 2H_2O;$$

where $\log k_1 = -3\cdot 0$; $\log k_2 = -3\cdot 3$ and $\log k_{2,2} = -2\cdot 9$. In calculations only the overall formation constants will be used:

$$Fe^{3+} + OH^- \underset{}{\overset{\beta_1}{\rightleftharpoons}} Fe(OH)^{2+}$$

$$Fe^{3+} + 2OH^- \underset{}{\overset{\beta_2}{\rightleftharpoons}} Fe(OH)_2^+$$

$$2Fe^{3+} + 2OH^- \underset{}{\overset{\beta_{2,2}}{\rightleftharpoons}} Fe_2(OH)_2^{4+}$$

where, since $\beta_1 = [Fe(OH)^{2+}]/[Fe^{3+}][OH^-] = k_1/[H^+][OH^-]$, etc.,

$\log \beta_1 = \log k_1 + 14 = 11\cdot 0$;

$\log \beta_2 = \log k_2 + \log \beta_1 + 14 = 21\cdot 7$,

$\log \beta_{2,2} = \log k_{2,2} + 2 \times 14 = 25\cdot 1$.

In cases where only the formation of mononuclear complexes is to be considered, the hydroxo-complex formation can be taken into account simply by using the α function in equilibrium calculations. According to literature data, virtually only mononuclear complexes are formed in dilute solutions of bivalent transition metal ions. However, this is not the case with other metal ions. When polynuclear complexes are also formed, the calculation becomes complicated, since the extent of formation depends not only on the concentration of the ligand but also on that of the metal ion. Equation (1.70) cannot be applied (see Example 40).

Fortunately, in analytical procedures it frequently occurs that the hydroxo-complex is formed in the presence of an excess of an effective complexant, which reduces the free metal ion to a concentration below that corresponding to the mononuclear wall and only the mononuclear complex need be considered (see Example 9). Another favourable factor is that polynuclear complexes usually form within a certain pH range, and in strongly acidic or alkaline solutions only mononuclear complexes are present.

The pH below which practically no hydroxo-complexes are formed depends on the stability constant of the hydroxo-complex first formed, since the first stability constant is usually the greatest. For the function

$$\alpha_{M(OH)} = 1 + \beta_1[OH^-] + \ldots$$

to be smaller than 1·01, the second term must be smaller than 10^{-2}. That is,

$$\log \beta_1 + \log [OH] \leq -2$$

hence

$$pH \leq 14 - 2 - \log \beta_1 \qquad (1.81)$$

In formation of complexes with multidentate ligands, other pH-dependent side-reactions may occur. Multivalent metal ions tend to form protonated complexes, basic complexes and other mixed-ligand complexes. For example, mercury(II) forms complexes with EDTA of composition $HgHL^-$, $Hg(OH)L^{3-}$, and $Hg(NH_3)L^{2-}$, in addition to the 'regular' HgL^{2-} species.

In calculating the conditional stability constant, the protonated and mixed-ligand complexes can also be taken into consideration if the stability constants are known.

If we are interested in the normal metal-ligand complex only and all other reactions are considered as side-reactions, the formation of the protonated and mixed-ligand species are taken into account by using the functions [15]:

$$\alpha_{M(HL)} = 1 + [H][L]\beta_{MHL} \qquad (1.82)$$

and

$$\alpha_{M(YL)} = 1 + [Y][L]\beta_{MYL} \quad (1.83)$$

where Y is the second ligand (hydroxide ion, ammonia, etc.). The α values are added to the α values for the other side-reactions of the metal ion [see Eq. (1.74)]. Such a case is encountered in a spectrophotometric determination when only the normal metal-ligand complex absorbs at the wavelengths used.

Another case frequently encountered is where the actual composition of the complex does not matter so long as M and L are present in it in the same molar ratio, e.g. 1 : 1. In this case, the value of the conditional stability constant depends on the degree of formation of all the 1 : 1 complexes, irrespective of their composition. The ratio of the total concentration of 1 : 1 complexes to that of the normal metal-ligand complex is:

$$\frac{[ML']}{[ML]} = \frac{[ML] + [MHL] + \ldots + [MYL]}{[ML]}$$

$$= 1 + [H]K_{MHL} + \ldots + [Y]K_{MYL} = \alpha_{ML} \quad (1.84)$$

The following relationships exist between the stability constants used in Eq. (1.84) and in Eqs. (1.82) and (1.83):

$$K_{MHL} = \frac{\beta_{MHL}}{K_{ML}} = \frac{[MHL]}{[ML][H]} \quad (1.85)$$

$$K_{MYL} = \frac{\beta_{MYL}}{K_{ML}} = \frac{[MYL]}{[ML][Y]} \quad (1.86)$$

In the calculation, the conditional stability constant is multiplied and not divided by α_{ML}:

$$K'_{ML} = \frac{\alpha_{ML}}{\alpha_M \alpha_L} K_{ML} \quad (1.87)$$

or

$$\log K'_{ML} = \log K_{ML} + \log \alpha_{ML} - \log \alpha_M - \log \alpha_L$$

In Fig. 1.15 the conditional stability constants calculated for the complexes of various metal ions with EDTA are plotted against pH. All the curves in the figure have a maximum. Initially, protonation of the ligand decreases with pH, favouring formation of the complex, and thus in-

Sec. 1.3] Conditional equilibrium constants 63

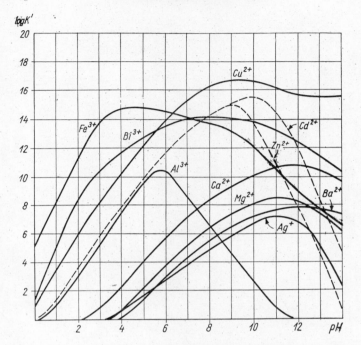

Fig. 1.15. Conditional stability constants calculated for the EDTA complexes of various metal ions as functions of pH [15]

creasing the conditional stability constant. At higher pH values the formation of hydroxo-complexes of the metals, as a side-reaction, reduces the values of the conditional stability constants of the EDTA complexes. The effect of the formation of hydroxo-complexes is especially great in the case of aluminium(III) and mercury(II). The maxima of the curves are lower than the value of the thermodynamic stability constants, except for calcium, magnesium and barium for which the conditional stability constant at pH 12 is practically equal to the thermodynamic value, as both α_M and $\alpha_{L(H)}$ approach 1. The shape of the curve for copper(II) is due to the formation of basic complexes.

1.3.2. WORKED EXAMPLES

11. Calculate the conditional stability constant of the EDTA-nickel(II) complex in a solution which is $0\cdot 1M$ in ammonia and $0\cdot 1M$ in ammonium chloride, given that the stability constant of EDTA-nickel(II) is

$\log K = 18{\cdot}6$ [25], and the basic dissociation constant of ammonium hydroxide is $\log k_b = -4{\cdot}65$ ($I = 0{\cdot}1$) [26].

In the calculation, the protonation of the ligand and the formation of nickel ammine complexes are taken into account [see Eqs. (1.70), (1.72) and (1.73)]. Assuming that the ammonia–ammonium chloride buffer is present in excess compared with the metal ion, the concentration of free ammonia, [NH$_3$], is approximately $10^{-1}\,M$. The pH of the solution is calculated from the basic dissociation constant of ammonium hydroxide or the protonation constant of ammonia. The protonation constant according to Eq. (1.37) is

$$\log K = 14 - 4{\cdot}65 = 9{\cdot}35;$$

i.e.

$$\frac{[\text{NH}_4^+]}{[\text{NH}_3][\text{H}^+]} = \frac{10^{-1}}{10^{-1}[\text{H}^+]} = 10^{9\cdot35};$$

$$[\text{H}^+] = 10^{-9\cdot35}, \quad \text{pH} = 9{\cdot}35.$$

The concentration of free ammonia and the pH being known, the necessary $\log \alpha$ values can be taken from Figs. 1.10, 1.11 and 1.14: $\log \alpha_{\text{Ni(NH}_3)} = 4{\cdot}2$; $\log \alpha_{\text{Ni(OH)}} = 0{\cdot}3$. The latter may be neglected compared with $4{\cdot}2$; thus

$$\log \alpha_{\text{Ni}} = 4{\cdot}2.$$

At pH $= 9{\cdot}35$, $\log \alpha_{\text{EDTA(H)}} = 1{\cdot}0$.

Thus the logarithm of the conditional stability constant

$$\log K'_{\text{Ni(EDTA)}} = 18{\cdot}6 - 4{\cdot}2 - 1{\cdot}0 = 13{\cdot}4$$

12. Calculate the conditional stability constant of the diaminocyclohexanetetra-acetate (DCTA) barium(II) complex at pH 6, taking into account the formation of the protonated complex BaHL ($\log K_{\text{BaL}} = 8{\cdot}0$ and $\log K_{\text{BaHL}} = 6{\cdot}7$ [25]). The calculation is done by using Eq. (1.87). According to Eq. (1.84):

$$\alpha_{\text{BaL}} = 1 + 10^{-6} \times 10^{6\cdot7} = 6; \quad \log \alpha_{\text{BaL}} = 0{\cdot}8$$

From Fig. 1.14: $\log \alpha_{L(H)} = 6{\cdot}2$. The conditional stability constant is given by

$$\log K'_{\text{BaL}} = 8{\cdot}0 + 0{\cdot}8 - 6{\cdot}2 = 2{\cdot}6$$

Sec. 1.3] Conditional equilibrium constants

13. Calculate the percentage of a trace amount of erbium(III) that will be present in the form of the ErL_4^- species, at pH 2 and pH 4, in $0.1 M$ α-hydroxyisobutyric acid solution. The dissociation constant of α-hydroxyisobutyric acid is $\log k = -3.8$, and the overall complex stability constants are $\log \beta_1 = 3.0; \log \beta_2 = 5.7; \log \beta_3 = 7.57, \log \beta_4 = 9.0$ [27].

First calculate the concentration of free ligand by using Eq. (1.77), for pH 2 and pH 4, remembering that $C_{Er} \ll C_L$.

At pH 2: $\alpha_{L(H)} = 1 + \dfrac{1}{k}[H^+] = 1 + 10^{3.8} \times 10^{-2} = 64.1$

At pH 4: $\alpha_{L(H)} = 1 + 10^{3.8} \times 10^{-4} = 1.63$

$$\log [L]_{pH2} = \log C_L - \log \alpha_{L(H)} = -1 - 1.81 = -2.82$$
$$\log [L]_{pH4} = -1 - 0.21 = -1.21$$

The calculation is continued by using Eq. (1.20).

$$\Phi_{4(pH\,2)} =$$
$$\dfrac{10^9 \times (10^{-2.82})^4}{1 + 10^3 \times 10^{-2.82} + 10^{5.7} \times (10^{-2.82})^2 + 10^{7.57} \times (10^{-2.82})^3 + 10^9 \times (10^{-2.82})^4}$$
$$= \dfrac{0.005}{1 + 1.51 + 1.15 + 0.13 + 0.005} = 0.0013$$

$$\Phi_{4(pH\,4)} =$$
$$\dfrac{10^9 \times (10^{-1.21})^4}{1 + 10^3 \times 10^{-1.21} + 10^{5.7} \times (10^{-1.21})^2 + 10^{7.57} \times (10^{-1.21})^3 + 10^9 \times (10^{-1.21})^4}$$
$$= \dfrac{1.45 \times 10^4}{2.51 \times 10^4} = 0.58$$

Thus at pH 2, 0.13%, and pH 4, 58% of the erbium(III) is present in the form of the four-coordinated species ErL_4^-.

14. Calculate the percentage of barium (II) present in the form of BaHL and BaL at pH 6 in a solution in which the total concentration of both barium(II) and the complexant DCTA is $10^{-2} M$. For the stability constants see Example 12.

The free ligand concentration is calculated first. As the total metal concentration cannot be neglected in comparison with the total ligand concentration, [L] is obtained by solving the following equation [see Eqs. (1.19) and (1.80)]:

$$\frac{K[L] + \beta[L][H^+]}{1 + K[L] + \beta[L][H^+]} = \frac{C_L - [L]\alpha_{L(H)}}{C_{Ba}}$$

where $K = K_{BaL} = 10^{8 \cdot 0}$ and $\beta = K_{BaL}K_{BaHL} = 10^{14 \cdot 7}$ [see Eq. (1.85)]. At pH 6 log $\alpha_{L(H)} = 6 \cdot 2$ (see Example 12).

Then on substitution and rearrangement,

$$C_{Ba}K[L] + C_{Ba}\beta[L][H] = C_L + C_L K[L] + C_L \beta[L][H]$$
$$- [L]\alpha_{L(H)} - K[L]^2\alpha_{L(H)} - \beta[L]^2[H]\alpha_{L(H)}$$

$10^{-2} \times 10^8[L] + 10^{-2} \times 10^{14 \cdot 7} \times 10^{-6}[L] = 10^{-2} + 10^{-2} \times 10^8[L] +$
$10^{-2} \times 10^{14 \cdot 7} \times 10^{-6}[L] - 10^{6 \cdot 2}[L] - 10^8 \times 10^{6 \cdot 2}[L]^2 - 10^{14 \cdot 7} \times 10^{-6} \times 10^{6 \cdot 2}[L]^2$

which simplifies to

$$9 \cdot 5 \times 10^{14}[L]^2 + 1 \cdot 58 \times 10^6[L] - 10^{-2} = 0$$

so

$$[L] = 2 \cdot 5 \times 10^{-9} M$$

The mole-fractions of the two complex species can be obtained as follows:

$$\Phi_{BaL} = \frac{[L]K}{1 + [L]K + [L][H]\beta}$$

$$= \frac{2 \cdot 5 \times 10^{-9} \times 10^8}{1 + 2 \cdot 5 \times 10^{-9} \times 10^8 + 2 \cdot 5 \times 10^{-9} \times 10^{-6} \times 10^{14 \cdot 7}}$$

$$= \frac{0 \cdot 25}{1 + 0 \cdot 25 + 1 \cdot 25} = 0 \cdot 10$$

$$\Phi_{BaHL} = \frac{1 \cdot 25}{1 + 0 \cdot 25 + 1 \cdot 25} = 0 \cdot 50$$

Thus 10% of the barium(II) is complexed as BaL, 50% as BaHL and 40% is uncomplexed.

15. Calculate the pH above which diethylenetriamine is not more than 1% protonated. The protonation constants are log $K_1 = 9 \cdot 94$;

$\log K_2 = 9\cdot 13$; $\log K_3 = 4\cdot 33$ [28]. The calculation is done by using inequality (1.78):
$$pH \geq 9\cdot 94 + 2$$
i.e.
$$pH \geq 11\cdot 94.$$

16. Calculate the maximum pH at which a $0\cdot 02M$ cadmium nitrate solution has at least 99% of its cadmium present as the aquo-complex only, i.e. <1% is present as a hydroxo-complex. The stability constant of the $CdOH^+$ complex is $\log \beta_1 = 4\cdot 3$ [29]. The calculation is done by using condition (1.81)
$$pH \leq 14 - 2 - 4\cdot 3$$
i.e.
$$pH \leq 7\cdot 7.$$

1.4. Factors Affecting Complex Formation

Metal ions form complexes of different composition, structure and stability with various complexing ligands. It is important for an analytical chemist to know what kind of complexes of what stability may be expected. As the selectivity of complex formation reactions depends on the stability of the complex formed, it is important to know the factors which affect the formation of complexes.

The solubility, absorption of light and the magnetic properties of complexes depend on the metal ion and ligand, the structure of the complex and the nature of the chemical bonds. It is usual to classify metal ions and ligands according to the nature of the complex they form. This classification and the properties of the complexes can be interpreted in terms of modern bond theories.

According to *Pauling's valence-bond theory*, bonds in complexes are formed by the incorporation of free electron pairs of the ligands into orbitals of the metal ion. Hybridization of the orbitals occupied makes the bonds to the ligands equivalent in complexes containing more than one ligand. The geometry, magnetic properties and reactivity of complexes may be explained by means of the theory. Compounds containing unpaired electrons are paramagnetic, whereas those containing paired electrons only are diamagnetic.

Theories by which the behaviour of complexes can be explained and some physico-chemical properties can be described quantitatively are the *crystal-field* and *ligand-field theories*. The crystal-field theory is suitable for describing complexes in which the interaction between the ligand and metal is mainly electrostatic, whereas the ligand-field theory considers bonds of greater covalency, using the molecular orbital method. The method of linear combination of atomic orbitals to form molecular orbitals (LCAO-MO) (although it can be considered as the most exact bond theory available at present) is not suitable for the quantitative treatment of the structure and properties of complicated complexes.

According to the crystal-field theory an electrostatic bond is formed between the positively charged metal ion and the negatively charged ligands or ligands with negative polarity. The chemical and physical properties of the complex formed and the absorption spectrum of the compound in solution depend mainly on the change caused by the ligands in the electronic energy levels of the metal ion. In the case of transition metals, splitting of the d orbitals is accompanied by an energy change which stabilizes the complex formed. The wavelength of the absorption maximum for the complex can be calculated from this energy change, or conversely the orbital-splitting and bond energies can be calculated from the shifts in the spectrum. The ligands may be arranged in the following order of increasing electric field strength and orbital-splitting ability:

$$I^- < Br^- < Cl^- < F^- < OH^- < H_2O < SCN^- < NH_3 < CN^-$$

As the splitting ability increases, the spectral bands are shifted to higher energies, i.e. to shorter wavelengths. The magnetic properties of the complexes can be explained by the crystal-field theory similarly to the valence-bond theory. The 'outer and inner-sphere' electron configurations in the valence-bond theory correspond to the 'high spin and low spin' states in the crystal- and ligand-field theory. In some cases the coordinate bonds are especially strong and cannot be accounted for by simple electrostatic interaction. In these cases either the filled orbitals of the ligand penetrate the p or d orbitals of the metal, or the filled orbitals of the metal overlap unfilled or partially filled orbitals of the ligand. By this back donation a π-bond can be formed.

The bond theory of complex formation will not be treated here in further detail. For a more detailed study see refs. [30, 31]. Factors affecting the stability of complexes are dealt with in the book by Beck [32].

1.4.1. The nature of the central ion

Metal ions may be classified into three groups according to their complex-forming ability.

The first group consists of metal ions with a noble-gas electronic configuration, such as alkali metal, alkaline earth metal and lanthanide and actinide ions which are of similar behaviour. All are generally poor complex-forming ions, with mainly electrostatic bonding. As the metal ions have no low-level d orbitals or only filled d orbitals, crystal-field stabilization cannot occur. These metal ions mainly bind small anions, particularly fluoride, and form complexes with ligands containing oxygen donor atoms. They thus have a strong tendency to form aquo-complexes and do not form ammine complexes in aqueous solution. Similarly, they form neither complexes nor precipitates with sulphide. In their complexes with EDTA-type chelating agents they also bind the nitrogen-containing groups. As the bonds are mainly ionic, the higher the charge density in the metal ion the more stable the complex. The orders of stabilities of the complexes of metals in the same group of the periodic table with the same ligand are generally as follows:

$$Li > Na > K > Rb > Cs; \quad Mg > Ca > Sr > Ba;$$

$$Lu > Tm > Er > Ho > Dy > Tb > Gd > Sm > Ce > La,$$

although there are exceptions in the case of the alkaline earth metals, e.g. $Ca > Sr > Ba \simeq Mg$ with EDTA as ligand.

The second, smaller, group contains those transition metal ions having d^{10} or d^8 configurations. This includes copper(I), silver(I), gold(I), mercury(II), platinum(II) and palladium(II), the ions of which are readily deformable and capable of forming covalent bonds. In contrast to the ions of the first group, they form highly stable complexes. In the first group the charge and ionic radius affect complex formation, whereas in the second one the electronegativity is of importance.

The lower the electronegativity of the donor atom in the ligand, the stronger is the bond in the complex. Metals of this group form stable complexes with ligands containing sulphur, arsenic and phosphorus. With the halogens the bond with fluoride is the weakest. In complexes of high stability, π-bonding may occur. If the d orbitals of the metal ion are full, the π-bonding electrons come from the metal ion. For the bond to be formed the ligand must have free p or d orbitals capable of accepting the electrons. Fluoride has no d orbitals available and its p orbitals are full, so cannot form a π-donor bond. For example π-donor bonds occur in the cyano- and chloro-complexes of copper(I), zinc(II) and gallium(III).

The third, large group contains transition metal ions with partially filled orbitals. Depending on the number of d electrons the properties of these ions lie between those of the two preceding groups. From the point of view of complex stability, the charge and radius of the ion as well as the stabilization due to orbital splitting are of importance. The covalency of the bond and the stability of the complexes formed with the same ligand generally increase with increasing oxidation number of the metal ion. Thus the [hexacyanoferrate(III)] complex is more stable than the similar [hexacyanoferrate(II)] complex. Exceptions occur when the lower oxidation state is spin-paired and the higher state is not, e.g. the 1,10-phenanthroline iron(II) complex (low-spin) is more stable than the high-spin iron(III) complex. The metal ions of this group mainly bind nitrogen- and oxygen-containing ligands. According to the investigations of Irving and Williams [33] the order of complex-forming ability in the series of similar, divalent metal ions is as follows:

$$Mn < Fe < Co < Ni < Cu > Zn$$

The order can be accounted for by the decrease in metal-ion radius from Mn to Zn, and by the crystal-field stabilization energy increasing from Fe to Cu. The d orbitals of zinc(II) are full, so no stabilization energy is released in its complex formation. This is why the order changes after copper. The order is mainly valid if the donor atom is nitrogen or oxygen and the coordination number is four but is upset in the case of ligands which may form π-donor bonds with zinc(II), e.g. chloride.

If complex formation is considered as a Lewis acid–base reaction, then, according to Pearson [34, 35], the metal ions in the first group are 'hard' acids whereas those in the second one are 'soft' acids. Hard acids are characterized by high electronegativity and low polarizability. The reverse is true for soft acids. Complexes of high stability are formed between hard acids and hard bases and soft acids and soft bases. The properties of the third group are intermediate between those of the first and second groups.

1.4.2. THE PROPERTIES OF THE LIGAND

The nature of the donor atom is of importance in determining the complex stability. The donor atoms in the ligands are elements on the right-hand side of the periodic table. These are, in order of

Sec. 1.4] Factors affecting complex formation

increasing electronegativity,

$$\text{As, P} < \text{C, Se, S, I} < \text{Br} < \text{N, Cl} < \text{O} < \text{F}$$

The metal ions of the first group (hard acids according to Pearson) are attracted mainly by the donor atoms at the right-hand end of the series, and those of the second group by atoms at the left-hand end of the series. The order of stabilities of the halogen complexes of ions in the first group is: $F > Cl > Br > I$, the reverse order being the case for the second group. According to Pearson metal ions of the first group form stable complexes with hard bases (F^-, OH^-, PO_4^{3-}, SO_4^{2-}, Cl^-, CO_3^{2-}, NO_3^-, NH_3, N_2H_4, R_2O, ROH, etc.), and those of the second group with soft bases (R_2S, RS^-, I^-, SCN^-, R_3P, CN^-, CO, C_2H_4, etc.).

There is a correlation between the complexing ability and Brønsted base strength of ligands of similar structure and containing the same donor atom. The smaller the dissociation constant of the conjugated acid, the stronger is the ligand as a base and generally the greater is the stability of the complex formed. Perchlorate is a very weak base and thus a poor complexant. The correlation between basic strength and complexing ability does not hold in the case of ligands of different structures containing different donor atoms. For example, copper(II) forms a more stable complex with phthalate than with the less basic 3-nitrophthalate of similar structure. Aluminium(III) forms a more stable complex with weakly basic fluoride than with strongly basic cyanide.

The complexes formed are especially stable if the ligand is of chelating character. Chelates can be formed by ligands which contain at least two donor atoms, thus forming a ring by the coordination at two sites. It is worth mentioning, however, that in the case of silver(I) with coordination number two, complexes with unidentate ligands are more stable than those with multidentate ones. The reason for this is the linear configuration of the complexes with unidentate ligands.

Aminopolycarboxylic acids are especially good chelating agents. A good example is ethylenediaminetetra-acetic acid (EDTA), which is a hexadentate ligand containing oxygen and nitrogen as donor atoms:

$$\begin{array}{c} \text{HOOC}-\text{CH}_2 \\ \text{HOOC}-\text{CH}_2 \end{array} \!\!\!\! N-\text{CH}_2-\text{CH}_2-N \!\!\!\! \begin{array}{c} \text{CH}_2-\text{COOH} \\ \text{CH}_2-\text{COOH} \end{array}$$

EDTA forms octahedral chelates with di- and trivalent metal ions while

DCTA forms even more stable complexes and TTHA is a decadentate complexant but for all ten donor groups of TTHA to be utilized

1,2-Diaminocyclohexanetetra-acetic acid (DCTA)

Triethylenetetraminehexa-acetic acid (TTHA)

the solution has to be satisfactorily alkaline, since owing to the greater number of acid groups the proton affinity of this compound is greater than that of the other compounds mentioned.

If sulphur is also present the chelating agent will be selective for the metal ions of the second group.

The stability of chelates also depends on the number of atoms in the ring formed. Chelates with 5- or 6-membered rings are the most stable. The stability also depends on the number of rings. The ethylenediamine

Sec. 1.4] Factors affecting complex formation

copper(II) complex is less stable than that with diethylenetriamine, and the latter is again less stable than the complex with triethylenetetramine. The number of rings formed per ligand molecule increases from ethylenediamine to triethylenetetramine.

Complex formation also depends on steric factors. It may happen that, although the ligand has the necessary number of donor atoms, chelate formation does not occur at all or only to a small extent, since ring formation is completely or partly hindered. Although alkylated ethylenediamine is a stronger base than ethylenediamine itself, it is a weaker complexing agent since the alkyl groups hinder ring formation.

Chelating agents may be classified according to whether the skeleton of the molecule is flexible or rigid. For example EDTA is flexible and 1,10-phenanthroline is rigid. Rigid chelating agents are more selective. There is also an entropy effect with rigid chelating agents, since there is less loss of configurational entropy when the complex is formed. The stability of chelates formed with multidentate ligands usually exceeds that of complexes formed with similar unidentate ligands by some orders of magnitude. However, care is required when comparing the stability constants of complexes formed with unidentate and multidentate ligands. It follows from the difference in the nature of complex formation that the relationship between the extent of complex formation and the free ligand concentration is different for the two cases.

Let us assume that the overall stability constants of a complex of a metal ion with four unidentate ligands, L, are $\log \beta_1 = 6$, $\log \beta_2 = 8$, $\log \beta_3 = 10$ and $\log \beta_4 = 12$, and that the stability constant of the chelate of the same metal ion with a quadridentate ligand Y is $\log K = 12$. Calculation of the ratios of the concentration of the complexed metal to the free metal ion concentration for different free ligand concentrations, and plotting of a logarithmic diagram gives the curves shown in Fig. 1.16. The broken line represents the ratio of the total concentration of the metal ion in the species ML, ML_2, ML_3 and ML_4 to that of the free metal, for given free ligand concentrations

$$\frac{\sum_{i=1}^{i=4} [ML_i]}{[M]} = [L]\beta_1 + [L]^2\beta_2 + [L]^3\beta_3 + [L]^4\beta_4$$

One of the continuous lines represents the ratio of the concentration of the metal in the species ML_4 to that of the free metal ion:

$$\frac{[ML_4]}{[M]} = [L]^4\beta_4$$

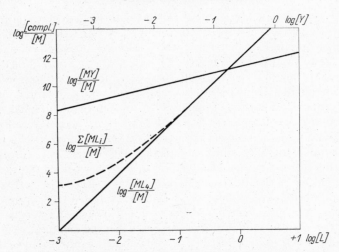

Fig. 1.16. The degree of complex formation as a function of the logarithm of the ligand concentration for a simple complex and a chelate. L: monodentate ligand; Y: tetradentate ligand; $\log \beta_{ML_4} = \log K_{MY} = 12$

As the complex formed with the tetradentate ligand has coordination number four, i.e. one molecule of Y is equivalent to four molecules of L, the molar ratio of the concentration of the complex MY to that of the free metal ion,

$$\frac{[MY]}{[M]} = [Y]K$$

was plotted with an abscissa scale corresponding to this stoichiometry, viz. $\log [Y] = \log [L] + 0.6$. By comparing the curves we can establish that more metal is bound in the chelate than in the complexes with unidentate ligand only when the free ligand concentration is less than *ca.* $10^{-1}M$ (i.e. $\log [L]$ and $\log [Y] < -1$).

1.5. Kinetics of the Reactions of Complexes

The majority of complexation calculations used in analytical chemistry are based on equilibrium relationships. Therefore it is important for the analytical chemist to know the factors which affect the rate of complex formation and the cases in which equilibrium data may safely be used in calculations.

Sec. 1.5]   Kinetics of the reactions of complexes

According to the rate at which a complex can exchange its ligand or possibly its central ion, we distinguish between inert and labile complexes. The distinction is only of practical importance, since no sharp boundary can be drawn between the two groups. It is important to note that lability is not equivalent to instability, the first being a kinetic and the second a thermodynamic term.

For example, according to equilibrium calculations [hexacyanoferrate(II)] should decompose to form hydrogen cyanide and iron(II) in acidic solution. However, the acidic solution of the complex can be preserved for hours without appreciable decomposition, because the complex is inert. On the other hand, [tetra-ammineiron(II)] is very labile, and on acidification immediately decomposes into iron(II) and ammonium ions.

In analytical procedures the rate of complex-formation is of great importance. If, for example a metal ion is titrated with a chelating agent directly, the reaction should be almost instantaneous, otherwise the endpoint is not readily observable. In other cases the inertness of the complex is advantageous to the analytical chemist. For example potassium [hexacyanoferrate(II)] can be used as a standard solution for the precipitation titration of zinc(II) in acidic solution.

A new complex may be formed by exchanging the central metal ion, or one or more ligands. The latter type of reaction generally occurs more often. The ligands are exchanged by a dissociation or a replacement mechanism. In the case of dissociation the first, slow, rate-controlling process is a monomolecular dissociation reaction which is followed by the fast binding of the new ligand:

$$ML_6 \rightarrow ML_5 + L \tag{1.88}$$

$$ML_5 + Y \rightarrow ML_5Y \tag{1.89}$$

In a displacement reaction, however, the rate-controlling process is bimolecular, which is followed by the fast transformation of the intermediate formed:

$$ML_6 + Y \rightarrow L_5M \begin{smallmatrix} \nearrow L \\ \searrow Y \end{smallmatrix} \rightarrow L_5MY + L \tag{1.90}$$

The dissociation mechanism can be expected for high-spin complexes, whereas the displacement mechanism operates in the case of complexes of metal ions having d orbitals not completely filled and which can thus increase their coordination number.

Any method which is suitable for monitoring the concentration of the reactants or products can be used for rate studies of complex formation or decomposition. Radioactive tracer techniques are particularly advantageous, as they enable the progress of the reaction to be followed as well as the location of the labelled atoms in the complex to be determined.

The mechanism of complexation reactions is treated in detail in the monograph by Basolo and Pearson [36]. Only some general statements concerning the reactivity of complexes will be given here.

(i) The complexes of the transition metals are more inert than the corresponding complexes of other metals, e.g. the EDTA-nickel(II) complex is more inert than the EDTA-calcium(II) complex.

(ii) The most inert complexes of the transition metals are those with the electronic configuration d^3, d^8, low-spin d^5 and d^6, e.g., [hexathiocyanatochromate(III)], [tetrachloroplatinate(II)], [hexacyanoferrate(III)], [hexacyanoferrate(II)].

(iii) The inert behaviour of the complexes of transition metals with the same electronic configuration increases with the group number in the periodic system. For example, the inert nature of 1,10-phenanthroline complexes increases in the order: iron(II) < ruthenium(II) < osmium(II). The electronic configurations are $3d^6$, $4d^6$ and $5d^6$.

(iv) Generally the higher the coordination number the more inert is the complex. For example [hexacyanonickelate(II)] is more inert than [tetracyanonickelate(II)].

(v) Metal chelates are more inert than complexes formed with unidentate ligands. [Bis-ethylenediaminecopper(II)] is more inert than tetraammine-copper(II).

(vi) Neutral, uncharged complexes generally react more slowly than complex ions.

(vii) Polynuclear complexes are usually much more inert than the corresponding mononuclear ones. Hydroxo-complexes, being generally polynuclear, react slowly. This is why aluminium(III) reacts slowly with EDTA in solutions of pH > 3. In titrations the reaction must be accelerated by heating, and only the indirect determination by back titration can be used.

(viii) Reactions in which a metal ion bound in a chelate is replaced, are generally slow:

$$M + NL \rightarrow ML + N \tag{1.91}$$

The reaction is particularly slow if two transition metal ions exchange place

(ix) Redox reactions of a complexed metal ion are generally faster than those of the corresponding aquo-complex. For example, the reduction of iron(III) becomes faster in the presence of complex-forming chloride or hydroxide ions.

1.6. Energy Changes Involved in Complex Formation Reactions

Thermodynamic relationships can be applied to complexation reactions. The equation

$$\Delta G^0 = -2{\cdot}303\, RT \log K \tag{1.92}$$

gives the relationship between the thermodynamic equilibrium constant and the standard free-energy change accompanying complex formation [38].

ΔG^0 is the difference between the sums of the standard free energies of the reactants and of the products; R is the universal gas constant, $1{\cdot}987$ cal.mole^{-1}.deg^{-1}; T is the absolute temperature; K is the equilibrium constant. The greater the value of K the greater the negative value of ΔG^0, i.e. the greater is the driving force of the complex formation reaction.

The standard free energy change is given by

$$\Delta G^0 = \Delta H^0 - T\Delta S^0 \tag{1.93}$$

where ΔH^0 is the enthalpy change (kcal/mole) and ΔS^0 is the entropy change (cal. mole^{-1}. deg^{-1}).

The enthalpy change accompanying complex formation can be determined by direct calorimetry or calculated from the values of the equilibrium constant at different temperatures by using Eq. (1.98).

The formation reactions of chelates have been studied extensively. Metal chelates are very stable, and accordingly, the standard free energy changes calculated by using Eq. (1.92) are very large.

In the case of complex formation with a unidentate ligand, ΔG^0 is mainly determined by the enthalpy change, as the exchange of the unidentate coordinated water molecules for other ligands does not involve a large entropy change. When, however, chelates are formed, the driving force (the so-called 'chelate effect') can be accounted for by the large increase in entropy [37].

In a statistical interpretation of the entropy change in the case of a complex formation reaction with EDTA, we can state that the release of the ordered water molecules from the aquo-complexes results in an increase of the number of free particles, i.e. in an increase of the degree of disorder. Accordingly, the entropy increases

$$M(H_2O)_6^{2+} + L^{4-} = ML^{2-} + 6H_2O \qquad (1.94)$$
$$\phantom{M(H_2O)_6^{2+}}1\quad 1\phantom{{}^{4-}}\quad 1\phantom{{}^{2-}}\quad\ 6$$

The large entropy increase in chelate formation can also be interpreted in terms of probability theory. The binding of further sites of a multidentate ligand which is already bound onto one coordination site is more probable than the binding of a new unidentate ligand particle.

1.6.1. Calculation of Equilibrium Constants from Thermodynamic Data

The equilibrium constants of complex formation or other reactions can be calculated by using Eq. (1.92) if the standard free energies of the starting materials and products are known.

The standard free energy change ΔG^0 of the reaction can be obtained by subtracting the sum of the standard free energies of formation of the reactants from the weighted sum of the standard free energies of formation of the products. (See Example 17.)

The standard free energy of formation is the free energy change accompanying the formation of a compound from its elements in the standard state. The free energy of formation of elements, the hydrogen ion and the electron in the standard state, is, by convention, zero. If only the heat of formation of a compound is known, the entropy of formation can be estimated by Latimer's method [11], and the approximate free energy of formation calculated by using Eq. (1.93).

If a chemical reaction takes place at reversible electrodes, the following relationship between the electromotive force measured between the electrodes and the free energy change of the chemical reaction is valid:

$$\Delta G = -zFE \qquad (1.95)$$

where E is the electromotive force of the cell, in volts; F the Faraday (96, 500 coulombs; 23·06 kcal V^{-1}) and z the number of electrons in-

Sec. 1.6] Energy changes in complex formation reactions

volved in the reaction. If there is a positive cell voltage ΔG is negative, and the cell reaction proceeds spontaneously.

If E is the difference between the standard potentials, i.e. $E = E_1^0 - E_2^0$, Eq. (1.95) gives the standard free energy change. If, in addition, one of the electrodes of the cell is the standard hydrogen electrode, i.e. $E_2^0 = 0.0$, then $E = E_1^0$, and by combining Eqs. (1.92) and (1.95) the relationship between the equilibrium constant of the redox reaction proceeding at the electrode and the standard potential is obtained:

$$-\Delta G^0 = zFE_1^0 = 2 \cdot 3\, RT \log K_{r1} \qquad (1.96)$$

This equation is equivalent to Eq. (1.55).

From the thermodynamic relationships given, the equilibrium constants of redox reactions can be calculated from standard redox potentials, or conversely, standard potentials or standard potential differences can be calculated from equilibrium constants (see Examples 20, 21 and 22).

From the equilibrium constants of simple chemical reactions those of composite reactions can be calculated by simple mathematical procedures (see p. 43 and Example 24). If the standard potentials of the free metal ion and complexed metal ion are known, the equilibrium constant of complex formation can also be calculated (see Example 23).

1.6.2. Temperature dependence of the equilibrium constant

Although, electrolyte equilibria in aqueous solutions at room temperature are mainly important in analytical chemistry, it is useful to know the direction and extent of the shift of equilibrium with temperature.

The relationship between the equilibrium constant and temperature is given by the van't Hoff equation,

$$\frac{d(\log K)}{dT} = \frac{\Delta H^0}{2 \cdot 3 RT^2} \qquad (1.97)$$

Assuming that ΔH^0 is constant in the temperature interval in question, Eq. (1.97) can be solved by integration

$$\log K = -\frac{\Delta H^0}{2 \cdot 3 RT} + \text{const.} \qquad (1.98)$$

Accordingly, the equilibrium constant of a heat-absorbing (endothermic) reaction increases, that of a heat-liberating (exothermic) reaction decreases with increasing temperature.

The enthalpy changes involved in the reactions of complexes are usually of the order of a few kcal/mole, thus the shift of equilibrium due to temperature changes is not very significant in aqueous solutions. It should be mentioned, however, that the enthalpy change is generally not constant even over a narrow temperature interval, and the enthalpy change for dissociation of some organic acids even changes sign at around room temperature. The ionic product of water, the stability constant of the Mg-EDTA complex, and the first and second dissociation constants of tartaric acid are shown as functions of temperature in Figs. 1.17, 1.18 and 1.19 respectively.

As well as thermodynamic equilibrium constants, activity coefficients change with temperature, and thus so do stability, protonation and other

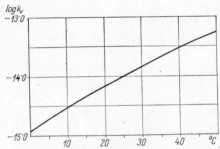

Fig. 1.17. Dependence of the ionic product of water on temperature [39]

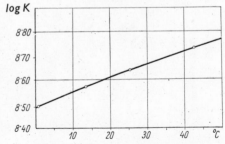

Fig. 1.18. Dependence of the stability constant of ethylenediaminetetra-acetatomagnesium(II) on temperature, according to Bohigian and Martell [40]

Sec. 1.6]   Energy changes in complex formation reactions   81

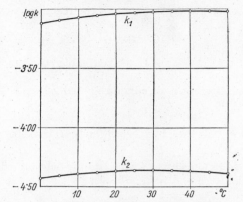

Fig. 1.19. Temperature dependence of the first and second dissociation constants of tartaric acid, according to Bates and Canham [41]

constants at a given ionic strength. For temperatures not far from room temperature, the logarithm of the activity coefficient can be calculated by using the Davies equation, the values of A in Eq. (1.4) being modified according to the temperature.

Values of A: at 15°C   0·50
             25°C   0·51
             45°C   0·53
             65°C   0·56

1.6.3. Worked examples

17. Calculate the protonation constant of ammonia at 25°C given the free energies of formation [42]:

	ΔG_f^0, kcal/mole
$NH_{3(aq)}$	− 6·30
H^+	0·0
NH_4^+	−18·93

The reaction equation is written first, viz.

$$NH_3 + H^+ \rightleftharpoons NH_4^+$$

6 Inczédy

then the standard free energy change of the reaction is calculated

$$\Delta G^0 = \Delta G^0_{fNH_4} - (\Delta G^0_{fNH_3} + \Delta G^0_{fH}) = -18.93 + 6.30$$
$$= -12.63 \text{ kcal/mole}$$

The equilibrium constant is calculated by using Eq. (1.92)

$$\log K = -\frac{\Delta G^0}{2.3\,RT} = \frac{12.63}{2.3 \times 1.987 \times 10^{-3} \times 298} = 9.26.$$

The result is in good agreement with the value determined experimentally (see Section 4.1 at the end of this book).

18. Calculate the first and second dissociation constants of carbonic acid at 25°C, using the free energies of formation [42]:

	ΔG^0_f, kcal/mole
$H_2CO_{3(aq)}$	-148.81
HCO_3^-	-140.00
CO_3^{2-}	-125.76
H^+	0.0

The equation for the first dissociation is

$$H_2CO_3 \rightleftharpoons H^+ + HCO_3^-$$

The standard free energy change is calculated as follows:

$$\Delta G^0 = (0 - 140.00) - (-148.81) = 8.81 \text{ kcal/mole}$$

Then the dissociation constant is determined by using Eq. (1.92):

$$\log k_1 = -\frac{8.81}{1.364} = -6.46.$$

The second dissociation constant is calculated similarly:

$$HCO_3^- \rightleftharpoons H^+ + CO_3^{2-}$$

$$\Delta G^0 = (0 - 125.76) - (-140.00) = 14.24 \text{ kcal/mole}$$

$$\log k_2 = -\frac{14.24}{1.364} = -10.43$$

Sec. 1.6] Energy changes in complex formation reactions

The calculated values agree well with the experimental constants (see Section 4.1).

19. Calculate the overall stability constant of [diacetatolead(II)] from thermodynamic data. The necessary free energies of formation are as follows [42, 43]:

	ΔG_f^0, kcal/mole
Pb^{2+}	-5.55
CH_3COO^-	-87.3
$Pb(CH_3COO)_2$	-184.4

The reaction equation is

$$Pb^{2+} + 2\, CH_3COO^- \rightleftharpoons Pb(CH_3COO)_2$$

The standard free energy change is given by

$$\Delta G^0 = -184.4 - (-5.55 - 2 \times 87.3) = -4.25 \text{ kcal/mole}$$

The overall stability constant is calculated by using Eq. (1.92):

$$\log \beta_2 = \frac{4.25}{1.364} = 3.12$$

The calculated value approximates to the experimental constant (see Section 4.1).

20. Calculate the equilibrium constants of the redox equations below from thermodynamic data

$$Fe^{3+} + e^- \rightleftharpoons Fe^{2+} \qquad (i)$$

$$BrO_3^- + 6H^+ + 6e^- \rightleftharpoons Br^- + 3H_2O \qquad (ii)$$

The free energies of formation necessary are as follows [42]:

	ΔG_f^0, kcal/mole
Fe^{3+}	-3.12
Fe^{2+}	-20.24
BrO_3^-	$+2.30$
Br^-	-24.6
$H_2O_{(liq)}$	-56.56
$Br_{2(aq)}$	$+0.98$
e^-	0.0

The free energy change in reaction (i) is given by

$$\Delta G^0_{Fe} = -20.24 + 3.12 = -17.12 \text{ kcal/mole}$$

The equilibrium constant [see Eq. (1.92)] is

$$\log K_{rFe(III)/Fe(II)} = \frac{17.12}{1.364} = 12.56$$

The standard redox potential of the redox couple (i), according to Eq. (1.55), is

$$E^0_{Fe} = 0.059 \times 12.56 = +0.74 \text{ V}$$

The standard free energy change in reaction (ii) is given by

$$\Delta G^0 = -24.6 - 3 \times 56.56 - 2.30 = -196.58 \text{ kcal/mole}$$

and the equilibrium constant is

$$\log K_{rBrO_3^-/Br^-} = \frac{196.58}{1.364} = 144.1$$

The standard redox potential of the redox system (ii) is

$$E^0_{BrO_3^-/\bar{Br}} = \frac{0.059}{6} \times 144.1 = 1.42 \text{ V}$$

The calculated equilibrium constants agree well with the experimentally determined values (see Section 1.4.3).

21. Calculate the equilibrium constant of the following redox reaction in aqueous solution at 25°C.

$$BrO_3^- + 6H^+ + 5Br^- \rightleftharpoons 3Br_{2(aq)} + 3H_2O$$

For the free energies of formation see Example 20.
The standard free energy change of the reaction is given by

$$\Delta G^0 = (3 \times 0.98 - 3 \times 56.56) - (2.3 - 5 \times 24.6) = -46.04 \text{ kcal/mole}$$

and the equilibrium constant according to Eq. (1.92) is

$$\log K = \frac{46.04}{1.364} = 33.7$$

Sec. 1.6]   **Energy changes in complex formation reactions**   85

The calculated value agrees fairly well with the literature value of 33·14 kcal/mole [44].

22. Calculate the equilibrium constant of the reaction given in Example 21 from the following standard redox potentials:

$$BrO_3^- + 6H^+ + 5e^- \rightleftharpoons 1/2Br_{2(aq)} + 3H_2O \quad E_1^0 = +1\cdot48 \text{ V}$$

$$1/2Br_{2(aq)} + e^- \rightleftharpoons Br^- \quad\quad\quad\quad\quad E_2^0 = +1\cdot085 \text{ V}$$

The calculation is done by using Eq. (1.95). The difference between the two standard redox potentials is the electromotive force of the standard cell in which the two electrode reactions proceed:

$$E_1^0 - E_2^0 = 1\cdot48 - 1\cdot085 = 0\cdot395 \text{ V}$$

$$-\varDelta G^0 = zF(E_1^0 - E_2^0) = 5 \times 23\cdot06 \times 0\cdot395 = 45\cdot54 \text{ kcal/mole}$$

$$\log K = \frac{45\cdot54}{1\cdot364} = 33\cdot4$$

The result obtained by using standard potentials agrees well with the value calculated in Example 21.

23. Calculate the overall stability constant of [tetracyanozincate(II)] from the following standard redox potentials:

$$Zn^{2+} + 2e^- \rightleftharpoons Zn \quad\quad\quad\quad E_1^0 = -0\cdot763 \text{ V}$$

$$Zn(CN)_4^{2-} + 2e^- \rightleftharpoons Zn + 4CN^- \quad E_2^0 = -1\cdot26 \text{ V}$$

The standard free energy change and the equilibrium constant can be calculated from the difference between the two standard potentials by using Eqs. (1.92) and (1.95):

$$-\varDelta G^0 = 2 \times 23\cdot06(-0\cdot763 + 1\cdot26) = 22\cdot92 \text{ kcal/mole}$$

$$\log \beta_4 = \frac{22\cdot92}{1\cdot364} = 16\cdot8$$

The calculated value agrees well with the measured one, given in Section 4.1 at the end of this book.

Similar results are obtained by calculating the $\log K_r$ values from the standard redox potentials by using Eq. (1.96) or (1.55) and $\log \beta_4$ from the difference of the two values (cf. p. 43).

$$\log \beta_4 = \log K_{rZn} - \log K_{rZn(CN)_4} = -25\cdot8 + 42\cdot6 = 16\cdot8$$

24. Calculate the stability constant of [tetrachloroaurate(III)], given the following redox equilibrium constants:

$$[AuCl_4]^- + 2e^- \rightleftharpoons [AuCl_2]^- + 2Cl^- \qquad \log K_{r_1} = 31\cdot 6$$
$$[AuCl_2]^- + e^- \rightleftharpoons Au_{(solid)} + 2Cl^- \qquad \log K_{r_2} = 18\cdot 8$$
$$Au^{3+} + 3e^- \rightleftharpoons Au_{(solid)} \qquad \log K_{r_3} = 76\cdot 0$$

The reaction equation for the complex formation is obtained by subtracting the first and second equations from the third:

$$Au^{3+} + 4Cl^- \rightleftharpoons AuCl_4^-$$

The logarithm of the equilibrium constant of the resultant reaction can be obtained similarly:

$$\log \beta_4 = \log K_{r_3} - (\log K_{r_1} + \log K_{r_2}) = 76\cdot 0 - 31\cdot 6 - 18\cdot 8 = 25\cdot 6$$

Compare the result with that given in Section 4.1.

1.7. Dependence of Equilibrium Constants on the Solvent

The statements made so far are true for electrolyte equilibria in aqueous solutions. Recently the equilibria in mixed and other non-aqueous solvents have become important in connection with analytical procedures. It is, however, not the purpose of this book to deal in detail with chemical reactions proceeding in solvents other than water. Here, and in Section 3.8, the author only wishes to draw the attention of the reader to some important questions which arise in connection with electrolyte equilibria in non-aqueous solutions.

Reactions taking place in non-aqueous solvents which are important in analytical chemistry are dealt with in detail in the books by Gyenes [46], and Charlot and Trémillon [47]. On addition of a solvent miscible with water to an aqueous solution, the equilibrium is shifted. Depending on the nature of the solvent, new chemical equilibria may be set up, or at least the activity of water is reduced by the addition of the solvent. Also a change in the dielectric constant of the medium causes a change in the activity coefficients of the ions. The value of A in Eq. (1.4) is closely related to the dielectric constant of the medium.

In aqueous solution the proton uptake proceeds according to the following equation:

$$A^- + H_3O^+ \xrightleftharpoons{K} HA + H_2O$$

The reaction can be broken down into the following partial reactions:

$$A^- + H^+ \xrightleftharpoons{K_A} HA$$

$$H_2O + H^+ \xrightleftharpoons{K_{H_2O}} H_3O^+$$

The protonation constant $K = \dfrac{K_A}{K_{H_2O}}$. When a solvent other than water is used, which is also capable of taking up protons but has a proton affinity greater than that of water, K will be smaller and the strength of the acid HA greater. The situation is the reverse in the case of solvents with lower proton affinity than that of water.

If the behaviour of acids and bases in various solvents is investigated from the point of view of the change in dielectric constant only, the following relationship can be derived according to Brønsted for describing the relationship between the protonation constant and the dielectric constant of the medium:

$$\log K = \log K_{aq} + z_A a \left(\frac{1}{D_{aq}} - \frac{1}{D} \right)$$

where D is the dielectric constant of the medium, z is the charge of the base, a is a constant characteristic of the compound (in the case of simple aliphatic carboxylic acids a is about 140), and D_{aq} is the dielectric constant of water at 25°C, 78·5.

As the dielectric constants of solvents used in practice are much smaller than that of water, and the sign of z is negative in the case of acids, the acid strength is noticeably reduced by adding such a solvent to an aqueous solution.

In complex formation processes, since the reaction consists of substitution for the solvent molecules coordinated to the central ion, the nature and composition of the solvent is of particularly great importance. Very often the change in stability constant with change of solvent is difficult to predict.

Electrolyte equilibria in a proton-containing non-aqueous solvent can, in principle, be treated according to the Brønsted theory, similarly to equilibria in water. In view of the great difference in dielectric constants, the solubilities, complex-formation and dissociation conditions differ markedly from those in aqueous solutions. The calculations are made more difficult by the fact that in solvents of low dielectric constant the degree of ion-pair formation is high. In calculations ion-association equilibria have to be considered in addition to the complex-formation, protonation equilibria, etc.

Polar and non-polar solvents immiscible with water are important in extraction methods of analysis. The great difference in the behaviour of water and organic solvents is very important in determining the distribution of electrolyte species between the two solvents (see Sections 3.8 and 3.9.4). The operation of liquid ion-exchangers is based on ion-pair formation in non-polar solvents.

Chapter 2

DETERMINATION OF EQUILIBRIUM CONSTANTS

As has been mentioned in the first chapter, chemical reactions can be divided into various groups. The methods for determining the equilibrium constants differ according to the type of reaction in question. Separate sections will be devoted to the determination of protonation constants and of the formation constants for mononuclear, mixed-ligand and polynuclear complexes. The determination of redox equilibrium constants will not be treated here.

2.1. Determination of Protonation Constants

The protonation constant of a ligand base and the dissociation constant of its conjugate acid differ only in the sign of the exponent (see p. 35). Protonation constants can thus be determined by using methods available for the measurement of acid dissociation constants. In practice mainly potentiometric, spectrophotometric and conductometric methods are used [49]. Of these only the first two will be treated here.

2.1.1. POTENTIOMETRIC METHODS

The principle of the method is that a solution of known concentration of the base (or acid) to be studied is titrated with a strong acid (or strong base) and the reaction followed potentiometrically with glass and calomel electrodes. A pH-titration curve is constructed from the results and the protonation constant determined from the curve.

The method can be used for studying the protonation equilibria of ligands which, in the protonated and non-protonated form, are sufficiently soluble to form at least $10^{-3}M$ solutions, and which do not decompose during the titration. The determination of the protonation constant becomes uncertain in aqueous solutions if $\log K > 11$, or if the protonation constant is smaller than the reciprocal of the concentration of the substance studied, i.e. if $\log K < -\log C_L$.

In the course of the titration of a monoacidic weak base with a strong acid a protonation reaction takes place. For example, in the case of ammonia,

$$NH_3 + H^+ \rightleftharpoons NH_4^+$$

In the titration with hydrochloric acid, ammonium chloride is formed which is completely dissociated to NH_4^+ (a Brønsted acid) and Cl^-. The equilibrium constant of the reaction is the protonation constant:

$$K_{NH_3} = \frac{[NH_4^+]}{[NH_3][H^+]} \tag{2.1}$$

The reaction is similar if a solution of an alkali metal salt of a monobasic weak acid is titrated with a strong acid. In the case of the titration of sodium acetate (NaAc):

$$Ac^- + H^+ \rightleftharpoons HAc,$$

the negatively charged acetate (Brønsted base) taking up a proton to form acetic acid. The equilibrium constant of the reaction is the protonation constant of the anion.

$$K_{Ac} = \frac{[HAc]}{[Ac^-][H^+]} \tag{2.2}$$

The logarithmic forms of Eqs. (2.1) and (2.2) are

$$\log K_{NH_3} = \log \frac{[NH_4^+]}{[NH_3]} + pH \tag{2.3}$$

$$\log K_{Ac} = \log \frac{[HAc]}{[Ac^-]} + pH \tag{2.4}$$

If the pH of a solution containing both the protonated and non-protonated form is known as well as the concentration ratio of the two species, the protonation constant can be calculated by using Eq. (2.3) or (2.4).

In the titration of ammonia with hydrochloric acid the following relationships hold for the species in solution.

Sec. 2.1] Protonation constants

Charge balance (the electroneutrality principle) gives

$$[NH_4^+] + [H^+] = [OH^-] + [Cl^-] \qquad (2.5)$$

Mass balance gives

$$C_{NH_3} = [NH_3] + [NH_4^+] \qquad (2.6)$$

where C_{NH_3} is the total concentration of the base.

The 'degree of neutralization' a is used for characterizing the degree of titration, defined as follows:

$$a = \frac{\text{total amount of titrant added}}{\text{total amount of the titrand}} \qquad (2.7)$$

Then a is zero at the beginning of the titration, and gradually increases to an integer at the equivalence point, the value of the integer depending on the stoichiometry of the reaction. In the case cited,

$$a = \frac{[Cl^-]}{C_{NH_3}} \qquad (2.8)$$

and $a = 1$ at equivalence.

By combination of Eqs. (2.5), (2.6) and (2.8), the concentrations of the protonated and non-protonated forms can be expressed as:

$$[NH_4^+] = aC_{NH_3} - [H^+] + [OH^-] \qquad (2.9)$$

$$[NH_3] = (1 - a)C_{NH_3} + [H^+] - [OH^-] \qquad (2.10)$$

In the case of the titration of sodium acetate with hydrochloric acid the concentrations of the protonated and non-protonated forms can be similarly expressed, by the following equations.

Charge balance:

$$[Na^+] + [H^+] = [OH^-] + [Cl^-] + [Ac^-] \qquad (2.11)$$

Mass balance:

$$C_{Ac} = [HAc] + [Ac^-] \qquad (2.12)$$

Taking into account that

$$[Cl^-] = aC_{Ac}$$

and

$$[Na^+] = C_{Ac}$$

[HAc] and [Ac$^-$] can be expressed from Eqs. (2.11) and (2.12) as follows:

$$[HAc] = aC_{Ac} - [H^+] + [OH^-] \qquad (2.13)$$

$$[Ac^-] = (1 - a)C_{Ac} + [H^+] - [OH^-] \qquad (2.14)$$

The forms of Eqs. (2.13) and (2.14) are the same as those of Eqs. (2.9) and (2.10), respectively. That is, irrespective of whether the titrand is a neutral molecule or an anion, Eqs. (2.3) and (2.4) can be given in the following general form for formation of the conjugate and of the base:

$$\log K = \log \frac{[HL]}{[L]} + pH \qquad (2.15)$$

and

$$\log K = \log \frac{aC_L - [H^+] + [OH^-]}{(1-a)C_L + [H^+] - [OH^-]} + pH \qquad (2.16)$$

where L is the base capable of proton uptake and C_L is its total concentration, irrespective of whether it is charged or uncharged.

The protonation constant of a base can be calculated by using Eq. (2.16) and the results of the potentiometric titration of the base with hydrochloric acid (from the degree of neutralization and corresponding hydrogen and hydroxyl ion concentrations), if the total concentration of the base titrated is known.

The concentrations of the protonated and non-protonated forms can be similarly deduced in the case when the conjugate acid of a base is titrated with a strong base:

$$[HL] = (1-a)C_{HL} - [H^+] + [OH^-];$$

$$[L] = aC_{HL} + [H^+] - [OH^-]$$

The protonation constant of the base can be calculated from the titration curve by using the following equation:

$$\log K = \log \frac{(1-a)C_{HL} - [H^+] + [OH^-]}{aC_{HL} + [H^+] - [OH^-]} + pH \qquad (2.17)$$

where C_{HL} is the total concentration of the titrand. Equation (2.17) can be applied if the substance to be titrated is present in the protonated form, as compared with Eq. (2.16) which is to be used if the base is titrated with a strong acid. (See Figs. 2.1 and 2.2 and Example 25.)

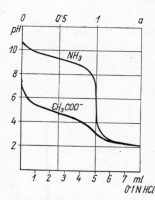
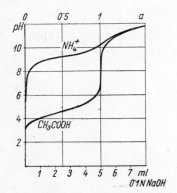

Fig. 2.1. Titration curves of ammonia and sodium acetate. The concentration of the titrated base in the half-titrated solution is $C_L = 10^{-2}M$

Fig. 2.2. Titration curves of acetic acid and ammonium chloride. The concentration of the titrated acid in the half-titrated solution is $C_{HL} = 10^{-2}M$

It is to be noted that in Eqs. (2.16) and (2.17), if C_L or $C_{HL} > 10^{-3}M$, [H⁺] can be neglected both in the numerator and denominator if pH > 5, and [OH⁻] can be neglected if pH < 9. In the pH range 5–9 both [H⁺] and [OH⁻] can be neglected, and the first terms on the right-hand side of the equations will be

$$\log \frac{a}{1-a} \quad \text{and} \quad \log \frac{1-a}{a}$$

These terms depend only on the degree of neutralization, and are equal to zero at $a = 0.5$. Therefore in the cases mentioned the pH of the half-titrated solution is equal to log K.

The first term in Eqs. (2.16) and (2.17) – which is the logarithm of the concentration ratio of the protonated and non-protonated forms – if calculated from the titration curve for various degrees of neutralization and plotted against pH, gives a straight line with slope -1 and an intercept on the abscissa which gives the value of log K.

$$\log \frac{[HL]}{[L]} = \log K - \text{pH} \tag{2.18}$$

(See Fig. 2.7, p. 107.)

When bases capable of taking up more than one proton are titrated with a strong acid, or polybasic acids are titrated with a strong base, the successive protonation constants can be calculated similarly to those

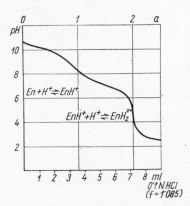

Fig. 2.3. Titration of ethylenediamine with standard hydrochloric acid solution. The initial volume of the titrated solution is 50 ml

of bases taking up one proton if the difference in the logarithms of the successive protonation constants is greater than 2·8. In this case the titration curve can be divided by vertical lines drawn at the inflection points, into independent parts in which practically only two species need to be considered. The equilibria within these sections involve the uptake or release of only one proton. Ethylenediamine (En) can take up two protons when titrated with a strong acid. The difference in the logarithms of the two protonation constants is about 2·8. During the titration, up to the first equivalence point practically only the En and EnH^+ species exist, and between the first and second equivalence points only the EnH^+ and EnH_2^{2+} species are present. That is, if $0 < a < 1$, then

$$C_{En} \simeq [En] + [EnH^+]$$

and if $1 < a < 2$

$$C_{En} \simeq [EnH^+] + [EnH_2^{2+}].$$

The first protonation constant

$$K_1 = \frac{[EnH^+]}{[En][H^+]}$$

can be calculated from an a value taken from the part of the titration curve where $a < 1$ and from the corresponding pH, $[H^+]$ and $[OH^-]$ concentrations; and the second protonation constant

$$K_2 = \frac{[EnH_2^{2+}]}{[EnH^+][H^+]}$$

from similar data taken from the part where $1 < a < 2$, by using Eq. (2.16). When the second protonation constant is calculated, 1 is to be subtracted from a (since $a > 1$) and this value used in the equation (see Fig. 2.3).

Sec. 2.1] **Protonation constants** 95

The general forms of Eqs. (2.15) and (2.16) which can be used in the case of the titration of bases which can take up more than one proton, are as follows:

$$\log K_i = \log \frac{[H_i L]}{[H_{i-1}]L} + \text{pH} \qquad (2.19)$$

and

$$\log K_i = \log \frac{(a - i + 1)C_L - [H^+] + [OH^-]}{(i - a)C_L + [H^+] - [OH^-]} + \text{pH} \qquad (2.20)$$

where K_i is the ith protonation constant of the base.

When the titration is done with a strong base and the titrand (the protonated base) is a polybasic acid, the protonation constants can be calculated similarly if the difference in the logarithms of the successive constants is greater than 2·8. The general form of Eq. (2.17) is

$$\log K_i = \log \frac{(1 - a + n - i)C_{H_nL} - [H^+] + [OH^-]}{(a - n + i)C_{H_nL} + [H^+] - [OH^-]} + \text{pH} \qquad (2.21)$$

where K_i is the ith protonation constant of the completely deprotonated L^{n-}-species of the acid H_nL. If the values of log ($[H_iL]/[H_{i-1}L]$) calculated from the points on neutralization portions $[i < a < (i + 1)]$ of the titration curve are plotted against pH, straight lines are obtained as in the case of monobasic acids, and intersect the abscissa at the values of log K_i. (See Fig. 2.7, p. 107.)

The protonation constants of uncharged bases which can take up more than one proton can also be determined by adding a known amount of a strong acid in excess of that required for the base and back-titrating with a standard solution of strong base. In this case the protonation constants can be calculated by using Eq. (2.21), if the difference in the logarithms of the protonation constants is large enough. In constructing the titration curve care should be taken that in the calculations no confusion arises because of the increase in the amount of the standard solution equivalent to the excess of strong acid added, that is the very first part of the titration curve (see Fig. 2.14, p. 130).

The protonation constants of polyacidic bases can be determined by Schwarzenbach's graphical method [51] even if the difference between the logarithms of the successive constants is smaller than 2·8.

Let us assume that the protonation constants of the base L^{2-} are determined from the results of the potentiometric titration of the acid

H_2L with standard potassium hydroxide solution, and the difference between the logarithms of the protonation constants is smaller than 2·8.

The following equations can be written for the concentrations of the various species present during the titration in the range $0 < a < 2$.

Charge balance:

$$[K^+] + [H^+] = [HL^-] + 2[L^{2-}] + [OH^-] \qquad (2.22)$$

Mass balance:

$$C_{H_2L} = [H_2L] + [HL^-] + [L^{2-}] \qquad (2.23)$$

[In this case the three species can be present simultaneously in significant amounts; when $\log (K_2/K_1)$ is $>2·8$, although all three species can co-exist, the concentration of at least one of them is negligible.] From the definition of the protonation constants:

$$K_1 = \frac{[HL^-]}{[L^{2-}][H^+]} \qquad (2.24)$$

$$K_2 = \frac{[H_2L]}{[HL^-][H^+]} \qquad (2.25)$$

and the facts that $[K^+] = aC_{H_2L}$, and if $[H^+] > [OH^-]$ the latter can be neglected, the following equation is obtained after eliminating $[L^{2-}]$, $[HL^-]$ and $[H_2L]$ from Eqs. (2.22) and (2.23):

$$\frac{1}{K_1}\left[\frac{(2-a)C_{H_2L}}{[H^+]} - 1\right] - K_2[H^+][aC_{H_2L} + [H^+]]$$
$$= (a-1)C_{H_2L} + [H^+]$$

For the purposes of graphical evaluation it is useful to rearrange the equation in the following form:

$$\frac{1}{K_1} = \left[\frac{[H^+](aC_{H_2L} + [H^+])}{\frac{(2-a)C_{H_2L}}{[H^+]} - 1}\right]K_2 + \frac{(a-1)C_{H_2L} + [H^+]}{\frac{(2-a)C_{H_2L}}{[H^+]} - 1} \qquad (2.26)$$

This corresponds to

$$\frac{1}{K_1} = \frac{B}{A}K_2 + B \qquad (2.27)$$

where

$$A = \frac{(a-1)C_{H_2L} + [H^+]}{[H^+](aC_{H_2L} + [H^+])}$$

$$B = \frac{(a-1)C_{H_2L} + [H^+]}{\dfrac{(2-a)C_{H_2L}}{[H^+]} - 1}$$

If A and B are calculated from several sets of values of a, $[H^+]$ taken from the titration curve, and C_{H_2L}, the set of straight lines drawn through corresponding values of A and B will all intersect at one point. (See Fig. 2.4.) From the coordinates of the point of intersection K_1 and K_2 can be calculated, as $y = 1/K$, and $x = K_2$ (see Example 27).

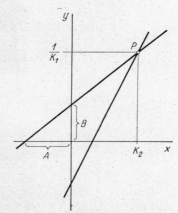

Fig. 2.4. Determination of two protonation constants, which are very close to one another, by Schwarzenbach's graphical method

In addition to the methods described, protonation constants can also be determined by the methods suitable for determining complex stability constants. From the data of the titration curve the 'formation curve of the protonated products' can be constructed, similarly to the formation curves of complexes. The protonation constants can be determined from the formation curves by one of the methods used for estimating complex stability constants (see p. 119).

For simplicity, the method of calculating $\bar{n}$, the 'average number of protons bound' and of constructing the 'formation curve' from the data of a titration curve will be shown by the example of the titration of the dibasic acid H_2L with standard potassium hydroxide solution. Equations (2.22) and (2.23) are valid for the titration. By combining them we get:

$$2C_{H_2L} - [K^+] + [OH^-] - [H^+] = [HL^-] + 2[H_2L]$$

The right-hand side of this equation gives the concentration of protons bound. Taking into account that $[K^+] = aC_{H_2L}$, the number of protons bound to one ligand is given by

$$\bar{n} = \frac{(2-a)C_{H_2L} + [OH^-] - [H^+]}{C_{H_2L}} \qquad (2.28)$$

It can be similarly deduced that in the case of the titration of an m-basic acid, $\bar{n}$ can be calculated as follows:

$$\bar{n} = \frac{(m-a)C_{H_mL} + [OH^-] - [H^+]}{C_{H_mL}} \qquad (2.29)$$

All the data required in Eq. (2.29) can be calculated from the titration curve. In cases where $C_{H_mL} > 10^{-2}M$, in the pH range 5–9, $[H^+]$ and $[OH^-]$ can be neglected in Eq. (2.29) and

$$\bar{n} = m - a.$$

By plotting the values of $\bar{n}$, calculated for various degrees of neutralization, against the pH, we obtain the 'formation curve'.

As $\bar{n}$ is a function of the hydrogen ion concentration and of the protonation constants:

$$\bar{n} = \frac{[H^+]K_1 + 2[H^+]^2 K_1 K_2 + \ldots}{1 + [H^+]K_1 + [H^+]^2 K_1 K_2 + \ldots}$$

[see Eq. (1.41) in Chapter 1, p. 37], K_1 and K_2 can be calculated by the numerical or graphical methods given in connection with the calculation of stability constants (see Example 31).

It should be noted that the concentration of inert electrolyte used for adjusting to constant ionic strength is not included in the equations. The added electrolyte should not, of course, take part in the reaction studied (i.e. it must not directly affect the position of the equilibria).

2.1.1.1. *Practical instructions*

A pH-meter which measures the pH with good reproducibility to ± 0.01 pH is required. The Radiometer (Copenhagen, Denmark) and Metrohm (Herisau, Switzerland) instruments are particularly suitable. Of Hungarian instruments the Radelkis Universal pH-meter can be recommended.

A calomel electrode and a good glass electrode (possibly a combined calomel–glass electrode) may be used. The instrument is calibrated as follows.

After switching on, the instrument is allowed to warm up for half an hour, then the electrode pair is immersed in a $0.05M$ potassium hydrogen phthalate solution prepared by direct weighing, at 20°C. The pH-meter is switched to the standardization position, and the needle of the pH-meter is set to read pH = 4·00 by means of the appropriate control. The electrodes are rinsed and then immersed in 0·01M sodium tetraborate solution at 20° C, prepared by weighing the salt and dissolving it in carbon dioxide-free water. The pH-meter should indicate a pH value of 9·22 with good reproducibility (see Table 2.1).

Table 2.1

The pH of buffer solutions (National Bureau of Standards, USA), and the ionic product of water [39] at different temperatures

Temperature °C	0·05M Potassium hydrogen phthalate solution pH	0·01M borax ($Na_2B_4O_7$) solution pH	$-\log k_w$
10	4·00	9·33	14·535
15	4·00	9·27	14·346
20	4·00	9·22	14·167
25	4·01	9·18	13·997
30	4·01	9·14	13·833
35	4·02	9·10	13·680
40	4·03	9·07	13·535
50	4·06	9·01	13·272
60	4·09	8·96	13·017

If a cell with ground-glass joints and gas flush is not available, it is advisable to make a titration device similar to that shown in Fig. 2.5. The titration vessel is a tall 100-ml beaker with a plastic lid with four inlets for the nitrogen inlet tube, electrodes and burette. Nitrogen from a cylinder is passed through a wash-flask containing alkaline sodium dithionite solution, then through another containing water. The titration vessel is placed on a magnetic stirrer and the temperature of the solution measured if necessary. If the titrant is an alkali, it is protected from the carbon dioxide in the air by means of a tube filled with soda-lime.

Hydrochloric or perchloric acid (0·1M) is the standard acid solution used. For its preparation see ref. [52]. Carbonate-free 0·1M potassium hydroxide solution can be made from a standard solution low in carbonate as follows: 500 ml of 0·2M potassium hydroxide standard solution are passed through an OH-form anion-exchange column, and the carbonate-free solution is collected in a 1-litre volumetric flask, and diluted to the mark with water passed through the column to wash all

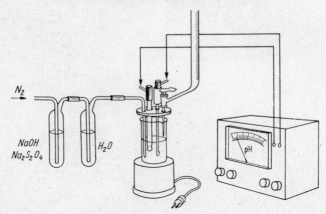

Fig. 2.5. Apparatus for pH titrations

the alkali through [53]. For ensuring constant ionic strength $1M$ potassium chloride (or nitrate) stock solution is used.

The substance under investigation is thoroughly purified by recrystallization or distillation. In the case of a crystalline, non-decomposing compound, it is dried to constant weight. The protonated or non-protonated form is chosen depending on which form is better for handling and weighing. If a free base or acid is necessary it can be prepared from the corresponding salts by ion-exchange [53].

Dissolution of the sample can be aided by adding some alcohol or dioxan. Care should be taken, however, that the amount of the organic solvent in the solution titrated is not greater than 1–2%. Otherwise a correction should be used in measuring the pH [54].

If the solubilities permit, it is expedient to titrate a 50-ml aliquot of $10^{-2}M$ solution (prepared by accurate weighing and containing potassium chloride to adjust the ionic strength), with $0.1M$ or more concentrated standard solution. The titration is done under a continuous nitrogen stream with stirring, the titrant being added in small portions. After the needle of the instrument has settled the burette reading and the pH are noted.

The titration curve is obtained by plotting pH against the volume of standard solution (see Figs. 2.1, 2.2 and 2.3). If some parts of the titration curves appear uncertain, the titration should be repeated, the standard solution being added in smaller portions (dropwise) for the pH interval in question. The point at which a stoichiometric amount of standard solution has been added is marked on the graph. (If the molecular weight is unknown, the abscissa of the inflection point is marked.) In the case of polyfunctional acids or bases all equivalence points are marked and denoted by $a = 1, 2 \ldots$ The starting point of titration is denoted by $a = 0$. If a strong acid or a strong base has been added in excess to the titrand, the point $a = 0$ is at the point where the amount of standard solution added is equivalent to the excess.

Dilution of the solution during titration may be neglected in the range $0 < a < 1$ (or possibly $0 < a < 2$), if the titrant is ten times as concentrated as the titrand and the volume of the half-titrated solution is used for the calculations. From the volume of the half-titrated solution, the concentrations C_L and C_{H_nL} corresponding to it are calculated. When Eqs. (2.16) and (2.17) or (2.20) and (2.21) are used, the calculation is done for at least 6 points taken from the neutralization interval from $a = 0·1$ to $a = 0·9$. The difference in log K values should not exceed $\pm 0·02$, if the measurements and calculations have been done accurately.

As the values of log ($[H_iL]/[H_{i-1}L]$) plotted *vs.* pH yield a straight line, log K can be advantageously determined from the related pairs of values by the method of least squares (see p. 44).

2.1.2. Spectrophotometric methods

The principle of the method is as follows. In a solution at a pH at which both the base and its protonated product are present, the ratio of the two species is determined by spectrophotometry and the protonation constant calculated by using the basic equation (2.15):

$$\log K = \log \frac{[HL]}{[L]} + \text{pH}$$

The advantage of the method as compared to the potentiometric one is that it is suited to the investigation of very dilute solutions of light-absorbing substances which are only slightly soluble in water. Furthermore, it can be used even if the logarithm of the protonation constant is greater than 11 or smaller than 1.

By use of Hammett's acidity function, the protonation constants of very weak bases which take up protons below pH 1, in strongly acidic solutions only, can be determined in sulphuric acid solutions of known H^0 values. The extension of the pH scale for concentrated sulphuric acid solutions by Hammett by introducing an H^0 scale analogous to the pH scale arises from the consideration that the ratio of the activities of two similarly charged ions (LH^+ and H^+) is, according to the Debye-Hückel theory, to a first approximation independent of the dielectric constant and ionic strength of the medium. The concentration ratio of the protonated and non-protonated forms of uncharged indicator bases is related to the pH, or, in strongly acidic medium to H^0,

$$H^0 = \log K + \log \frac{[L]}{[LH^+]} \qquad (2.30)$$

where K is the protonation constant of an indicator base. If K and $[L]/[LH^+]$ are known, the value of H^0 can be calculated (or conversely, the value of K from H^0 and $[L]/[LH^+]$).

In order to select a suitable wavelength the absorption spectra of the protonated and non-protonated species are recorded. A 10^{-4} or $10^{-5} M$ solution is prepared from the substance in question, one part of which is acidified with a strong acid, and another is rendered alkaline by addition of a strong base. For practically all the base to be present in the non-protonated form it is necessary that $pH > \log K + 2$ [see Eq. (1.78)], and in the protonated form, $pH < \log K - 2$. On the basis of the spectra the wavelength is selected at which the difference in the absorbances of the two species is the greatest and at which the difference does not change much with the wavelength. In the ideal case only one form absorbs light (see Fig. 2.6).

At the selected wavelength the absorbances of equimolar solutions in one of which all the substance is present in the protonated form and in

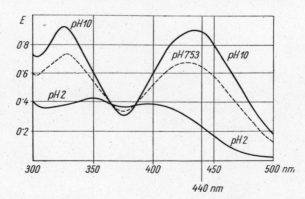

Fig. 2.6. Absorption spectrum of β-nitroso-α-naphthol in solutions of different pH but the same concentration, $10^{-2} M$, according to the measurements by Dyrssen and Johansson [59]

the other in the non-protonated form, are determined. Let these values be denoted by E_{HL} and E_L. Then a measurement is made for a solution of the same concentration but at a pH where the two species are present in commensurable amount. Log K can be approximated by either of the following equations:

$$\log K = \log \frac{E_L - E}{E - E_{HL}} + \text{pH} \qquad (2.31a)$$

$$\log K = \log \frac{E - E_L}{E_{HL} - E} + \text{pH} \qquad (2.31b)$$

where E stands for the absorbance of the solution containing both species.

If $E_L > E_{HL}$, then Eq. (2.31a) is used, otherwise Eq. (2.31b). For the calculation of the accurate value of log K the measurements are made on solutions of pH within the interval pH $= \log K \pm 0.5$.

As all the measurements are made on solutions of the same concentration, the molar absorptivities need not be calculated.

The method can be used similarly for substances taking up or releasing more than one proton, if $\log K_1 - \log K_2 > 3$ and suitable wavelengths can be found for the determination of the different species. If the protonation constants are closer to each other, the determination and calculation can still be done, but this will not be treated here in detail (see [50, 55, 56]).

If neither the protonated nor the non-protonated species has absorption in the visible or ultraviolet range of the spectrum, then log K can be determined by an indirect method, by means of an acid–base indicator of similar protonation constant. To a very dilute solution ($10^{-5}M$) of the indicator known amounts of the protonated and non-protonated species are added ($10^{-2}M$) and the absorbance of the solution is measured at two wavelengths best suited to the determination of the two forms of the indicator; log K can be calculated by using the following equation:

$$\log K = \log K_I + \log \frac{[HL]}{[L]} + \log \frac{[I]}{[HI]} \qquad (2.32)$$

where K_I is the protonation constant of the acid–base indicator, and [I] and [HI] are the concentrations of the non-protonated and protonated forms of the indicator, the ratio of which is calculated from the absorbances [57].

2.1.2.1. *Practical advice*

Although it is easier to obtain spectra with an automatic, recording spectrophotometer, equilibrium constants can be determined with a greater accuracy by using a single-beam, compensation-type spectrophotometer [e.g. Spectromom 201 or 202 (MOM, Hungary)].

A good pH meter must be used to obtain accurate results [e.g. Radiometer (Denmark) or Radelkis (Hungary) Universal pH-meter].

As stock buffer solutions, $0.1M$ solutions of the substances tabulated in Table 2.2 are recommended. The test solutions are made so that they are at least $0.01M$ in a buffer. After addition of the stock solution, the pH is adjusted with sodium hydroxide or hydrochloric acid.

For preparing solutions of pH higher than 11, sodium hydroxide is used instead of buffer. If the measurements are to be made with solutions of pH lower than 2, then the acidity scale can be extended on the basis of Hammett's acidity function by means of sulphuric acid solutions of appropriate concentration (see Table 2.3).

The concentration of the substance is chosen so that the absorbance measured is in the range 0·2–0·6. The same concentration should be used in every experimental solution. Care should be taken that the reference solution contains the same reagents other than the test-substance

Table 2.2

Buffer substances for spectrophotometric measurements, according to Albert and Serjeant [50]

Substance	pH range
Oxalic acid	1·0– 3·0
Formic acid	3·2– 4·4
Acetic acid	4·2– 5·4
Potassium dihydrogen phosphate	6·5– 7·7
Boric acid	8·6– 9·8
Ethylamine	10·1–11·3

and has the same pH as the solution to be measured. Constant ionic strength can be ensured by means of the buffer or by adding potassium chloride or nitrate.

If the temperature is also controlled, good results can be expected.

Table 2.3

pH values of strongly alkaline solutions and Hammett function (H^0) values of sulphuric acid solutions of different concentrations, at 25°C [50, 57]

NaOH, M	pH	H_2SO_4, M	H^0
0·01	11·95	0·005	2·08
0·02	12·23	0·065	1·06
0·05	12·61	0·54	0·09
0·10	12·88	0·90	−0·23
		1·18	−0·42
0·20	13·16	1·54	−0·64
0·50	13·53	2·04	−0·90
1·00	13·82	2·61	−1·20
2·00	14·14	3·55	−1·68
		4·29	−2·03

2.1.3. Worked examples

25. Calculate the protonation constant of acetate from the titration curves in Figs. 2.1 and 2.2.

Two points are selected from the curve in Fig. 2.1, one to the left, one to the right of the point where $a = 0·5$.

For example, at

$$a = 0·3 \quad \text{pH} = 5·02 \quad [H^+] = 9·55 \times 10^{-6} M$$

and at

$$a = 0·7 \quad \text{pH} = 4·31 \quad [H^+] = 4·90 \times 10^{-5} M$$

The calculation is done for both points by using Eq. (2.16).

The hydroxide ion concentration, $10^{-8·98}$ and $10^{-9·69}$ can be neglected in both cases. The acetate concentration in the half-titrated solution is $10^{-2} M$.

$$\log K = \log \frac{0·3 \times 10^{-2} - 9·55 \times 10^{-6}}{(1 - 0·3) \times 10^{-2} + 9·55 \times 10^{-6}} + 5·02$$

$$= \log \frac{0·00299}{0·00701} + 5·02 = -0·37 + 5·02 = 4·65.$$

$$\log K = \log \frac{0{\cdot}7 \times 10^{-2} - 4{\cdot}90 \times 10^{-5}}{(1 - 0{\cdot}7) \times 10^{-2} + 4{\cdot}90 \times 10^{-5}} + 4{\cdot}31$$

$$= \log \frac{0{\cdot}00695}{0{\cdot}00305} + 4{\cdot}31 = 0{\cdot}34 + 4{\cdot}32 = 4{\cdot}65.$$

The results obtained from the two points agree.

Now, two points are selected from the titration curve in Fig. 2.2, where $a = 0{\cdot}4$ and $a = 0{\cdot}7$.

At $a = 0{\cdot}4$ pH $= 4{\cdot}49$ $[H^+] = 3{\cdot}24 \times 10^{-5} M$
At $a = 0{\cdot}7$ pH $= 5{\cdot}02$ $[H^+] = 9{\cdot}55 \times 10^{-6} M$

Hydroxide ion concentrations can be neglected in both cases. $C_{AcH} = 10^{-2} M$. The calculation is done by using Eq. (2.17).

$$\log K = \log \frac{(1 - 0{\cdot}4) \times 10^{-2} - 3{\cdot}24 \times 10^{-5}}{0{\cdot}4 \times 10^{-2} + 3{\cdot}24 \times 10^{-5}} + 4{\cdot}49$$

$$= \log \frac{0{\cdot}00597}{0{\cdot}00403} + 4{\cdot}49 = 0{\cdot}17 + 4{\cdot}49 = 4{\cdot}66.$$

$$\log K = \log \frac{(1 - 0{\cdot}7) \times 10^{-2} - 9{\cdot}55 \times 10^{-6}}{0{\cdot}7 \times 10^{-2} + 9{\cdot}55 \times 10^{-6}} + 5{\cdot}02$$

$$= \log \frac{0{\cdot}00299}{0{\cdot}00701} + 5{\cdot}02 = -0{\cdot}37 + 5{\cdot}02 = 4{\cdot}65.$$

The protonation constant of acetate is given by

$$\log K = 4{\cdot}65 \qquad (20°\ C;\ I = 10^{-1}).$$

Calculated log [HL]/[L] values are given as a function of pH in Fig. 2.7.

26. Calculate the protonation constants of ethylenediamine by using data taken from the titration curve in Fig. 2.3.

For the calculation of the first protonation constant the points where $a = 0{\cdot}4$ and $a = 0{\cdot}6$ are chosen.

At $a = 0{\cdot}4$ the volume of standard solution consumed is $0{\cdot}4 \times 3{\cdot}6 = 1{\cdot}44$ ml; pH $= 10{\cdot}08$; $[OH^-] = 10^{-3{\cdot}92} = 1{\cdot}2 \times 10^{-4} M$.
At $a = 0{\cdot}6$ the volume of standard solution consumed is $0{\cdot}6 \times 3{\cdot}6 = 2{\cdot}16$ ml; pH $= 9{\cdot}76$; $[OH^-] = 10^{-4{\cdot}24} = 5{\cdot}76 \times 10^{-5} M$.

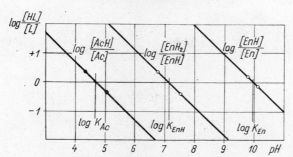

Fig. 2.7. Graphical estimation of protonation constants. See Examples 25 and 26

The ethylenediamine concentration in the half-titrated solution can be calculated from the volume and concentration ($0.1085M$) of the standard solution consumed, and the volume of the half-titrated solution.

The volume of the half-titrated solution is

$$50 + 1.8 = 51.8 \text{ ml}$$

$$C_{En} = \frac{3.6 \times 0.1085}{51.8} = 7.54 \times 10^{-3} M.$$

The protonation constant is calculated by using Eq. (2.20). Since in both cases pH > 5, [H$^+$] can be neglected.

$$\log K_1 = \log \frac{(0.4 - 1 + 1) \times 7.54 \times 10^{-3} + 1.2 \times 10^{-4}}{(1 - 0.4) \times 7.54 \times 10^{-3} - 1.2 \times 10^{-4}} + 10.08$$

$$= \log \frac{0.00314}{0.00440} + 10.08 = -0.15 + 10.08 = 9.93$$

$$\log K_1 = \log \frac{(0.6 - 1 + 1) \times 7.54 \times 10^{-3} + 5.76 \times 10^{-5}}{(1 - 0.6) \times 7.54 \times 10^{-3} - 5.76 \times 10^{-5}} + 9.76$$

$$= \log \frac{0.00458}{0.00296} + 9.76 = 0.19 + 9.76 = 9.95$$

For the calculation of the second protonation constant the points corresponding to $a = 1.3$ and $a = 1.7$ should be used.

At $a = 1.3$ the volume of standard solution used is $3.6 \times 1.3 = 4.68$ ml and the pH is 7.50.

At $a = 1.7$ the volume of standard solution consumed is $3.6 \times 1.7 = 6.12$ ml and the pH is 6.78.

The $[H^+]$ and $[OH^-]$ values can be neglected in both cases. The concentration of ethylenediamine in the half-titrated solution is

$$C_{En} = \frac{3.6 \times 0.1085}{55.4} = 7.05 \times 10^{-3} M.$$

This value, however, is not necessary for the calculation.

$$\log K_2 = \log \frac{(1.3 - 2 + 1) \times 7.05 \times 10^{-3}}{(2 - 1.3) \times 7.05 \times 10^{-3}} + 7.50$$

$$= \log \frac{0.3}{0.7} + 7.50 = -0.37 + 7.50 = 7.13$$

$$\log K_2 = \log \frac{1.7 - 2 + 1}{2 - 1.7} + 6.78$$

$$= \log \frac{0.7}{0.3} + 6.78 = 0.37 + 6.78 = 7.15$$

The protonation constants, each calculated from two points, are

$$\log K_1 = 9.94; \quad \log K_2 = 7.14 \quad (25°C, I \simeq 0.1)$$

(see also Fig. 2.7).

27. The sulphosalicylate ion

$$\begin{array}{c} -O_3S \diagdown \diagup COO^- \\ \bigcirc \\ \diagdown O^- \end{array}$$

can take up three protons. Calculate the three protonation constants from titration curve I in Fig. 2.12 (p. 127).

No inflection point can be observed on the titration curve between the first and second neutralization sections. The third and second protonation constants (the first and second dissociation constants of sulphosalicylic acid) are very close to each other. Hence, these are calculated by using Schwarzenbach's graphical method [see Eqs. (2.26) and (2.27), and Fig. 2.4].

Sec. 2.1]   Protonation constants

The value of the first protonation constant, estimated from the pH at $a = 2\cdot5$ in the third section of the curve, is much greater than $\log K_2 + 2\cdot 8$. Therefore it can be calculated from the third neutralization section, by using Eq. (2.21).

For the calculation of K_2 and K_3, the points where $a_1 = 0\cdot6$, $a_2 = 1\cdot2$ and $a_3 = 1\cdot6$ are chosen in the range $0 < a < 2$.

At $a = 0\cdot6$ the standard solution added is $0\cdot6 \times 4\cdot2 = 2\cdot52$ ml, the pH is $2\cdot22$, therefore $[H^+] = 6\cdot03 \times 10^{-3} M$. The total concentration of sulphosalicylic acid at the mid-point of the two neutralization sections, where $a = 1$, is given by

$$C_{H_3SS} = \frac{50}{54\cdot 2} \times 10^{-2} = 9\cdot 22 \times 10^{-3} M.$$

The parameters necessary for the construction of the straight line are

$$A_1 = \frac{(0\cdot6 - 1) \times 9\cdot22 \times 10^{-3} + 6\cdot03 \times 10^{-3}}{(0\cdot6 \times 9\cdot22 \times 10^{-3} + 6\cdot03 \times 10^{-3}) \times 6\cdot03 \times 10^{-3}} = \frac{2\cdot34 \times 10^{-3}}{69\cdot7 \times 10^{-6}} = 33\cdot6;$$

$$B_1 = \frac{(0\cdot6 - 1) \times 9\cdot22 \times 10^{-3} + 6\cdot03 \times 10^{-3}}{\dfrac{(2 - 0\cdot6) \times 9\cdot22 \times 10^{-3}}{6\cdot03 \times 10^{-3}} - 1} = \frac{2\cdot34 \times 10^{-3}}{2\cdot14 - 1} = 2\cdot05 \times 10^{-3}$$

At the degree of neutralization $a = 1\cdot2$, the standard solution added is $1\cdot2 \times 4\cdot2 = 5\cdot04$ ml; the pH is $2\cdot60$, therefore $[H^+] = 2\cdot51 \times 10^{-3} M$.

$$A_2 = \frac{(1\cdot2 - 1) \times 9\cdot22 \times 10^{-3} + 2\cdot51 \times 10^{-3}}{(1.2 \times 9\cdot22 \times 10^{-3} + 2\cdot51 \times 10^{-3}) \times 2\cdot51 \times 10^{-3}}$$

$$= \frac{4\cdot35 \times 10^{-3}}{34\cdot06 \times 10^{-6}} = 127;$$

$$B_2 = \frac{(1.2 - 1) \times 9\cdot22 \times 10^{-3} + 2\cdot51 \times 10^{-3}}{\dfrac{(2 - 1\cdot2) \times 9\cdot22 \times 10^{-3}}{2\cdot51 \times 10^{-3}} - 1}$$

$$= \frac{4\cdot35 \times 10^{-3}}{2\cdot94 - 1} = 2\cdot24 \times 10^{-3}$$

At the degree of neutralization $a = 1\cdot6$, the pH is $3\cdot02$ and therefore $[H^+] = 0\cdot96 \times 10^{-3} M$.

$$A_3 = \frac{(1\cdot6 - 1) \times 9\cdot22 \times 10^{-3} + 0\cdot96 \times 10^{-3}}{(1\cdot6 \times 9\cdot22 \times 10^{-3} + 0\cdot96 \times 10^{-3}) \times 0\cdot96 \times 10^{-3}}$$

$$= \frac{6\cdot49 \times 10^{-3}}{15\cdot08 \times 10^{-6}} = 430$$

$$B_3 = \frac{(1\cdot6 - 1) \times 9\cdot22 \times 10^{-3} + 0\cdot96 \times 10^{-3}}{\dfrac{(2 - 1\cdot6) \times 9\cdot22 \times 10^{-3}}{0\cdot96 \times 10^{-3}} - 1}$$

$$= \frac{6\cdot49 \times 10^{-3}}{3\cdot84 - 1} = 2\cdot28 \times 10^{-3}$$

The values of A and B being known, three straight lines can be plotted, as seen in Fig. 2.8. The coordinates of the intersection of the three lines are as follows:

$$y = \frac{1}{K_2} = 2\cdot3 \times 10^{-3}$$

and

$$x = K_3 \approx 2.$$

The third protonation constant is very small and its determination is uncertain. The logarithms of the constants are

$$\log K_3 < 1; \quad \log K_2 = 2\cdot64 \quad (I = 0\cdot1).$$

The first protonation constant is calculated from the coordinates of the point corresponding to a degree of neutralization $a = 2\cdot4$.

At $a = 2\cdot4$ the volume of standard solution consumed is $2\cdot4 \times 4\cdot2 = 10\cdot08$ ml; the pH is $11\cdot14$ and therefore $[OH^-] = 10^{-2\cdot86} = 1\cdot38 \times 10^{-3} M$.

$$\log K_1 = \log \frac{(1 - 2\cdot4 + 3 - 1) \times 9\cdot22 \times 10^{-3} + 1\cdot38 \times 10^{-3}}{(2\cdot4 - 3 + 1) \times 9\cdot22 \times 10^{-3} - 1\cdot38 \times 10^{-3}} + 11\cdot14$$

$$= \log \frac{6\cdot91 \times 10^{-3}}{2\cdot31 \times 10^{-3}} + 11\cdot14 = 0\cdot48 + 11\cdot14 = 11\cdot62 \quad (I = 0\cdot1).$$

The reliability and accuracy of the results can be increased by further calculations from more points and a statistical evaluation of the results obtained.

28. Dyrssen and Johansson [59] have studied the absorption spectra of β-nitroso-α-naphthol in solutions of different pH. The absorbances of $10^{-4} M$ solutions at pH 2 and 10 are plotted against wavelength in Fig. 2.6. The difference in absorbance is greatest at 440 nm: $E_{HL} = 0\cdot235$ at pH 2 and $E_L = 0\cdot890$ at pH 10.
The absorbance of a solution of the same concentration at pH $= 7\cdot53$ was found to be $E = 0\cdot651$.

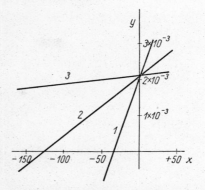

Fig. 2.8. Determination of the second and third protonation constant of sulphosalicylate by Schwarzenbach's graphical method. (See Example 27)

Calculate the protonation constant of β-nitroso-α-naphtholate from the data given.

The calculation is done by using Eq. (2.31a):

$$\log K = \log \frac{0\cdot890 - 0\cdot651}{0\cdot651 - 0\cdot235} + 7\cdot53 = -0\cdot24 + 7\cdot53 = 7\cdot29.$$

2.2. Determination of the Stability Constants of Mononuclear Complexes

In the case of mononuclear complexes, correlations exist between the concentrations C_L, C_M, [L], [M], [ML$_n$] etc., the average ligand number $\bar{n}$ and the mole fractions Φ_i (see pages 27 and 29) and it is usually sufficient to establish the relationship between two variables experimentally for the determination of stability constants. The methods available for determining stability constants are mainly based on the preparation of a series of solutions containing known amounts of the complex-forming components, in which the concentration of one component is gradually varied and the concentration of one of the reactants or products is followed directly or indirectly by a suitable analytical method. Usually the concentration of the metal ion is kept constant and that of the ligand varies within wide limits. In order to make the stability constants calculated from analytical concentrations unambiguous, the ionic strength should be the same in each solution. Hence, in general the concentration of the complex-forming substances should be low ($\leq 10^{-2}M$) and a large concentration ($\geq 10^{-1}M$) of inert electrolyte should be present to ensure constant ionic strength.

The most suitable analytical methods are those which do not disturb the equilibria existing in solution and by which low concentrations can be selectively determined with high accuracy.

Instrumental methods of analysis are mostly based on the measurement of an intensive physico-chemical property which is proportional to the concentration of a substance. If the proportionality constant is the same for all species of a substance, then the method is suitable for the determination of the total concentration of the complex-forming substance present (C_M or C_L). In the investigation of complex formation, and especially of successive complex formation, when several similar species are present in the solution, it is very important to have a selective analytical method by which the concentration of a single species can be determined in the presence of the others, independently of the composition of the solution. Such is the case, for instance, when the concentration of the uncomplexed, free metal ion is measured directly by potentiometry, or indirectly through ion-exchange.

In the following sections, only the principles of the most generally used potentiometric, spectrophotometric, polarographic, extraction and ion-exchange methods are given, together with information necessary for the calculations and measurements.

For other methods and details of the methods described here see refs. [32, 60, 61, 62].

2.2.1. Potentiometric methods

Potentiometric methods of analysis are often applicable to the investigation of solutions containing complexes, since very often the activity of the free metal and also of the free ligand can be directly determined.

The concentration of the free ligand can also be determined indirectly on the basis of Eq. (1.77) through measuring the pH of the solution if the ligand can be protonated and the value of protonation constant(s) is known.

2.2.1.1. *Direct methods*

Measurement of free metal ion concentration. A series of solutions is prepared in which the total concentration of the metal ion is constant and the ligand concentration gradually increases. Constant ionic strength

Sec. 2.2] Stability constants of mononuclear complexes

is ensured by adding an inert electrolyte to each of the solutions. The concentration of the uncomplexed metal ion is measured by direct potentiometry, and the stability constants are calculated from the related free metal and free ligand concentration data.

If only one complex species is formed, with the composition ML_n, the overall complex formation constant can be calculated by using the basic Eq. (1.11) and the [M], C_L and C_M values:

$$[ML_n] = C_M - [M];$$
$$[L] = C_L - n[ML_n]$$

Plotting $\log([ML_n]/[M])$ against $\log[L]$ (both calculated from experimental data) gives a straight line the intercept of which is $\log \beta_n$ and the slope n, i.e., from Eq. (1.11):

$$\log \frac{[ML_n]}{[M]} = n \log [L] + \log \beta_n \qquad (2.33)$$

To obtain accurate results, in preparing the solutions care should be taken that [M] and $[ML_n]$ are of similar magnitude.

The calculation is more complicated if more complex species are formed. Consider the ratio of total and free metal ion concentration, taking Eq. (1.14) into account.

$$\frac{C_M}{[M]} = \frac{[M] + [M][L]\beta_1 + [M][L]^2\beta_2 + \ldots [M][L]^N\beta_N}{[M]}$$

$$= 1 + \sum_{1}^{N} \beta_i [L]^i = A_{M(L)} \qquad (2.34)$$

The function $A_{M(L)}$ obtained is the reciprocal of the function Φ for the free metal ion [see Eq. (1.21)]. It is similar in form to the α function described earlier [see Eq. (1.70)], but a different symbol is used in order to stress that function A is related to the main reaction, and α to the disturbing side-reactions of the components taking part in the main (complex, acid–base or redox) reaction.

If the potential of a reversible M^{z+}/M electrode is measured in a solution of concentration C_M in the absence and also in the presence of a complexing agent, the value of the function A can be calculated directly from the difference between the two potentials measured. In the first case $[M] = C_M$, whereas in the second $[M] = C_M/A_{M(L)}$. The electrode potentials are

$$E_1 = E_M^0 + \frac{0\cdot 059}{z} \log C_M f_M,$$

$$E_2 = E_M^0 + \frac{0\cdot 059}{z} \log \frac{C_M}{A_{M(L)}} f_M.$$

The difference is:

$$\Delta E = \frac{0\cdot 059}{z} \log A_{M(L)} \qquad (2.35)$$

Calculation of complex products from related ligand concentration and $A_{M(L)}$ data. In the case where $C_L \gg C_M$ throughout the series of experiments and the ligand does not take part in any side-reaction, the free ligand concentration is practically equal to the total ligand concentration, $C_L \sim [L]$, and the overall stability constants can be calculated by using Eq. (2.34) for $A_{M(L)}$ values determined from measurements and the related [L] values.

For the calculation of N different β values N measurements, or N equations are sufficient in principle. The accuracy of the β values obtained in this way is not satisfactory, but can be increased remarkably by recalculating the real value of [L] from the calculated approximate values of β by using Eqs. (1.19) and (1.80), and repeating the calculation of the β values with the new [L] values, instead of C_L, and by doing the calculation on the basis of a large number of experimental results, using the method of least squares.

The calculations become more and more involved as the number of β values to be determined increases. The calculations are vastly facilitated by the application of a computer [63, 64]. It should be emphasized that it is not enough to use a computer to get reliable results, even if the determinations have been done with great accuracy. The concentration range for which the data were obtained and the equation used are also very important. β_i, for example, can be calculated only from data for a concentration range in which the ith complex really exists.

The calculation can also be facilitated by using the so-called relaxation method [65] (see later).

Graphical methods are particularly advantageous in studying complex equilibria, and usually draw the attention of the researchers to trends which are hidden or appear as errors in the course of numerical calculations.

One of the oldest and most widely used methods for the graphical resolution of the function $A_{M(L)}$ is due to Leden [66].

Sec. 2.2] Stability constants of mononuclear complexes

The principle of the method is as follows.

$$\frac{A_{M(L)} - 1}{[L]} = \beta_1 + \beta_2[L] + \ldots + \beta_N[L]^{(N-1)} \tag{2.36}$$

On calculating the fraction on the left-hand side of the equation from experimental [M] and [L] results and plotting it against [L], we get β_1, which is the extrapolated intercept of the initial linear part of the curve. After determination of β_1, the left-hand side of the equation

$$\frac{\dfrac{A_{M(L)} - 1}{[L]} - \beta_1}{[L]} = \beta_2 + \beta_3[L] + \ldots + \beta_N[L]^{(N-2)} \tag{2.37}$$

is plotted against [L]; the intercept of the curve is equal to β_2. The successive extrapolation procedure is continued until all β values are obtained (see Example 37).

It is useful, before determining the overall complex stability constants, to plot the experimental and calculated values of $A_{M(L)}$ vs. [L]. The curve should intersect the ordinate, where [L] = 0, at $A_{M(L)} = 1$ [see Eq. (2.34)]. By this means the points of the curve, i.e. $A_{M(L)}$ values in the range of small [L] values, which are most important from the point of view of further calculations can be evaluated more correctly.

Some important practical aspects of the potentiometric determination of the concentration of the uncomplexed, free metal ion are as follows. It is very important that an appropriate reversible electrode should be available. In some cases (Cu, Hg, Cd, Zn) metal electrodes of the first kind or amalgam electrodes can be used for the direct determination of the metal ion. In principle, ion-selective membrane electrodes are also suitable for this purpose. Unfortunately, the number of metal ions which can be determined directly is rather limited.

A calomel electrode may be used as the reference electrode. According to the Nernst equation [38], the potential difference between the two electrodes when a zero current is passed is proportional to the logarithm of the activity of the free metal ion. It is expedient to calibrate the voltage as a function of log [M] with solutions which do not contain the complexing agent. As the measuring instrument a Radiometer or Radelkis Universal millivoltmeter can be used.

Care should be taken that the determination is not disturbed by a side-reaction taking place in addition to the complex formation investigated. If the ligand is easily protonated, either a pH range is chosen in which practically no protonation occurs (pH $\geq \log K + 2$) or the measurements are made on solutions at the same pH, and the true free ligand concentration is calculated by using Eq. (1.77) or (1.79), provided that the protonation constant is known. If the protonation is not taken into account, then the calculations will yield the conditional stability constant for the pH used.

In the choice of pH, protonation of the ligand as well as the formation of hydroxo-complexes of the metal ion must be considered.

Determination of the free ligand concentration. A series of solutions is prepared, in which the total concentration of the central metal ion is the same but the concentration of the ligand is varied within wide limits. The concentration of the free ligand is determined potentiometrically in each solution by using an ion-specific electrode. The average ligand number $\bar{n}$ is calculated for each solution by using Eq. (1.17) with the total metal, total ligand and free ligand concentrations. From the calculated $\bar{n}$ and related [L] data the overall stability constants are determined by using Eq. (1.18).

The calculation of the stability constants will be dealt with later (see the subsection below, and especially p. 119).

In contrast to the method based on determination of the concentration of the free metal ion, it is important from the point of view of correct calculations that the concentration of bound ligand is not negligible compared with the total ligand concentration. More reliable results are obtained if the concentrations of the bound and free ligand are of the same order of magnitude.

For the determination of the concentration of free ligand, electrodes of the second kind (possibly specific ion-selective electrodes) can be used. Chloride, bromide or iodide ions can be determined by means of the corresponding silver–silver halide electrode. A calomel electrode which is connected with the solution studied, by means of a tube filled with potassium nitrate or sodium perchlorate in agar gel, may be used as a reference. A Radiometer or Radelkis Universal millivolt meter can be used. The correctness of the potential *vs.* log [L] relationship is checked with solutions of the complexing agent in the absence of the metal ion but of the same ionic strength.

2.2.1.2. *Methods based on measurement of pH*

As the majority of complexing ligands used in analytical chemistry are moderately strong bases and become protonated in the pH range mostly applied in practice, methods based on pH measurement are very

Sec. 2.2]   Stability constants of mononuclear complexes

often applicable to the determination of stability constants. The great advantage of these methods is that the measurements are not time-consuming and the instruments necessary are not expensive.

The basis of the methods is that during the complex formation between the metal ion and protonated ligand

$$M + HL \rightleftharpoons ML + H^+ \qquad (2.38)$$

protons are liberated. By determination of the hydrogen-ion concentration the degree of complex formation or the position of the equilibrium can be established. Only the titration methods will be treated here in detail.

The complex-forming agent, the weak acid (protonated base) is titrated potentiometrically with a standard solution of a base, in the presence of an inert electrolyte of suitable concentration and a pH vs. a curve is plotted. Then a solution of similar composition but also containing metal ion is titrated similarly and the titration curve again constructed. If complex-formation [see Eq. (2.38)] occurs extensively in the pH range 3–10 (i.e. the complex is rather stable), and C_L is not much greater than C_M, then there is an appreciable difference between the two curves. If we know the protonation constant(s) of the ligand, then by using the values of the coordinates, from points on the curves we can calculate the stability constants.

If the protonation constants of the ligand are not known, they can be determined from the data of the first titration curve (see p. 94).

If the complexing agent is a weak dibasic acid, H_2L, and the metal ion is a bivalent ion M^{2+}, and the concentrations are chosen so as to ensure the formation of the complex containing the maximum number of ligands, i.e. $C_{H_2L} > NC_M$, then the following equations hold for the concentrations of the constituents of the solution titrated with potassium hydroxide:

Charge balance:

$$[K^+] + [H^+] + 2[M^{2+}] = 2C_M + [OH^-] + 2[L^{2-}]$$
$$+ [HL^-] + 2[ML_2^{2-}] + \ldots \qquad (2.39)$$

(The first term on the right-hand side represents the concentration of anions introduced with the metal salt, and is stoichiometric with the metal ion.)

Mass balance:

$$C_M = [M^{2+}] + [ML] + [ML_2^{2-}] + \ldots, \qquad (2.40)$$

$$C_{H_2L} = [L^{2-}] + [HL^-] + [H_2L] + [ML] + 2[ML_2^{2-}] + \ldots \quad (2.41)$$

By substituting C_M from Eq. (2.40) and $[L^{2-}]$ from Eq. (2.41) into Eq. (2.39), we get

$$[K^+] + [H^+] = [OH^-] + 2C_{H_2L} - [HL^-] - 2[H_2L].$$

Rewriting $[HL^-]$ and $[H_2L]$ in terms of the protonation constants, and $[H^+]$ and $[L^-]$ we obtain

$$[K^+] + [H^+] = [OH^-] + 2C_{H_2L} - K_1[H^+][L^{2-}] - 2K_1K_2[H^+]^2[L^{2-}]$$

Taking into account that during the titration

$$[K^+] = aC_{H_2L},$$

the free ligand concentration may be expressed as

$$[L] = \frac{(2-a)C_{H_2L} - [H^+] + [OH^-]}{[H^+]K_1 + 2[H^+]^2 K_1 K_2} \tag{2.42}$$

It can be deduced that if the complexant is an acid containing m protons (protonated base), the following general equation is valid:

$$[L] = \frac{(m-a)C_{H_mL} - [H^+] + [OH^-]}{[H^+]K_1 + 2[H^+]^2 K_1 K_2 + \ldots + m[H^+]^m K_1 \ldots K_m} \tag{2.43}$$

It should be noted that the function in the denominator of the fraction in Eqs. (2.42) and (2.43) is not equal to the $\alpha_{L(H)}$ function.

The equations deduced are suitable for the calculation of the free ligand concentration from $[H^+]$ and $[OH^-]$ at different degrees of neutralization, if C_{H_mL} and $K_1 \ldots K_m$ are known.

Since, owing to the gradual dissociation of the weak acid H_mL, the free ligand concentration may change by several orders of magnitude during the titration, the degree of complex formation can be studied over a wide range of ligand concentration without the need for the tedious preparation of a long series of solutions. This is the great advantage of the titration method, compared with others.

The average ligand number for the free ligand concentration calculated by means of Eq. (2.42) or (2.43) can be obtained as follows, if the total metal and ligand concentrations are known [see Eq. (1.80)]:

$$\bar{n} = \frac{C_{H_mL} - [L]\alpha_{L(H)}}{C_M} \tag{2.44}$$

Sec. 2.2] Stability constants of mononuclear complexes

From the related pairs of $\bar{n}$ and [L] calculated from the coordinates of points on the titration curve, the formation curve can be constructed and the stability constants calculated.

If at some stage of the titration of the acid practically only the species HL and L are present in the solution, and there is an appreciable shift of the titration curve in the pH range 5–9 in the presence of the metal ion, the concentration of bound ligand can be read directly from a graph of both titration curves in the same system of coordinates by Calvin and Melchior's method [67]. The horizontal distances between the two titration curves give the amounts of standard solution equivalent to the hydrogen ions liberated during the formation of the complex according to Eq. (2.38). If the degree of neutralization of an acid to a given pH is denoted by a^0, and in the presence of the metal ion by a, then the concentration of the bound ligand is given by $(a - a^0)C_{HL}$, and the average ligand number by the following fraction (see p. 27).

$$\bar{n} = \frac{[L_{bound}]}{C_M} = \frac{(a - a_0)C_{HL}}{C_M} \quad (2.45)$$

The free ligand concentration can be calculated according to Eq. (1.79) as follows:

$$[L] = \frac{C_{HL} - (a - a^0)C_{HL}}{\alpha_{L(H)}} \quad (2.46)$$

If the conditions for applicability of the method are fulfilled – that is, the species HL and L are dominant in the range of neutralization studied – then the denominator will be given by $\alpha_{L(H)} \simeq 1 + [H^+]K_1$ even in the case of polybasic acids, i.e. further terms corresponding to species containing more protons can be neglected (see Example 29).

Calculation of stability constants from related pairs of [L] *and* $\bar{n}$. When only one complex species is formed of the composition 1 : 1, the calculation or graphical evaluation can be done in the same way as described in connection with the determination of protonation constants. Since in this case $[ML]/C_M = \bar{n}$ and $[M]/C_M = (1 - \bar{n})$, from Eq. (1.11) defining the stability constant,

$$\log \frac{[ML]}{[M]} = \log \frac{\bar{n}}{1 - \bar{n}} = \log K + \log [L] \quad (2.47)$$

On plotting log $[\bar{n}/(1 - \bar{n})]$ against log [L], we get a straight line, the intercept of which is log K. The calculation is more involved if $N > 1$.

The complex formation function defined by Eq. (1.18) gives a relationship between the average ligand number, free ligand concentration and the constants β_i. By rearranging the equation we obtain

$$\bar{n} + (\bar{n} - 1)\beta_1[L] + (\bar{n} - 2)\beta_2[L]^2 + \ldots + (\bar{n} - N)\beta_N[L]^N = 0$$

or

$$\bar{n} + \sum_1^N (\bar{n} - i)\beta_i[L]^i = 0 \qquad (2.48)$$

From equations given for N related $\bar{n}$ and [L] pairs, the various β values can be calculated. The accuracy of the constants obtained can be increased by calculating them by using the method of least squares and a large number of related $\bar{n}$ and [L] pairs. Since the statistical evaluation becomes more complicated as the number of β constants increases, it is advisable to use a computer. For computer procedures see refs. [68]–[74]. Statements similar to those made in connection with the resolution of the function A hold also in this case for the numerical calculations (see p. 114).

Of the many graphical methods of evaluation only some will be dealt with here. The first and simplest method is due to Bjerrum [16] and is similar to the method of determination of protonation constants. It is based on the principle that the formation curve $\bar{n}$ vs. log [L] obtained experimentally has inflection points where $\bar{n}$ is an integer, if the difference between the logarithms of successive stability constants is greater than 2·8. In this case, only two complex species are present in significant amount within a section of $\bar{n}$, and the negative logarithms of the successive stability constants can be directly read from the curve by means of the 'half-value points'. The reciprocal value of the ligand concentration belonging to the $\bar{n} - 0.5$ points is equal to the stability constant.

Since $K_n = [ML_n]/[ML_{n-1}][L]$, then at the half-value point, $\bar{n} - 0.5$

$$[ML_n] = [ML_{n-1}] \qquad (2.49)$$

$$K_n = \frac{1}{[L]_{n-0.5}} \qquad (2.50)$$

In general, the following relationship holds for any $\bar{n}$ points between n and $n - 1$:

$$K_n = \frac{\bar{n} + 1 - n}{(n - \bar{n})[L]} \qquad \text{(if } n > \bar{n} > n - 1\text{)} \quad (2.51)$$

Sec. 2.2] Stability constants of mononuclear complexes

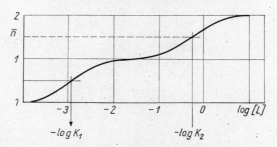

Fig. 2.9. Graphical determination of complex stability constants from the half-value points, using Bjerrum's method

For the simplest case, if the highest complex is ML_2 the determination of the stability constants by graphical method is shown in Fig. 2.9.

If the difference in the logarithms of two consecutive stability constants is smaller than 2·8 the inflection points become indistinct, and the condition given by Eq. (2.49) is met at ligand concentrations other than those belonging to the half-value points.

If complexes higher than ML_2 can be formed, the difference in the logarithms of successive equilibrium constants is often smaller than 2·8. In such cases more than two complex species are present in the solution simultaneously, and obviously the half-value points or graphical methods based on Eq. (2.51) yield only the 'mean' value of stability constants.

Nevertheless, the great advantage of the method is that it gives information rapidly and simply.

Another graphical method can be used if ML and ML_2 are formed. The formation function can be rewritten in linear form as follows:

$$\frac{\bar{n}}{(1-\bar{n})[L]} = \beta_1 + \beta_2 \left(\frac{2-\bar{n}}{1-\bar{n}}\right)[L] \qquad (2.52)$$

If the left-hand side of Eq. (2.52) is plotted against $(2-\bar{n})[L]/(1-\bar{n})$, then a straight line is obtained, the intercept of which is β_1 and the slope β_2. Accurate evaluation can be greatly facilitated by the application of the method of least squares.

The third graphical method, which is due to Fronaeus [75], can be used even in the case of multistep complex formation.

In Eq. (2.34) differentiate the function $A_{M(L)}$ with respect to [L].

$$\frac{dA}{d[L]} = \beta_1 + 2\beta_2[L] + \ldots + N\beta_N[L]^{N-1} \qquad (2.53)$$

A comparison of the formation function defined by Eq. (1.18) and the functions given by Eqs. (2.34) and (2.53) clearly shows that the ratio $\bar{n}/[L]$ is just the ratio of the two functions.

That is

$$\frac{\bar{n}}{[L]} = \frac{\dfrac{dA}{d[L]}}{A} \qquad (2.54)$$

By rearranging the equation we get

$$\frac{\bar{n}}{[L]} d[L] = \frac{1}{A} dA \qquad (2.55)$$

Taking into account that $dA/A = d \ln A$, on integrating Eq. (2.55) between 0 and $[L]_i$ we obtain

$$\int_0^{[L]_i} \frac{\bar{n}}{[L]} d[L] = \ln A_i \qquad (2.56)$$

According to Eq. (2.56) if the $\bar{n}/[L]$ values calculated from experimental data are plotted against $[L]$, A_i corresponding to $[L]_i$ can be obtained by graphical integration (see Fig. 2.10). Provided that a number of related A and $[L]$ values are known, the complex products can be obtained one after the other by using the successive graphical extrapolation method due to Leden. The value of the term $(A-1)/[L]$ is plotted against $[L]$. The intercept will give β_1 etc. (see p. 115).

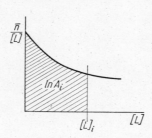

Fig. 2.10. Determination of the value of the function A by a graphical method, due to Fronaeus

The modified correction-term method due to Gergely, Nagypál and Mojzes can be used successfully in the case of multi-step complex formation and yields very accurate results even if the ratio of consecutive constants is small [76].

The next method is a general method of evaluation which can be used when related values of two variables can be determined experimentally, the correlation of the two variables is known, and the constants of the function are to be calculated.

Stability constants of mononuclear complexes

The method is due to Sillén [77] and is called the curve-fitting method. Its principle is as follows. If the basic function containing several constant parameters, $y = f(x, p_1, p_2 \ldots)$, is known, then a 'normalized' function $Y = F(X)$ is to be looked for, which gives the relationships $Y = (y + P_1)$ and $X = (x + P_2)$ between the normalized variables and the variables of the original function. The parameters p_1, p_2 etc. of the original function need not be equal to the parameters P_1, P_2 etc. of the normalized function, but should be unambiguously related to them. If the normalized function is plotted with coordinates X and Y and the curve constructed from x and y values based on experimental data is fitted to it, in the case of a good fit the parameters P_1 and P_2 can be obtained from the x value corresponding to $X = 0$ and from the y value corresponding to $Y = 0$. Namely, if $X = 0$, $x = -P_2$ and if $Y = 0$ then $y = -P_1$. As the relationship between the parameters p_1, p_2 etc. of the original function and the parameters P_1 and P_2 is known, they can also be calculated.

One of the constant parameters can also be found by introducing a parameter into the normalized function, calculating the corresponding X and Y values of the function $Y = F(X, P)$ for different values of P, then plotting the series of curves obtained, with coordinates X and Y. The basic curve constructed from experimental data with x and y coordinates is fitted to the series of the normalized curves, and the appropriate constant of the original function is calculated from the P value of the best fitting normalized curve.

Accordingly, not only two, but even three constants can be determined by the curve-fitting method.

Finding the appropriate normalized function can be greatly facilitated by using logarithmic coordinates. In such cases additivity of the logarithms is the condition of normalization instead of the original condition (i.e. the additive correlation between the basic variable and the constant).

The application of the curve-fitting method will be shown in the case where two constants are to be calculated by using Eq. (1.18), that is, $N = 2$, and the corresponding $\bar{n}$ and [L] values have been calculated from experimental data. Only one variable is normalized. One parameter is determined by shifting the basic diagram along one axis of the normalized diagram, the other by finding the normalized curve which fits best.

The basic function which contains the constants sought is [Eq. (1.18)]

$$y = f(x); \quad \bar{n} = \frac{\beta_1[L] + 2\beta_2[L]^2}{1 + \beta_1[L] + \beta_2[L]^2} \quad (2.57)$$

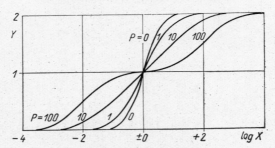

Fig. 2.11. Series of normalized curves for the application of Sillén's curve-fitting method; i.e. representation of function (2.59) for different P values [77]

The variable $x = [L]$ is normalized:

$$X = [L]\sqrt{\beta_2}; \quad Y = \bar{n} \tag{2.58}$$

The normalized function will be as follows:

$$Y = F(X, P); \quad Y = \frac{PX + 2X^2}{1 + PX + X^2} \tag{2.59}$$

where

$$P = \frac{\beta_1}{\sqrt{\beta_2}} = \sqrt{\frac{K_1}{K_2}} \tag{2.60}$$

The original function is plotted with coordinates $\bar{n}$ and log [L] on the basis of experimental data. Similarly, the normalized function is also plotted on a graph of the same scale with different P values. The series of curves shown in Fig. 2.11 is obtained in this way.

As one coordinate of both the original and the normalized functions is logarithmic, the condition of normalization refers to the variables log x and log X, namely:

$$\log X = \log [L] + \frac{1}{2} \log \beta_2 \tag{2.61}$$

[see Eq. (2.58)].

It is expedient to draw the normalized curves on transparent paper, then to shift this along the log [L] axis on the graph of the basic function until the best fit is reached. The value of P can be obtained directly from the best fitting curve. β_2 is calculated by using the relationship:

$$\log [L]_0 = -\frac{1}{2} \log \beta_2,$$

Sec. 2.2]   Stability constants of mononuclear complexes

where log $[L]_0$ is the abscissa value on the original graph which coincides with the value $\log X = 0$ on the fitted normalized graph. Once β_2 and P are known, β_1 is calculated from Eq. (2.60).

A comparison of the basic (2.57) and the normalized (2.59) functions shows that with a suitable modification the equations can be used for calculating the overall stability constants if the values of pairs of [L] and Φ_1 or Φ_2 are known. If, for example, the mole-fraction of the ML_2 species is known at different free ligand concentrations, then the basic and normalized functions will be as follows [see Eq. (1.20)]:

$$y = f(x); \quad \Phi_2 = \frac{\beta_2[L]^2}{1 + \beta_1[L] + \beta_2[L]^2} \quad (2.62)$$

$$Y = F(X, P); \quad Y = \frac{X^2}{1 + PX + X^2} \quad (2.63)$$

Obviously, Eqs. (2.58) and (2.60) are valid also in this case. The evaluation can be made similarly, but the shape and position of the curves will be different (see Example 33).

It should be noted that Eqs. (2.57) and (2.62) can also be normalized as follows

$$Y = \frac{X + 2PX^2}{1 + X + PX^2} \quad (2.64)$$

and

$$Y = \frac{PX^2}{1 + X + PX^2} \quad (2.65)$$

where

$$X = \beta_1[L] \quad \text{and} \quad P = \frac{\beta_2}{\beta_1^2} = \frac{K_2}{K_1}$$

Practical information for pH-titrations. The titrations can be done with apparatus similar to that described in detail in connection with the determination of protonation constants (see p. 98).

As the alkali error of a glass electrode is greater in sodium hydroxide than in potassium hydroxide solution, the latter is recommended for use as titrant but if the maximum pH to be measured is <10, carbonate-free sodium hydroxide solution can also be used.

To ensure constant ionic strength during the titration an inert electrolyte is added which does not react with the metal ion and/or the ligand. Sodium perchlorate, sodium or potassium nitrate, and sometimes sodium chloride are used.

It is best to determine the protonation constant(s) of the ligand under conditions similar to those for determination of the complex formation constants (see p. 117). Therefore the protonation constants are always determined by titrating the protonated ligand with strong base and not *vice versa*.

It is necessary in the determination of stability constants that the metal ion and ligand be present in commensurable amounts in the solution titrated, and it is also important in the case of multistep complex formation that the amount of ligand be somewhat greater than NC_M.

Similarly to the determination of the protonation constants of the ligand, the error caused is not very great if the total ligand and metal concentrations C_{H_mL} and C_M are considered as constants in the range of neutralization from which the results necessary for the calculation are obtained. This is only justified when the concentration of the titrant is at least ten times that of the titrand, and the concentrations C_{H_mL} and C_M are taken as those corresponding to the midpoint of the range investigated. The results of the calculation cannot be accepted if the calculations are performed on results from a range in which the ligand is practically completely unprotonated, i.e. $pH > \log K_1 + 2$ or is protonated to such an extent that practically no complex formation occurs $[\beta_1 C_L < \alpha_{L(H)}]$ and $\bar{n}$ is smaller than 0·5. Nor can the method be used in cases where the complex is so stable that $\bar{n}$ is equal to the maximum number of ligands even in strongly acidic solution.

As the great majority of complex-forming metal ions also tend to form hydroxo-complexes, and it is difficult to take into account the simultaneous equilibria occurring during titration, the investigations should be made in a pH range where the formation of hydroxo-complexes is negligible. In these circumstances the correlation between $\bar{n}$ and log [L] is independent of the pH.

If turbidity or precipitation occurs, the titration should be stopped immediately and the phenomenon noted. If precipitation occurs, the relationships deduced are no longer valid, as the concentrations are controlled by the solubility conditions.

In the case of complicated systems the determination of the stability constants of the protonated or basic complexes of chelating agents can be facilitated if not only the pH but also pM is followed during the titration (for more details see [78]).

2.2.1.3. Worked examples

29. Two 50-ml solutions have been prepared: one is a mixture of 5 ml of $0.1M$ sulphosalicylic acid, 20 ml of $0.2M$ sodium chloride and 25 ml of water, the other is a mixture of 5 ml of $0.1M$ sulphosalicylic acid, 20 ml of $0.2M$ sodium chloride, 10 ml of $0.01M$ copper(II) chloride and 15 ml of water. Both solutions have been titrated potentiometrically with $0.1M$ sodium hydroxide. The titration curves are shown in Fig. 2.12. From the titration curves calculate the stability constants of the sulphosalicylate copper(II) complexes.

As the second and third dissociation constants of sulphosalicylic acid (or, in other words the first and second protonation constants of the sulphosalicylate ligand) differ by 9 orders of magnitude, the calculation can be done by using Calvin and Melchior's simple method, on the basis of the points taken from the third part of the titration curve.

The horizontal distances, that is, the differences in consumption of standard solution or in degree of neutralization between the titration curves obtained in the absence and in the presence of copper(II) are determined. The concentration of bound ligand is calculated, then the average ligand number and the corresponding free ligand concentration by using Eqs. (2.45) and (2.46).

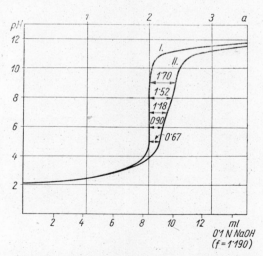

Fig. 2.12. I: Titration of 50 ml of $10^{-2}M$ sulphosalicylic acid with strong base standard solution; II: the same, in the presence of $2 \times 10^{-3}M$ copper(II) chloride ($I \simeq 0.1$)

The total concentrations of sulphosalicylic acid and copper are calculated by taking the solution volume as 59 ml, corresponding to consumption of 9 ml of titrant.

$$C_{SS} = \frac{50}{59} \times 10^{-2} = 8\cdot5 \times 10^{-3} M;$$

$$C_{Cu} = \frac{50}{59} \times 2 \times 10^{-3} = 1\cdot7 \times 10^{-3} M.$$

The pH, difference in standard solution consumption read directly from the titration curve and the calculated bound ligand concentration, the average ligand number and the free ligand concentration are summarized in Table 2.4. The log $\alpha_{SS(H)}$ values belonging to various pH values can be read from the curve in Fig. 1.13.

The formation curve constructed from the $\bar{n}$ and log [L] values calculated, is shown in Fig. 2.13. The approximate stability constants read from the diagram, according to Bjerrum's 'half-value point' method, are

$$\log K_1 = 9\cdot5, \quad \log K_2 = 6\cdot6 \quad (I = 0\cdot1)$$

More accurate results can be obtained by constructing the formation curve on the basis of a large number of experimental results and using Sillén's curve-fitting method for calculating the constants.

30. Calculate the stability constants of the [ethylenediaminenickel(II) complexes on the basis of titration curve II in Fig. 2.14.

Two 50-ml solutions have been prepared. Solution I contains 10 ml of $0\cdot078M$ ethylenediamine, 18 ml of $0\cdot0984M$ hydrochloric acid, 16 ml of $0\cdot2M$ sodium chloride and 6 ml of water. Solution II contains 10 ml of $0\cdot078M$ ethylenediamine, 18 ml of $0\cdot0984M$ hydrochloric acid, 16 ml of $0\cdot2M$ sodium chloride, 2 ml of $0\cdot0935M$ nickel(II) chloride and 4 ml of water.

Both solutions have been titrated with $0\cdot1982M$ sodium hydroxide. Chloride does not form complexes with nickel(II) in dilute solution.

The protonation constants of En, calculated in Example 26 (see p. 106), are

$$\log K_1 = 9\cdot94; \quad \log K_2 = 7\cdot14 \quad (I = 0\cdot1).$$

Hence, only titration curve II is necessary for the calculation of the stability constants.

Table 2.4

Calculation of the points of the formation curve for sulphosalicylate–copper(II) complexes, by using Calvin and Melchior's method, from the data of the titration curve

$C_{SS} = 8.5 \times 10^{-3}M$; $C_{Cu} = 1.7 \times 10^{-3}M$ (Example 29)

pH	0.1M NaOH ml	$(a-a^0)C_{SS} = [L_{bound}]$	$\bar{n} = \dfrac{[L_{bound}]}{C_{Cu}}$	$\log (C_{SS} - [L_{bound}]) - \log \alpha_{SS(H)} = \log [L]$
4.0	0.38	$0.090 \times 8.5 \times 10^{-3} = 0.76 \times 10^{-3}$	0.45	$-2.11 - 7.6 = -9.71$
5.0	0.67	$0.159 \times 8.5 \times 10^{-3} = 1.35 \times 10^{-3}$	0.79	$-2.15 - 6.6 = -8.75$
6.0	0.90	$0.214 \times 8.5 \times 10^{-3} = 1.82 \times 10^{-3}$	1.07	$-2.18 - 5.6 = -7.78$
7.0	1.18	$0.281 \times 8.5 \times 10^{-3} = 2.39 \times 10^{-3}$	1.41	$-2.21 - 4.6 = -6.81$
8.0	1.52	$0.362 \times 8.5 \times 10^{-3} = 3.08 \times 10^{-3}$	1.81	$-2.26 - 3.6 = -5.86$
9.0	1.70	$0.404 \times 8.5 \times 10^{-3} = 3.43 \times 10^{-3}$	2.0	$-2.30 - 2.6 = -4.9$

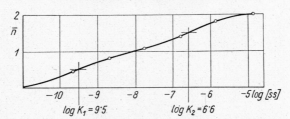

Fig. 2.13. Determination of the stability constants of [sulphosalicylatocopper(II)] complexes by using Bjerrum's half-value point method. (See Example 29)

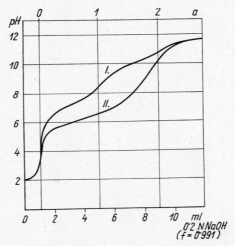

Fig. 2.14. I: Titration of 50 ml of $0.0156M$ ethylenediamine (made acid with hydrochloric acid) with standard sodium hydroxide solution; II: the same in the presence of $0.00374M$ nickel(II) chloride. (See Examples 30 and 31)

The free ligand concentrations are calculated from the total concentrations of ethylenediamine and nickel(II) and the pH values corresponding to different volumes of titrant consumed, by using Eq. (2.42), then the corresponding $\bar{n}$ values by using Eq. (2.44). Finally, the formation curve is constructed on the basis of the related log [L] and $\bar{n}$ values calculated.

The total ethylenediamine and total nickel(II) concentrations are considered as constants and their values are calculated by taking the volume

Sec. 2.2] Stability constants of mononuclear complexes

as 55 ml, corresponding to the addition of 5 ml of titrant.

$$C_{En} = \frac{10}{55} \times 0.078 = 1.42 \times 10^{-2} M;$$

$$C_{Ni} = \frac{2}{55} \times 0.0935 = 3.4 \times 10^{-3} M.$$

As $18 \times 0.0984 - 2 \times 10 \times 0.078 = 0.210$ mmole of hydrochloric acid was present in excess in the solution titrated, $0.210/0.1982 \simeq 1.06$ ml of $0.1982 M$ sodium hydroxide was necessary to neutralize the excess acid. This will be taken into consideration in determining the neutralization ranges.

The $a = 0$ point is at 1·06 ml, the $a = 1$ point at

$$1.06 + \frac{10 \times 0.078}{0.1982} = 1.06 + 3.94 = 5.00 \text{ ml},$$

and the $a = 2$ point at $1.06 + 2 \times 3.94 = 8.94$ ml of titrant added.

The calculation done on the basis of six points of the titration curve is shown in Tables 2.5 and 2.6. In calculating the free ligand concentration the [H$^+$] and [OH$^-$] concentrations in the numerator were neglected in each case. However, the [H$^+$] concentration may not be neglected in the denominator and in the calculation of the value of the $\alpha_{L(H)}$ function.

The formation curve constructed from the log [L] and $\bar{n}$ values calculated is shown in Fig. 2.15. The mean stability constants on the basis of the half-value points are

$$\log K_1 = 7.7; \quad \log K_2 = 6.5; \quad \log K_3 = 4.7.$$

As the difference between the logarithms of consecutive stability constants is less than 2·8, a correction is necessary to determine accurately the values of the constants. Consider the relaxation method [65]. The basic function is rewritten as follows [see Eq. (2.48)]:

$$\bar{n} + (\bar{n} - 1)\beta_1[L] + (\bar{n} - 2)\beta_2[L]^2 + (\bar{n} - 3)\beta_3[L]^3 = 0.$$

Select three $\bar{n}$, [L] pairs and insert them into this equation. In this way three equations are obtained, which will be set equal to A, B and C. For the corresponding pairs, the following data are taken from the diagram: $\bar{n} = 0.5$, $[L] = 2 \times 10^{-8} M$; $\bar{n} = 1.5$, $[L] = 3.16 \times 10^{-7} M$; $\bar{n} = 2.5$, $[L] = 2 \times 10^{-5} M$.

Table 2.5

Calculation of the free ligand concentration by using the data of the titration curve in Fig. 2.14. $C_{En} = 1.42 \times 10^{-2} M$; $C_{Ni} = 3.4 \times 10^{-3} M$ (Example 30)

Standard solution 0.2M NaOH ml	Degree of neutralization a	pH	$\dfrac{(2-a)C_{En}}{[H^+]K_1 + 2[H^+]^2 K_1 K_2} = [L]$
2	0.24	5.60	$\dfrac{1.76 \times 1.42 \times 10^{-2}}{10^{-5.6} \times 10^{9.94} + 2 \times 10^{-11.2} \times 10^{17.08}} = \dfrac{2.50 \times 10^{-2}}{1.53 \times 10^{6}} = 1.63 \times 10^{-8}$
3	0.49	5.95	$\dfrac{1.51 \times 1.42 \times 10^{-2}}{10^{-5.95} \times 10^{9.94} + 2 \times 10^{-11.9} \times 10^{17.08}} = \dfrac{2.14 \times 10^{-2}}{3.12 \times 10^{5}} = 6.85 \times 10^{-8}$
4	0.75	6.25	$\dfrac{1.25 \times 1.42 \times 10^{-2}}{10^{-6.25} \times 10^{9.94} + 2 \times 10^{-12.5} \times 10^{17.08}} = \dfrac{1.78 \times 10^{-2}}{8.08 \times 10^{4}} = 2.20 \times 10^{-7}$
5	1.00	6.57	$\dfrac{1.42 \times 10^{-2}}{10^{-6.57} \times 10^{9.94} + 2 \times 10^{-13.14} \times 10^{17.08}} = \dfrac{1.42 \times 10^{-2}}{1.97 \times 10^{4}} = 7.20 \times 10^{-7}$
6	1.25	6.95	$\dfrac{0.75 \times 1.42 \times 10^{-2}}{10^{-6.95} \times 10^{9.94} + 2 \times 10^{-13.9} \times 10^{17.08}} = \dfrac{1.07 \times 10^{-2}}{4.0 \times 10^{3}} = 2.67 \times 10^{-6}$
7	1.51	7.65	$\dfrac{0.49 \times 1.42 \times 10^{-2}}{10^{-7.65} \times 10^{9.94} + 2 \times 10^{-15.30} \times 10^{17.08}} = \dfrac{6.96 \times 10^{-3}}{3.15 \times 10^{2}} = 2.21 \times 10^{-5}$

Sec. 2.2] Stability constants of mononuclear complexes 133

Table 2.6

Calculation of the average ligand number belonging to the free ligand concentrations calculated in Table 2.5. $C_{En} = 1.42 \times 10^{-2} M$; $C_{Ni} = 3.4 \times 10^{-3} M$ (Example 30)

[L]	log [L]	$1 + [H^+]K_1 + [H^+]^2K_1K_2 = \alpha_{L(H)}$	$\dfrac{C_{En} - [L]\alpha_{L(H)}}{C_{Ni}} = \bar{n}$
1.63×10^{-8}	-7.79	$1 + 10^{-5.6} \times 10^{9.94} + 10^{-11.2} \times 10^{17.08} = 7.81 \times 10^{5}$	$\dfrac{1.42 \times 10^{-2} - 1.63 \times 10^{-8} \times 7.81 \times 10^{5}}{3.4 \times 10^{-3}} = 0.44$
6.85×10^{-8}	-7.16	$1 + 10^{-5.95} \times 10^{9.94} + 10^{-11.9} \times 10^{17.08} = 1.61 \times 10^{5}$	$\dfrac{1.42 \times 10^{-2} - 6.85 \times 10^{-8} \times 1.61 \times 10^{5}}{3.4 \times 10^{-3}} = 0.94$
2.20×10^{-7}	-6.66	$1 + 10^{-6.25} \times 10^{9.94} + 10^{-12.5} \times 10^{17.08} = 4.29 \times 10^{4}$	$\dfrac{1.42 \times 10^{-2} - 2.20 \times 10^{-7} \times 4.29 \times 10^{4}}{3.4 \times 10^{-3}} = 1.40$
7.20×10^{-7}	-6.14	$1 + 10^{-6.57} \times 10^{9.94} + 10^{-13.14} \times 10^{17.08} = 1.10 \times 10^{4}$	$\dfrac{1.42 \times 10^{-2} - 7.2 \times 10^{-7} \times 1.10 \times 10^{4}}{3.4 \times 10^{-3}} = 1.85$
2.67×10^{-6}	-5.57	$1 + 10^{-6.95} \times 10^{9.94} + 10^{-13.9} \times 10^{17.08} = 2.49 \times 10^{3}$	$\dfrac{1.42 \times 10^{-2} - 2.67 \times 10^{-6} \times 2.49 \times 10^{3}}{3.4 \times 10^{-3}} = 2.22$
2.21×10^{-5}	-4.66	$1 + 10^{-7.65} \times 10^{9.94} + 10^{-15.30} \times 10^{17.08} = 2.56 \times 10^{2}$	$\dfrac{1.42 \times 10^{-2} - 2.21 \times 10^{-5} \times 2.56 \times 10^{2}}{3.4 \times 10^{-3}} = 2.51$

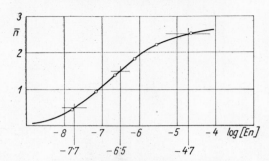

Fig. 2.15. Formation curve of [ethylenediamine-nickel(II)] complexes, calculated from a titration curve. (See Example 30 and Table 2.7)

After inserting these values into the equation, we obtain the three equations

$$0.5 - 10^{-8}\beta_1 - 6 \times 10^{-16}\beta_2 - 2 \times 10^{-23}\beta_3 = A;$$

$$1.5 + 1.58 \times 10^{-7}\beta_1 - 5 \times 10^{-14}\beta_2 - 4.74 \times 10^{-20}\beta_3 = B;$$

$$2.5 + 3 \times 10^{-5}\beta_1 + 2 \times 10^{-10}\beta_2 - 4 \times 10^{-15}\beta_3 = C.$$

The correct β values satisfy the simultaneous equations if $A = B = C = 0$.

Before finding the accurate β values consider the effect on A, B and C of increasing the different β values by 1. These considerations lead to the results shown in the upper part of Table 2.7. Substitute into all three equations the overall stability constants calculated from the stability constants found by the half-value point method:

$$\log \beta_1 = \log K_1 = 7.7; \quad \beta_1 = 5 \times 10^7;$$

$$\log \beta_2 = \log K_1 + \log K_2 = 14.2; \quad \beta_2 = 1.58 \times 10^{14};$$

$$\log \beta_3 = \log K_1 + \log K_2 + \log K_3 = 18.9; \quad \beta_3 = 7.94 \times 10^{18}.$$

The values of A, B and C obtained by inserting these data into the equations are shown in the first line of the lower part of Table 2.7. These values of A, B and C differ from 0. By taking into account the change in A, B and C on changing the β values by 1, the remainders are eliminated by trial and error. The partial results of further calculations are shown in Table 2.7. First β_1 is altered, by which the remainders are reduced. C can be markedly decreased by altering β_3, as this has very small

Sec. 2.2] Stability constants of mononuclear complexes

Table 2.7

Calculation of the overall stability constants of [ethylenediaminenickel(II)] complexes, by the relaxation method

$\pm \Delta\beta_i$	A	B	C
$\Delta\beta_1 = 1$	-10^{-8}	$+1.58\times 10^{-7}$	$+3\times 10^{-5}$
$\Delta\beta_2 = 1$	-6×10^{-16}	$-5\ \times 10^{-14}$	$+2\times 10^{-10}$
$\Delta\beta_3 = 1$	-2×10^{-23}	-4.74×10^{-20}	-4×10^{-15}
$\Delta\beta_1 = 5\ \times 10^7$	-0.0949	$+1.124$	$+1342.5$
$\Delta\beta_2 = 1.58\times 10^{14}$			
$\Delta\beta_3 = 7.94\times 10^{18}$			
$\Delta\beta_1 = -0.6\ \times 10^7$	-0.0349	$+0.176$	$+1162.5$
$\Delta\beta_3 = +0.29\ \times 10^{18}$	-0.0349	$+0.1623$	$+\quad 2.5$
$\Delta\beta_1 = -0.1\ \times 10^7$	-0.0249	$+0.004$	$-\quad 27.5$
$\Delta\beta_3 = -0.007\times 10^{18}$	-0.0249	$+0.004$	$+\quad 0.5$
$\Sigma\Delta\beta_1 = 4.3\ \times 10^7$			
$\Sigma\Delta\beta_2 = 1.58\ \times 10^{14}$			
$\Sigma\Delta\beta_3 = 8.223\times 10^{18}$			

influence on A and B. The remainders could be reduced to an acceptably low value by a further correction step. The improved estimates of the β values are given by the algebraic sum of the $\Delta\beta$ values used during the calculations. That is:

$$\beta_1 = \Sigma\Delta\beta_1 = 5\times 10^7 - 0.6\times 10^7 - 0.1\times 10^7 = 4.3\times 10^7;$$

$$\beta_2 = \Sigma\Delta\beta_2 = 1.58\times 10^{14};$$

$$\beta_3 = \Sigma\Delta\beta_3 = 7.94\times 10^{18} + 0.29\times 10^{18} - 0.007\times 10^{18} = 8.223\times 10^{18}.$$

The overall stability constants of the [ethylenediaminenickel(II)] complexes are as follows:

$$\log\beta_1 = 7.63;\quad \log\beta_2 = 14.2;\quad \log\beta_3 = 18.92.$$

31. Construct the formation curve for protonated ethylenediamine and determine the protonation constants by using the titration curve I in Fig. 2.14.

The calculation is done by using Eq. (2.28). As the important portion of the titration curve is in a pH range where the $[H^+]$ and $[OH^-]$ con-

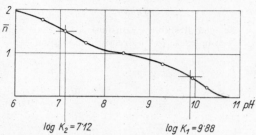

Fig. 2.16. Formation curve of the protonation of ethylenediamine, calculated from a titration curve. (See Example 31 and Table 2.8)

centrations can be neglected, the calculation is simplified. The pH and $\bar{n}$ values belonging to various degrees of neutralization (the latter having been calculated in the preceding Example) are given in Table 2.8 and the corresponding curve in Fig. 2.16.

The protonation constants read from the half-value points are:

$$\log K_1 = 7 \cdot 12; \quad \log K_2 = 9 \cdot 88 \quad (I = 0 \cdot 1).$$

(Compare with the results of Example 26.)

Table 2.8

Calculation of the average number of bound protons necessary for the construction of the formation curve of the protonation of ethylenediamine, by using Eq. (2.28).

$$C_{En} = 1 \cdot 42 \times 10^{-2} M$$

Standard solution consumption 0·2M NaOH ml	Degree of neutralization a	pH	Average proton number $\dfrac{(2-a)\,C_{En} + [OH^-] - [H^+]}{C_{En}} = \bar{n}$
2	0·24	6·60	$2 - 0·24 = 1·76$
3	0·49	7·10	$2 - 0·49 = 1·51$
4	0·75	7·45	$2 - 0·75 = 1·25$
5	1·00	8·40	$2 - 1·00 = 1·00$
6	1·25	9·35	$2 - 1·25 = 0·75$
7	1·51	9·90	$\dfrac{0·49 \times 1·42 \times 10^{-2} + 10^{-4 \cdot 1}}{1·42 \times 10^{-2}} = 0·49$
8	1·76	10·25	$\dfrac{0·24 \times 1·42 \times 10^{-2} + 10^{-3 \cdot 75}}{1·42 \times 10^{-2}} = 0·24$

2.2.2. Spectrophotometric methods

The optical properties of solutions containing complexes usually differ from those of the constituent ions or molecules. The change in the optical behaviour is closely related to the formation of coordinate bonds (see p. 67 and [30]). Analytical methods based on the measurement of light absorption can be used to advantage for studying complexation equilibria, since they are suited to the selective determination of very small concentrations of certain species without changing the composition of the solution. The determination can usually be rendered selective by an appropriate choice of the wavelength. The basic condition for application of all analytical methods based on the measurement of light absorption is that the Beer−Lambert law is obeyed by the constituents to be determined.

2.2.2.1. *Method of continuous variation*

This method is sometimes known as Job's method [79] although it is not originally due to him. The principle of the method is that the mole-ratio of the metal ion and ligand is varied between 0 and 1 at constant total concentration $C = C_L + C_M$ and the absorbance of the solutions of different composition is measured. The absorbances are then plotted against the mole-fraction, x_L, of the ligand. If only one complex species has been formed, with composition ML_n, and the absorbance is measured at a wavelength where neither the metal ion nor the ligand but only the complex absorbs, then n can be calculated from the abscissa of the maximum of the curve (x_{max}):

$$n = \frac{x_{max}}{1 - x_{max}} \qquad (2.66)$$

The stability constant can also be calculated from the curve, by drawing tangents to the initial and final parts of the curve, and by using the co-ordinates of certain points on the tangents and the curve, without knowing the molar absorptivity of the complex formed (see Fig. 2.17).

The points on the tangent drawn to the last part of the curve give the absorbances which would be measured if the metal were completely present in the form of the complex. Before the intersection of the two straight lines, the complex formation is limited by the ligand concentration, and the absorbances of the points on the initial part of the ascending

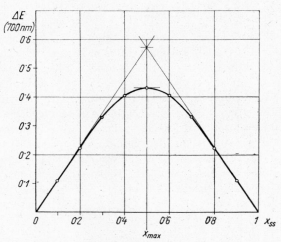

Fig. 2.17. Investigation of the copper(II)–sulphosalicylate system by the method of continuous variation. $C = C_{Cu} + C_{SS} = 0.045M$ at pH 5

line are proportional to the concentration of complex equivalent to the total ligand concentration. The ratio of the observed absorbance to that indicated by the tangent, for the same value of x, is equal to the mole-fraction of the metal ion in the complex, when $x > x_{max}$

$$\frac{E}{E_{ex}} = \frac{[ML_n]}{C_M} = \Phi_n \qquad (2.67a)$$

and the mole-fraction of the ligand in the complex when $x < x_{max}$

$$\frac{E}{E_{ex}} = \frac{n[ML_n]}{C_L} \qquad (2.67b)$$

The concentration of the complex species can be expressed from these equations as follows

$$[ML_n] = \frac{E}{E_{ex}} C_M$$

or

$$[ML_n] = \frac{E}{E_{ex}} \times \frac{C_L}{n}$$

Sec. 2.2] **Stability constants of mononuclear complexes**

If the multiplier of the ratio of absorbances is denoted by C_x,

if $x > x_{max}$, then $C_x = C_M$

if $x < x_{max}$, then $C_x = \dfrac{C_L}{n}$

if $x = x_{max}$, then $C_x = C_M = \dfrac{C_L}{n}$

From these, the concentration of the free metal ion and free ligand can be expressed as follows:

$$[M] = C_M - \frac{E}{E_{ex}} C_x;$$

$$[L] = C_L - n \frac{E}{E_{ex}} C_x$$

The overall stability constant is given by

$$\beta_n = \frac{[ML_n]}{[M][L]^n} = \frac{\dfrac{E}{E_{ex}} C_x}{\left[C_M - \dfrac{E}{E_{ex}} C_x\right]\left[C_L - n\dfrac{E}{E_{ex}} C_x\right]^n} \qquad (2.68)$$

and can thus be calculated from the ordinates of points with the same abscissa.

The greater the stability of the complex, the sharper is the peak of the curve and the greater is the error caused by incorrect reading of the ordinate values. If on the other hand, the stability of the complex is small, the curve becomes flat and the drawing of the tangents becomes uncertain.

If the free metal ion (or the ligand) also absorbs at the wavelength chosen, then the absorbance due to the metal (or ligand) has to be subtracted from the absorbance measured, and the difference in absorbances is plotted as the ordinate.

The method of continuous variation is the most widely used of photometric methods, because of its simplicity and rapidity. Its applicability is, however, rather limited. If the stability constant is small or very large, or n is greater than 3, the evaluation becomes uncertain. The method cannot normally be used when more than one complex is formed simultaneously, as the position of the maximum changes with the total concentration C chosen, and Eq. (2.66) yields only the average ligand number. An important condition for the applicability of the method is that the metal and ligand do not react with other constituents present in the solution, or that any such interfering equilibrium can be taken into account. As the majority of ligands are weak bases, the commonest side-reaction is protonation of the ligand. If the protonated form of the ligand is to be taken into account in the case of aqueous solutions, the pH must be kept constant during the experiments and therefore the free ligand concentration in Eq. (2.68) and hence the stability constant calculated will be conditional. The stability constant can be obtained from the relationship

$$\beta_n = \beta'_n \alpha_{L(H)}^n \qquad (2.69)$$

if $\alpha_{L(H)}$ has been calculated from the pH used in all the experiments and from the protonation constants.

Similarly, possible side-reactions of the metal (formation of hydroxocomplexes or complexes with buffer components) have to be taken into account if the position of the equilibrium is not to shift during the experiments [see Eq. (1.73)].

Important instructions for the measurement. As the validity of the Beer–Lambert law is a basic condition for the applicability of the method, it is expedient to use low concentrations ($<10^{-2}M$) of the complex-forming constituents. To ensure constant ionic strength an inert electrolyte is added in a concentration greater than that of the complexing constituents ($\sim 10^{-1}M$). If the protonation of the ligand is to be taken into account, the pH of the solutions studied should be adjusted carefully as the pH has a marked effect on the value of $\alpha_{L(H)}$. [See Eq. (1.72).] This can be done by using a buffer of sufficient capacity (if there are no side-reactions between the buffer and the metal ion) or by adding concentrated perchloric acid or sodium hydroxide solution (6–12M) from a dropper with a long capillary jet immersed in the vigorously stirred solution, until the desired pH is reached, as measured by a pH-meter with its electrodes also immersed in the solution.

The absorption spectrum of the complex is used to choose the optimum wavelength, where the absorption of the complex is high, and the constituents absorb only slightly or not at all, and where a small shift in the wavelength does not cause an appreciable change in the absorbance.

2.2.2.2. *The mole-ratio method*

The principle of the method is that a series of solutions is prepared in which the concentration of one component (usually C_M) is kept constant and that of the other varied. The absorbance of the solutions is measured at a suitable wavelength and plotted vs. the ratio of the variable and constant concentrations. If only one stable complex is formed, which has selective light absorption, then the absorbance increases approximately linearly with the mole-ratio and then becomes constant (see Fig. 2.18). The abscissa of the point of intersection of the two tangents gives the number of ligands in the complex, if it was the ligand concentration that was varied.

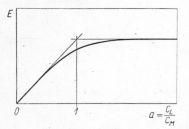

Fig. 2.18. Investigation of the formation of a metal complex by the mole-ratio method

The stability constant can be calculated from the coordinates of the points of the straight lines and the curve in a similar way to the case of the method of continuous variation [80, 81].

In the simplest case, if only one complex of composition ML is formed the mole-fraction of the complex $\Phi_{ML} = [ML]/C_x = E/E_{ex}$ and hence

$$K_{ML} = \frac{\dfrac{E}{E_{ex}} C_x}{\left[C_M - \dfrac{E}{E_{ex}} C_x \right] \left[C_L - \dfrac{E}{E_{ex}} C_x \right]} \qquad (2.70)$$

where E_{ex} stands for the extrapolated, and E for the actual absorbance at the same abscissa value. Before the intersection $C_x = C_L$, after that $C_x = C_M$.

In the case of successive complex formation, if the stability constants of the various complex species markedly differ, then several linear sections occur on the curve.

For the conditions for the applicability of the method, the choice of suitable wavelength, the role of the pH and the instruments necessary, the considerations given in connection with the method of continuous variation apply.

2.2.2.3. *Bjerrum's method*

This method can be used in the case of successive complex formation, if the ligand tends to protonate (and the protonation constants are known) and the complex with maximum number of ligands has selective light absorption at the wavelength chosen. When determining stability constants, it is also necessary to know the composition of the complex species formed.

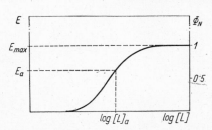

Fig. 2.19. Investigation of the formation of a metal complex by spectrophotometry, using Bjerrum's method. Only the complex with maximum ligand number absorbs

A series of solutions is prepared in which both C_M and C_L are constant, $C_L \gg C_M$, but the pH is varied. The absorbances of the solutions are measured and plotted vs. the logarithm of the free ligand concentration. The free ligand concentration of the solutions is calculated from the pH and the protonation constants, by using Eq. (1.77). If the overall stability constant of the light-absorbing complex with maximum coordination number is high enough, then the absorbance will reach a limiting value in the pH range investigated (see Fig. 2.19). As the concentration of the complex with maximum ligand number is practically equal to C_M when the limiting value is reached, the mole-fraction Φ_N can be calculated at any point of the curve by dividing the actual absorbance by the maximum absorbance:

$$\Phi_N = \frac{[ML_N]}{C_M} = \frac{E_a}{E_{max}} \qquad (2.71)$$

From the related [L] and Φ_N values, the constants can be calculated by using either Eq. (1.20) or Sillén's curve-fitting method (see p. 123).

If $N = 1$ (complex ML), then Eq. (1.20) can be written in the form

$$\log \frac{\Phi_1}{1 - \Phi_1} = \log \beta_1 + \log [L] \qquad (2.72)$$

Plotting the left-hand side of Eq. (2.72) against log [L] gives a straight line, the intercept of which is $\log \beta_1$ [82].

A great advantage of studies made by changing the pH is that the free ligand concentration can be changed sensitively over a wide range without changing the ionic strength appreciably. The drawback of the method is that above pH 7 the results may be falsified by competing formation of metal hydroxo-complexes.

The conditions for the applicability of the method are as follows: the maximum ligand number should not be greater than 2 (possibly 3), and only the complex species with maximum coordination number should have appreciable absorption at the wavelength chosen. If $N = 1$, another light-absorbing constituent is permissible. The value of β_N has to be high enough for E_{max} to be reached in the pH range investigated but not so high that the metal ion is present in the form of ML_N even in acid solution (pH < 3). It is also very important that the first protonation constant is large and that the metal ion is not involved in any side-reaction in the pH range studied.

The practical directions given for determination of protonation constants and application of the method of continuous variation apply (see pp. 104 and 140). The complete series of measurements can be made on one solution of known composition by adding to it small portions of standard alkali solution from a burette, and measuring the pH and absorbance of the solution simultaneously. Obviously, it is important that the dilution of the solution during the measurements is negligible or can be corrected for (without causing other errors).

2.2.2.4. *Other methods*

Several other methods are known in the literature for the determination of stability constants on the basis of the measurement of light absorption (only some of which will be mentioned here), which can be used in the case of multistep complex formation. One is the method of 'corresponding solutions' due to Bjerrum [83] and the other methods are the spectrophotometric methods due to Yatsimirskii [84] and Asmus [85]. A number of methods are described in detail in monographs [60–62]. For computer applications see [86, 87].

2.2.2.5. *Worked examples*

32. According to literature data, copper(II) forms only a 1 : 1 complex with sulphosalicylate at pH 5. For the purpose of studying the complex formation, a series of 11 solutions is prepared. The sum of the total concentration of the metal and ligand is constant, but the metal to ligand ratio is different for each solution. $C = C_{Cu} + C_{SS} = 0.045M$. The pH of the solutions is adjusted to 5·00 with sodium hydroxide and the absorbances are measured at 700 nm. The absorbance of the solution which contains copper(II) chloride is 0·330. The absorbance corresponding to the gradually decreasing amount of copper(II) (corrected for the amount of copper complexed) is subtracted from that of the other solutions. Plotting the corrected absorbances *vs.* the mole-fraction of sulphosalicylic acid gives the graph shown in Fig. 2.17. Determine the composition and the stability constant of the complex.

The mole-fraction of sulphosalicylic acid at the point of intersection of the tangents drawn to the initial and final sections of the curve is $x_{SS} = 0.5$, and, according to Eq. (2.66) $n = 1$, that is, practically only the 1:1 complex has been formed. For simplicity, the stability constant is calculated from the extrapolated and observed absorbances for $x_{SS} = 0.5$. At this point $C_x = C_{Cu} = C_{SS} = 2.22 \times 10^{-2} M$.

$$\frac{E}{E_{ex}} = \frac{0.435}{0.575} = 0.756.$$

According to Eq. (2.68):

$$K' = \frac{0.756 \times 2.22 \times 10^{-2}}{(2.22 \times 10^{-2} - 0.756 \times 2.22 \times 10^{-2})(2.22 \times 10^{-2} - 0.756 \times 2.22 \times 10^{-2})}$$

$$= \frac{1.68 \times 10^{-2}}{(5.4 \times 10^{-3})(5.4 \times 10^{-3})} = 5.75 \times 10^2$$

$\alpha_{SS(H)}$ is given by Fig. 1.13.

$$K = K'\alpha_{SS(H)} = 5.75 \times 10^2 \times 5 \times 10^6 = 2.88 \times 10^9$$

$$\log K = 9.45 \quad (I \simeq 0.1)$$

(Compare with the result obtained in Example 29.)

33. The formation of the cobalt(II) complexes of pyridine-2-aldoxime (P2A) has been studied by Burger et al. [88] by a spectrophotometric method. It has been found that only the complex species with maximum ligand number, the ML_2 complex, absorbs at 333 nm.

For the determination of the stability constants a series of solutions is prepared in which the total metal concentration and the pH are the same, but the total ligand concentration varies. $C_{Co} = 2 \times 10^{-5} M$; pH = 6.5; $[NaClO_4] = 0.3 M$. The absorbances measured at 333 nm are as follows:

C_{P2A}, M	E
3×10^{-5}	0.048
7×10^{-5}	0.096
1.4×10^{-4}	0.109
4.4×10^{-4}	0.116
10^{-3}	0.119

Sec. 2.2] Stability constants of mononuclear complexes

Calculate the stability constants of the cobalt complexes, given the protonation constants of P2A, log $K_1 = 10\cdot 0$; log $K_2 = 3\cdot 4$.

The stability constants are determined from the formation curve for the 1 : 2 complex by using Sillén's curve-fitting method (see p. 123).

For the construction of the absorbance vs. log [P2A] curve the free ligand concentrations are calculated first by using Eq. (1.77). As the metal and ligand were present in the solutions in commensurable amounts at the small ligand concentrations used, the estimated bound ligand concentration is subtracted from the total ligand concentration:

$$[\text{P2A}] = \frac{C_{\text{P2A}} - [\text{L}_{\text{bound}}]}{\alpha_{\text{P2A(H)}}}$$

At pH 6.5

$$\alpha_{\text{P2A(H)}} = 1 + 10^{-6\cdot 5} \times 10^{10} + 10^{-13} \times 10^{13\cdot 4} = 10^{3\cdot 5}$$

The logarithms of the free ligand concentrations in the solutions are

$$\log [\text{P2A}] = \log (3 \times 10^{-5} - 2 \times 10^{-5}) - 3\cdot 5 = -8\cdot 5;$$

$$\log (7 \times 10^{-5} - 3 \times 10^{-5}) - 3\cdot 5 = -7\cdot 9;$$

$$\log (1\cdot 4 \times 10^{-4} - 3\cdot 5 \times 10^{-5}) - 3\cdot 5 = -7\cdot 5;$$

$$\log (4\cdot 4 \times 10^{-4} - 3\cdot 5 \times 10^{-5}) - 3\cdot 5 = -6\cdot 9;$$

$$\log (10^{-3} - 4 \times 10^{-5}) - 3\cdot 5 = -6\cdot 5.$$

The absorbance vs. log [P2A] curve is shown in Fig. 2.20. Apparently the curve approximates to a limiting value of $E = 0\cdot 12$. Since in a solution with an absorbance of $0\cdot 12$ the metal ion is completely present in the form of the 1 : 2 complex, that is $\Phi_2 = 1$, the Φ_2 scale can be drawn on the right-hand side of the diagram.

For the purpose of estimating the stability constants the function described by Eq. (2.63) is plotted with $P = 0\cdot 5$, 1, 2, 5 and 10 with Y and log X as coordinates, on the same scale as that used in Fig. 2.20. The series of curves shown in Fig. 2.21 is obtained. If this diagram is drawn on transparent paper and placed over Fig. 2.20, the best fit is given by the curve for $P = 1$, and the point log $X = 0$ on the abscissa of the normalized diagram coincides with $-8\cdot 6$ on the log [P2A] axis

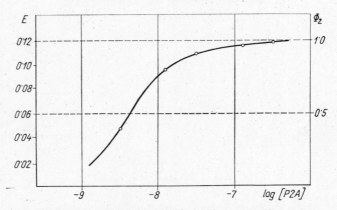

Fig. 2.20. Formation curve of the [1:2 pyridine-2-aldoxime–cobalt(II)] complex, obtained by spectrophotometry, according to the measurements of Burger, Egyed, Ruff and Ruff [88]. (See Example 33)

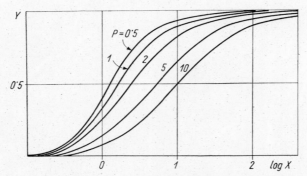

Fig. 2.21. Series of normalized curves for the determination of the overall stability constants of [pyridine-2-aldoxime–cobalt(II)] complexes. Representation of function (2.63) for different values of P. (See Example 33)

of the other diagram. Since, according to Eq. (2.61)

$$\log X = 1/2 \log \beta_2 + \log [P2A]$$

$$0 = 1/2 \log \beta_2 - 8 \cdot 6$$

it follows that

$$\log \beta_2 = 17 \cdot 2 \qquad (I = 0 \cdot 3).$$

Sec. 2]   Stability constants of mononuclear complexes   147

The constant β_1 can be calculated by using Eq. (2.60):

$$\log P = \log \beta_1 - \frac{1}{2}\log \beta_2,$$

$$0 = \log \beta_1 - \frac{1}{2} \times 17\cdot 2.$$

$$\log \beta_1 = 8\cdot 6 \qquad (I = 0\cdot 3).$$

2.2.3. Polarographic methods

If a metal ion can be reversibly reduced to the metallic state at a dropping mercury electrode, the metal ion may behave in the presence of a complexing agent in either of the following two ways.

1. If the complex-formation reaction is fast compared with the reduction of the metal ion, then one polarographic wave occurs, with a half-wave potential more negative than that of the uncomplexed metal ion. From the shift in half-wave potential with concentration of the complexant, the composition and stability constant of the complex can be determined.

2. If the complex-formation reaction is slow compared with the reduction of the metal ion, which occurs less frequently, then two polarographic waves are obtained, the first corresponding to the reduction of the free metal ion, the second to that of the complexed metal ion. The amount of free metal ion can be determined directly from the height of the first wave. From the dependence of the height of the waves on the ligand concentration, the composition and stability constant of the complex formed can be calculated.

In the first case, where the formation of the complex is fast compared with the reduction, the deposition potential of the metal ion reducible at the cathode is determined by the stationary free metal ion concentration due to the dissociation of the complex.

The reduction of a metal ion on a mercury cathode proceeds as follows:

$$M^{z+}_{(aq)} + ze \rightarrow M_{(Hg)} \qquad (2.73)$$

where M^{z+} is the metal ion, and $M_{(Hg)}$ the elemental metal dissolved in mercury. The equilibrium constant is given by

$$K_r = \frac{(M)}{[M][e]^z} \qquad (2.74)$$

The activity of the metal in mercury is denoted by the symbol in round brackets, and is practically equal to the concentration, which is very small; K_r is constant at given ionic strength and temperature.

By taking logarithms and rearranging:

$$-\log [e] = \frac{1}{z} \log K_r + \frac{1}{z} \log \frac{[M]}{(M)} \qquad (2.75)$$

If both sides of Eq. (2.75) are multiplied by 2·3 RT/F, the Nernst equation is obtained [see Eqs. (1.50) and (1.54)]:

$$E = E^0 + \frac{2 \cdot 3\, RT}{zF} \log \frac{[M]}{(M)} \qquad (2.76)$$

In the layer of solution in contact with the surface of the mercury drop the ratio of the total metal ion concentration to the potential-determining free metal ion concentration in the presence of a complexing ligand L, is determined by the complex-formation equilibrium [see Eq. (2.34)]:

$$\frac{C_M^0}{[M]} = 1 + [L]\beta_1 + [L]^2\beta_2 \ldots = A_{M(L)} \qquad (2.77)$$

Substituting for [M] from Eq. (2.77) in Eq. (2.76) gives

$$E = E^0 + \frac{2 \cdot 3\, RT}{zF} \log \frac{C_M^0}{(M)} \left[\times \frac{1}{A_{M(L)}} \right] \qquad (2.78)$$

According to the Ilkovič equation the intensity of the limiting diffusion current i_d is proportional to the concentration (c) of the reduced component under given experimental conditions

$$i_d = dc \qquad (2.79)$$

where

$$d = 6 \cdot 3\, zFD^{1/2}m^{2/3}t^{1/6} \qquad (2.80)$$

D being the diffusion coefficient of the ion, m the drop rate of the mercury and t the drop time. The value of d depends only on the diffusion coefficient if the head of mercury is kept constant [89].

That is

$$i_d = d_0[M] + d_1[ML] + d_2[ML_2] + \ldots = \bar{d}C_M \qquad (2.81)$$

Sec. 2.2] Stability constants of mononuclear complexes

where $\bar{d}$ is the proportionality factor calculated from the average diffusion constant, and C_M the total concentration of the metal ion in the bulk of the solution.

During the polarographic measurement the current is proportional to the concentration gradient in the solution:

$$i = \bar{d}(C_M - C_M^0) \qquad (2.82)$$

When i reaches the value of the limiting diffusion current i_d, C_M^0 becomes very small and can be neglected. That is

$$i_d = \bar{d} C_M \qquad (2.83)$$

From Eqs. (2.82) and (2.83)

$$C_M^0 = \frac{i_d - i}{\bar{d}} \qquad (2.84)$$

For the amalgam the relation between the current and concentration can be expressed by means of a proportionality factor δ

$$i = \delta(M) \qquad (2.85)$$

From Eqs. (2.84) and (2.85), Eq. (2.78) may be rewritten as

$$E = E^0 + \frac{2 \cdot 3\, RT}{zF} \log\left[\left(\frac{i_d - i}{\bar{d}}\right) \frac{\delta}{i}\right] - \frac{2 \cdot 3\, RT}{zF} \log A_{M(L)} \qquad (2.86)$$

At the half-wave potential

$$\frac{i_d - i}{i} = 1,$$

and hence

$$E_{1/2 M(L)} = E^0 + \frac{2 \cdot 3\, RT}{zF} \log \frac{\delta}{\bar{d}} - \frac{2 \cdot 3\, RT}{zF} \log A_{M(L)} \qquad (2.87)$$

In the absence of a complexing ligand,

$$C_L = 0, \quad \log A_{M(L)} = 0, \quad \text{and} \quad E_{1/2 M} = E^0 + \frac{2 \cdot 3\, RT}{zF} \log \frac{\delta}{d_0} \qquad (2.88)$$

Subtracting Eq. (2.87) from Eq. (2.88) gives the shift in half-wave potential due to the addition of a complexing agent:

$$\Delta E_{1/2} = \frac{2 \cdot 3\, RT}{zF} \log \frac{\bar{d}}{d_0} + \frac{2 \cdot 3\, RT}{zF} \log A_{M(L)} \qquad (2.89)$$

If C_M is the same in the absence and in the presence of the complexing agent, then

$$\frac{\bar{d}}{d_0} = \frac{i_{dM(L)}}{i_{dM}}, \qquad (2.90)$$

and $A_{M(L)}$ and $\log A_{M(L)}$ can be expressed from Eq. (2.89) as follows:

$$\log A_{M(L)} = \frac{zF}{2 \cdot 3\,RT} \Delta E_{1/2} + \log \frac{i_{dM}}{i_{dM(L)}}$$

At 25° C

$$A_{M(L)} = \text{antilog}\left[16 \cdot 9\, z\Delta E_{1/2} + \log \frac{i_{dM}}{i_{dM(L)}}\right] \qquad (2.91)$$

If a series of solutions is prepared in which the total metal ion concentration is the same [see Eq. (2.90)] but different amounts of ligand (including zero) are added to each solution, the $A_{M(L)}$ values for different ligand concentrations can be calculated from the polarographic data by using Eq. (2.91), $\Delta E_{1/2}$ being the difference between the half-wave potentials of the metal ion in the presence and absence of the complexing ligand, provided the polarograms have been run under same conditions.

The ratio $i_{dM}/i_{dM(L)}$ can be obtained by dividing the height of the polarographic wave for the metal in the absence of complexant by that in the presence of the complexing agent.

If $C_M \ll C_L$ then $C_L \simeq [L]$, and in the case of successive complex formation the corresponding stability constants can be calculated from corresponding $[L]$ and $A_{M(L)}$ values, by a computer or by Leden's successive extrapolation method.

The method which is described above was developed by DeFord and Hume [90] and can be used in many cases.

When only one complex ML_n predominates in a given ligand concentration range, the function A can be simplified as follows:

$$A_{M(L)} \simeq 1 + [L]^n \beta_n$$

Furthermore 1 may be neglected compared with $[L]^n\beta_n$. Combination of Eqs. (2.89) and (2.90) gives

$$\Delta E_{1/2} = \frac{2 \cdot 3\,RT}{zF} \log \frac{i_{dM(L)}}{i_{dM}} + \frac{2 \cdot 3\,RT}{zF} \log ([L]^n \beta_n) \qquad (2.92)$$

At 25° C

$$\Delta E_{1/2} + \frac{0 \cdot 059}{z} \log \frac{i_{dM}}{i_{dM(L)}} = \frac{0 \cdot 059}{z} n \log [L] + \frac{0 \cdot 059}{z} \log \beta_n \qquad (2.93)$$

If the shift in half-wave potential and the height of the wave are determined for a constant total metal ion concentration and varied ligand concentration, then a plot of the left-hand side of Eq. (2.93) vs. the logarithm of the ligand concentration produces a straight line. From the slope, $0.059\, n/z$, and intercept $(0.059/z)\log \beta_n$, of the line, n and β_n can be calculated [91].

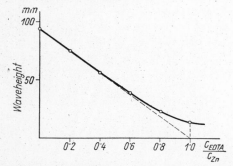

Fig. 2.22. Heights of the polarographic reduction waves of 'free zinc ion' in the presence of various amounts of EDTA. $C_{Zn} = 0.014M$; pH = 9.0 [92]

If the ligand is partly protonated and the pH of the solution is kept constant during the experiments and the calculations are done with the total ligand concentration, then the real overall formation constants can be calculated from the conditional constants obtained by using the $\alpha_{L(H)}$ function (see pp. 51 and 140).

The second case, i.e. the complex formation reaction is slow compared with the reduction of the metal ion, occurs much more rarely. In this case two polarographic waves occur in the presence of the complexant. If a series of solutions is prepared with constant C_M, but with C_L increasing from zero, then the free metal concentrations, determined from the height of the first wave, are plotted against C_L/C_M. A curve similar to that in Fig. 2.22 is obtained. The intersection of the abscissa with the tangent to the initial section of the curve gives information about the composition of the complex. From the related C_M, C_L and [M] values, the stability constant of the complex can be calculated. The method can be used only if one complex species is formed. From the conditional stability constant determined under the given experimental conditions the stability constant can be calculated by using the α-functions for the side-reactions (see Example 35).

2.2.3.1. Practical aspects

The fundamentals and practical realization of polarographic measurements are dealt with in detail in the literature [89, 93]. Here the reader's attention is drawn only to the most important problems in connection with the investigation of complex formation.

A Radiometer (Denmark) Type PO4 or a Radelkis (Hungary) Type

OH 102 polarograph is suitable for the measurements. Usually a pH meter is also necessary.

Stock solutions can be prepared by accurate weighing and dissolution or standardized if necessary. The following points should be noted when preparing the solutions to be studied.

(i) The concentration of the metal ion should be small (approx. $10^{-4}M$): the free ligand concentration may be considered as equal to the total ligand concentration only if $pH > 2 + \log K_1$ and $C_L > 20\, C_M$.

(ii) Although total ligand concentration is varied over a wide interval within a series of experiments, the ionic strength should be kept constant. To achieve this the concentration of inert electrolyte is varied so that the total ionic strength due to the complexing agent and the inert electrolyte is constant. If possible, only one kind of alkali metal (or possibly the ammonium) ion should be present in the solution.

(iii) Gelatin may be used for suppressing maxima.

(iv) If the side-reactions of the metal ion and ligand are also to be considered, the species involved in the side-reactions should be present in constant concentration in the whole series of experiments (ammonia, hydrogen ions, hydroxide ions, etc.).

In the polarographic measurements the following main points should be noted.

(i) The checking and adjustment of the polarograph before making measurements is very important. If the complex stability constants are to be calculated from the shift of the half-wave potential (in volts), the voltage scale should be checked.

(ii) Nitrogen rather than sulphite should be used to remove oxygen from the solution.

(iii) A calomel electrode should be used as a reference, either immersed in the solution to be measured or connected to the cell via a salt bridge.

(iv) In determining half-wave potentials it is important that the errors due to possible backlash of the cylinder carrying the chart paper are eliminated by using the same sequence of operations in starting the potential scan.

Half-wave potentials can be determined from the polarograms by graphical projection, or more accurately by plotting the calculated values of $\log i/(i_d - i)$ as a function of the voltage, to obtain a straight line which intersects the abscissa at a voltage equal to the half-wave potential. The values of the current i and limiting current i_d can be read from the polarogram directly in mm, as the units cancel in the ratio.

Sec. 2.2] Stability constants of mononuclear complexes

2.2.3.2. Worked examples

34. The formation of the 1,10-phenanthroline complex of cadmium is investigated by the polarographic method. The total concentration of cadmium and the concentration of the supporting electrolyte, potassium nitrate, are constant, whereas the concentration of 1,10-phenanthroline is varied. All the solutions contain 40% ethanol and the pH is 6·6. $C_{Cd} = 2 \times 10^{-4} M$, and $C_{KNO_3} = 10^{-1} M$.

The half-wave potentials measured against a calomel electrode and determined from the polarograms, and the wave heights in mm, are as follows:

C_{Phen} M	$E_{1/2}$ V	i_d mm
—	−0·591	170
10^{-3}	−0·745	135
2×10^{-3}	−0·790	122
4×10^{-3}	−0·822	120
10×10^{-3}	−0·862	113
20×10^{-3}	−0·895	113

Determine the composition and overall stability constant of the complex (or possibly complexes) formed.

As $C_{Phen} \gg C_{Cd}$, as a first approximation the shift of the half-wave potential is plotted against $\log C_{Phen}$, and the differences in wave heights due to the different diffusion coefficients of the species are disregarded. The ligand number determined from the slope of the curve obtained is about 3, and $\log \beta_3 \simeq 16$ is given by the intercept.

As, apparently, a highly stable complex containing three ligands is formed in the ligand concentration range studied, the amount of bound

Table 2.9

Data for the calculation of the overall stability constant of the 1,10-phenanthroline–cadmium(II) complex

[Phen]	log [Phen]	$\Delta E_{1/2}$	$\log \dfrac{i_{Cd}}{i_{Cd(Ph)}}$	$\Delta E_{1/2} + 0·03 \log \dfrac{i_{Cd}}{i_{Cd(Ph)}}$
4×10^{-4}	−3·4	0·154	0·10	0·157
$1·4 \times 10^{-3}$	−2·85	0·199	0·14	0·203
$3·4 \times 10^{-3}$	−2·47	0·231	0·15	0·236
$9·4 \times 10^{-3}$	−2·03	0·271	0·18	0·276
$1·94 \times 10^{-2}$	−1·71	0·304	0·18	0·309

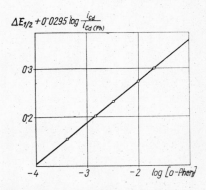

Fig. 2.23. Investigation of the formation of the 1,10-phenanthroline–cadmium complex by the polarographic method. $C_{Cd} = 2 \times 10^{-4} M$; pH = 6·6. (See Example 34)

ligand is taken into account when calculating the free ligand concentration more accurately:

$$[Phen] \simeq C_{Phen} - 3 C_{Cd}$$

The corresponding concentration, $\Delta E_{1/2}$ etc. data necessary for the further calculations are given in Table 2.9. The value of

$$\Delta E_{1/2} + \frac{0.059}{z} \log \frac{i_{Cd}}{i_{Cd(Ph)}}$$

is then plotted against the logarithm of the free ligand concentration to give the line shown in Fig. 2.23. As cadmium(II) is reduced to elemental cadmium at the dropping mercury cathode, $z = 2$.

As the data produce a straight line, it can safely be stated that one complex species is predominant in the ligand concentration range studied. From Eq. (2.93), the ligand number can be calculated from the slope of the line

$$\frac{0.059}{2} n = \frac{0.360 - 0.190}{2} = 0.085$$

$$n = \frac{0.085}{0.0295} = 2.9;$$

and the overall stability constant from the intercept:

$$\log \beta_3 = \frac{0.445}{0.0295} \simeq 15.1$$

Since the pH used is 6·6 and the protonation constant of 1,10-phenanthroline is $\log K = 4.95$ ($I = 0.1$), the value of the $\log \alpha_{L(H)}$ function is smaller than the error of the calculation and therefore it can be neglected:

$$\alpha_{L(H)} = 1 + 10^{-6.6} \times 10^{4.95} \simeq 1.02$$

$$3 \log \alpha_{L(H)} = 0.026$$

Sec. 2.2] Stability constants of mononuclear complexes

[See Eq. (2.69).] Thus the value calculated is the logarithm of the overall stability constant.

35. A solution that was $0.014M$ in both copper(II) and EDTA, $0.5M$ in ammonia, contained an ammonium salt and had a pH of 9·2, was investigated by polarography. The concentration of free, uncomplexed copper(II), as determined from the height of the first wave was found to be $7 \times 10^{-5}M$. Calculate the stability constant, knowing that copper(II) forms a 1:1 complex with EDTA.

First the conditional stability constant is calculated from the conditional concentrations. The 'conditional free' EDTA concentration is equal to the conditional free concentration of copper(II), and the concentration of the complex is the difference between the total copper(II) or EDTA concentration and the 'conditional free' copper(II) concentration.

$$[\text{Cu}'] = [\text{EDTA}'] = 7 \times 10^{-5} M$$

$$[\text{CuEDTA}] = 1.4 \times 10^{-2} - 7 \times 10^{-5} \simeq 1.4 \times 10^{-2} M.$$

The conditional stability constant is given by

$$K' = \frac{[\text{CuEDTA}]}{[\text{Cu}'][\text{EDTA}']} = \frac{1.4 \times 10^{-2}}{(7 \times 10^{-5})^2} = 2.9 \times 10^6$$

In calculating the true stability constant the side-reactions of copper(II) and the protonation of EDTA have also to be taken into account. As can be estimated on the basis of the diagrams in Figs. 1.10 and 1.11 the hydroxo-complex formation by copper(II) can be neglected compared with ammine-complex formation in the presence of $0.5M$ ammonia. The value of $\log \alpha_{\text{Cu(NH}_3)}$ is taken from Fig. 1.11, and of $\log \alpha_{\text{EDTA(H)}}$ from Fig. 1.14:

$$\log K = \log K' + \log \alpha_{\text{Cu(NH}_3)} + \log \alpha_{\text{EDTA(H)}} = 6.5 + 11.0 + 1.4 = 18.9$$

As the uncertainty of calculated log α values is at best ± 0.1, the stability constant calculated can be considered as a good approximation. (Compare with the value given in Section 4.1.)

2.2.4. EXTRACTION METHODS

The basis of the liquid–liquid extraction methods is Nernst's distribution law, according to which the ratio of the activities or (at constant ionic strength) of the concentrations of a substance in the two phases when it is distributed between two immiscible liquids (usually water and an organic solvent) is constant at constant temperature, i.e.

$$\frac{(A)}{[A]} = d_A \quad (p,T,I=\text{const.}) \tag{2.94}$$

where d_A is the distribution constant of the substance A. The round brackets mean the concentration of species A in the organic phase, the square brackets that in the aqueous phase. If $d_A > 1$, the substance is more soluble in the organic than in the aqueous phase.

The value of the distribution constant defined by Eq. (2.94) may differ greatly from the distribution coefficient even in the case of the same substance and solvents. The distribution coefficient, D, is the ratio of the total or analytical concentrations of the substance in the two phases.

$$\frac{G_A}{C_A} = D_A \tag{2.95}$$

where G_A is the total concentration in the organic phase, C_A that in the aqueous phase.

Obviously, the value of the distribution constant differs from that of the distribution coefficient if the solute enters into chemical reaction with other components present in either or both of the phases.

Generally, of the different complex species, the neutral (uncharged) complex is much less soluble in aqueous medium than in organic solvents less polar than water. As the charged species show just the opposite behaviour, the neutral complex can usually be selectively extracted.

If the distribution coefficient of the metal ion is determined by extracting aqueous solutions containing the metal ion and ligand in different concentrations with a suitable solvent, the results obtained provide information concerning the composition of the complex species present and the positions of the equilibria.

Let s denote the ligand number of the neutral complex which can be extracted by a solvent. Assuming that only the complex with a composition of ML_s is soluble in the organic solvent chosen, and also that the free ligand remains in the aqueous phase, the distribution coefficient of the metal ion M will be

$$D_M = \frac{(ML_s)}{C_M} \tag{2.96a}$$

Sec. 2.2] Stability constants of mononuclear complexes

Taking into account the distribution constant of the neutral complex

$$d_{ML_s} = \frac{(ML_s)}{[ML_s]}$$

and the concentrations of the different species present in the aqueous phase, Eq. (2.96a) can be rewritten as

$$D_M = \frac{d_{ML_s}[ML_s]}{[M] + [ML] + [ML_2] + \ldots + [ML_N]} \quad (2.96b)$$

On insertion of the overall stability constants and division of both the numerator and denominator by [M], the following simpler equation is obtained

$$D_M = \frac{d_{ML_s}[L]^s \beta_s}{1 + [L]\beta_1 + \ldots + [L]^N \beta_N} = d_{ML_s} \frac{[L]^s \beta_s}{\sum_0^N \beta_i [L]^i} \quad (2.97)$$

where N is the maximum ligand number. Comparing Eqs. (2.97) and (1.20) we obtain

$$D_M = d_{ML_s} \Phi_s \quad (2.98)$$

which means that the distribution coefficient is proportional to the mole-fraction of the extractable complex species.

If the metal ion M forms complexes with other ligands in addition to L (e.g. with hydroxide ion), and the complexes formed in the side-reaction are not extractable, then the denominator on the right-hand side of Eq. (2.97) should also contain terms corresponding to the formation of these species. For example, in the case of hydroxo-complex formation:

$$D_M = \frac{d_{ML_s}[L]^s \beta_s}{1 + [L]\beta_1 + [L]^2 \beta_2 + \ldots + [L]^N \beta_N + [OH]\gamma_1 + \ldots} \quad (2.99)$$

If the number of ligands in the extractable complex is equal to the maximum ligand number, i.e. $s = N$, and a series of solutions is prepared in which the total metal concentration is constant and the ligand concentration is varied, then if a constant phase-volume ratio is used and the distribution coefficient of the metal ion is determined on the basis of extraction experiments, a plot of D_M vs. log [L] gives a saturation-type curve, that is, the distribution coefficient approaches a limiting value at higher ligand concentrations.

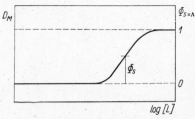

Fig. 2.24. Investigation of complex formation by extraction. Relationship between the distribution coefficient and logarithm of the ligand concentration, if the complex with maximum ligand number can be extracted selectively

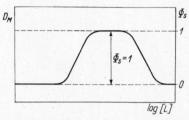

Fig. 2.25. Investigation of complex formation by extraction. Relationship between the distribution coefficient and logarithm of the free ligand concentration if $s < N$, and the extractable complex has high stability

As, according to Eq. (2.98), D_M is proportional to the mole-fraction of the complex extracted, at the ligand concentration where the curve reaches the limiting value, the mole-fraction Φ_s of the species ML_s is unity, and the Φ_s values for different free ligand concentration values can be obtained graphically (see Fig. 2.24). If N does not exceed 2 (or possibly 3), the β values can be obtained by Sillén's curve-fitting method described in connection with spectrophotometric methods (see Example 33).

It follows from Eq. (2.98) that when the limiting value is reached the distribution coefficient is equal to the distribution constant of the species extracted (namely, $\Phi_s = \Phi_N = 1$). In practice, the metal ion can be considered as being wholly in the form of the complex ML_N, if the amount of the metal ion in other forms does not exceed 1% of the total. Taking into account that the value of the overall stability constant increases with increasing ligand number, and therefore the concentrations of the species with ligand numbers $N-2$, $N-3$, etc. can be neglected compared with that of the species with ligand number $N-1$, the ligand concentration above which the metal ion is completely present in the form of the species ML_N can be deduced:

$$[ML_N] > 100 \, [ML_{N-1}];$$

$$[M][L]^N \beta_N > 100 \, [M][L]^{N-1} \beta_{N-1}$$

$$[L] > 100 \frac{\beta_{N-1}}{\beta_N} = 100 \frac{1}{K_N}$$

or $\log [L] > \log 2 - \log K_N$.

Sec. 2.2] Stability constants of mononuclear complexes

The procedure is similar in the rather rare case when $s < N$, but the extractable complex is particularly stable, and predominates in a fairly wide range of ligand concentrations. In such cases, the D_M vs. log [L] plot exhibits a plateau corresponding to $\Phi_s = 1$ (see Fig. 2.25).

If the extractable complex is not particularly stable and $s < N$, then a simple maximum-curve s obtained from which the Φ_s values cannot be read directly. In this case the experimental data are plotted in a log D_M vs. log [L] diagram.

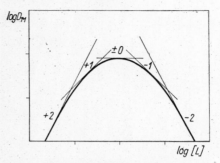

Fig. 2.26. Investigation of complex formation by extraction. Determination of the value of $(s - \bar{n})$ by graphical differentiation

The logarithmic form of Eq. (2.97) is

$$\log D_M = \log d_{ML_s} + s \log [L] + \log \beta_s - \log \sum_0^N \beta_i [L]^i \quad (2.100)$$

Differentiating this equation with respect to log [L] gives the simple relationship

$$\frac{d(\log D_M)}{d(\log [L])} = s - \bar{n} \quad (2.101)$$

where $\bar{n}$ is the average ligand number [94].

According to Eq. (2.101), the related [L] and $\bar{n}$ values can be obtained by the graphical differentiation of the log D_M vs. log [L] plot (see Fig. 2.26), if s is known. From a sufficiently large number of values taken from a wide ligand concentration range the overall stability constants can be calculated by using Eq. (1.18).

Another graphical evaluation of the log D_M vs. log [L] plot is due to Dyrssen and Sillén [95], according to which the equations of the asymptotes drawn to the initial and final part of the experimental curve possessing a maximum, are:

if $[L] \to 0$

$$\log D'_M = \log d_{ML_s} + \log \beta_s + s \log [L] \quad (2.102)$$

if $[L] \to \infty$

$$\log D''_M = \log d_{ML_s} + \log \beta_s - \log \beta_N - (N - s) \log [L] \quad (2.103)$$

When writing Eq. (2.102) it was assumed that, at small ligand concentrations, the metal ion is present practically entirely as the aquo-complex in the aqueous phase. Similarly, Eq. (2.103) indicates that in the high ligand concentration range practically only the species ML_N with maximum ligand number is present in the aqueous phase. The straight lines represent the two extremes of Eq. (2.100) (see Fig. 2.27).

At the intersection of the two asymptotes log D'_M is equal to log D''_M, and log β_N can be calculated from the resulting simple equation

$$\log \beta_N = -N \log [L]_0 \qquad (2.104)$$

where log $[L]_0$ is the abscissa of the intersection. The intercepts of the asymptotes [see Eqs. (2.102) and (2.103)] are: $A = \log d_{ML_s} + \log \beta_s$ and $B = \log d_{ML_s} + \log \beta_s - \log \beta_N$, and the difference between the values of A and B gives log β_N. The slopes of the asymptotes are s and $-(N-s)$, respectively. These authors assumed the following relationship always to be valid between the successive and overall stability constants

$$\frac{K_n}{K_{n+1}} = \frac{\beta_n^2}{\beta_{n-1}\beta_{n+1}} = P^2 \quad (\text{if } n \geq 1) \qquad (2.105)$$

where the value of P is constant in the case of a given metal ion and ligand, and can be calculated from the difference between the ordinates of the point of intersection of the asymptotes and that of the point on the experimental curve at log $[L]_0$ (see Fig. 2.27):

$$\Delta = \log \sum_0^N P^{i(N-i)} \qquad (2.106)$$

If P and β_N are known, the value of all the stability constants can be calculated by using Eq. (2.105).

The evaluation can also be made by a curve-fitting method if the experimental log D_M vs. log $[L]$ curve is fitted to the normalized set of curves drawn with log Y and log X coordinates, and calculated with different P values by using Eq. (2.107).

$$Y = \frac{P^{s(N-s)} X^s}{\sum_0^N P^{i(N-i)} X^i}, \qquad (2.107)$$

where

$$Y = \frac{D_M}{d_{ML_s}} \quad \text{and} \quad X = \beta_N^{1/N}[L]$$

Sec. 2.2] Stability constants of mononuclear complexes 161

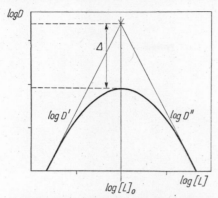

Fig. 2.27. Investigation of complex formation by extraction, using the two-parameter method due to Dyrssen and Sillén [95]

The investigation of extraction equilibria can be used also as an indirect method for studying complexes which are not soluble in organic solvents, and for the determination of their overall stability constants [96, 97].

Let Y be a complexing agent which forms practically only one complex of the composition MY_s with a metal ion M, the complex being extractable from the aqueous phase with an organic solvent. The distribution coefficient of the metal ion can be expressed similarly to Eq. (2.96b) as follows:

$$D_M = \frac{d_{MY_s}[MY_s]}{[M] + [MY_s]}$$

If a complexing agent L is also present and forms complexes which cannot be extracted by the organic solvent, then the distribution coefficient will be as follows:

$$D_{M(Y,L)} = \frac{d_{MY_s}[MY_s]}{[M] + [MY_s] + [ML] + [ML_2] + \ldots}$$

If we introduce the complex formation equilibrium equations and denote the overall stability constant of the complex MY_s by γ_s and the constants for $ML_1, ML_2 \ldots$ by $\beta_1, \beta_2 \ldots$ the equations can be written

11 Inczédy

in the following form [see also Eq. (2.97)]:

$$D_{M(Y)} = \frac{d_{MY_s}\gamma_s[Y]^s}{1 + \gamma_s[Y]^s} \qquad (2.108)$$

$$D_{M(Y,L)} = \frac{d_{MY_s}\gamma_s[Y]^s}{1 + \gamma_s[Y]^s + \beta_1[L] + \beta_2[L]^2 + \ldots} \qquad (2.109)$$

By dividing Eq. (2.108) by Eq. (2.109) we get:

$$\frac{D_{M(Y)}}{D_{M(Y,L)}} = \frac{1 + \gamma_s[Y]^s + \beta_1[L] + \beta_2[L]^2 + \ldots}{1 + \gamma_s[Y]^s} \qquad (2.110)$$

Because the concentration of the extractable complex MY_s is very low in the aqueous phase, the term $\gamma_s[Y]^s$ can be neglected both in the numerator and denominator. Thus the ratio of the distribution coefficients gives the $A_{M(L)}$ function:

$$\frac{D_{M(Y)}}{D_{M(Y,L)}} = 1 + [L]\beta_1 + [L]^2\beta_2 + \ldots = A_{M(L)} \qquad (2.111)$$

Thus, by determining the distribution coefficients of the metal ion when the ligand Y is used, in the absence and presence of the ligand L, respectively, we can find the value of $A_{M(L)}$ for the given ligand concentration [L].

A series of solutions is prepared in which the concentrations of the complex-forming agent Y and the metal ion M (and, for protonating ligands also the pH) are constant, whereas that of the ligand L is varied, starting from zero. The distribution coefficients of the metal ion and hence the values of the function $A_{M(L)}$ are determined for each member of the series for each ligand concentration [L]. (It is very important that the concentration of the ligand Y, the pH and the concentration of additives remain the same in the solutions which do and do not contain the ligand L.) From the related [L] and $A_{M(L)}$ values the overall stability constants can be calculated by Leden's successive extrapolation method (see p. 115).

For the computer evaluation of the results obtained by extraction methods see refs. [69, 98, 99].

When extraction experiments are done in the laboratory the following points should be noted.

In preparing the series of solutions care should be taken that the total metal ion concentration is low ($<10^{-3}M$), and the concentration of

Sec. 2.2]   Stability constants of mononuclear complexes   163

bound ligand is negligible compared with the total ligand concentration. As the ligand is usually a weak base and so tends to become protonated, and the metal ion may form hydroxo-complexes at higher pH values, the pH of the solutions must be measured and the protonation and hydroxo-complex constants must be known. In the direct methods it is expedient to keep either the pH or the total ligand concentration constant (and vary the other). If the concentration of bound ligand is negligible compared to the total concentration, the value of [L] can be calculated by using Eq. (1.77).

A suitable inert electrolyte is used to ensure constant ionic strength of the solutions.

A suitable organic solvent can only be selected on the basis of experiments. Care should be taken that the solvent does not give an interfering chemical reaction with the components of the aqueous solution and that the miscibility with water is as low as possible. It is advisable to make the ratio of the volume of the organic phase to that of the aqueous phase constant throughout the whole series of experiments.

It is very important that the distribution and chemical equilibria are fully attained during the period of shaking. As the rate of chemical reactions is generally higher than that of material transport between the two phases, the extraction should be done in a well-stoppered separating-funnel by vigorous shaking for a long enough period. The minimum time of shaking is determined by performing the extraction on samples of the same composition by shaking for different times and subsequently analysing the phases.

For calculating the distribution coefficient it is sufficient to determine the concentration in one of the phases after extraction if the total amount of metal present and the ratio of the volumes of the phases are known.

Good phase separation is very important from the point of view of the reliability of the result of chemical analysis; also the phases should not be turbid. An inert additive (cetyl alcohol) may be used to reduce foaming. The separation of the phases can be accelerated by centrifuging.

The concentration of the metal ion distributed can be determined in either phase or in both by spectrophotometry, radioactive tracer technique, polarography, etc. The distribution coefficient is calculated according to Eq. (2.95).

As it is a basic requirement of the calculations that the concentrations used are truly equilibrium data, the pH measured *after* the extraction, when distribution equilibrium has been attained, is to be used in calculating the free ligand concentrations or α-functions.

In the calculation of the free ligand concentration and consideration of possible side-reactions the statements made in connection with pH and spectrophotometric methods are valid (see pp. 98, 119 and 140).

2.2.4.1. *Worked example*

36. The hydrolysis of vanadium(V) has been studied by Dyrssen and Sekine [100] by the extraction method. The non-dissociated HVO_3 molecule was found to be extractable with methyl isobutyl ketone from a $0.5M$ $NaClO_4$–$HClO_4$ medium. The extraction was performed with solutions of different pH values, and the equilibrium concentrations were determined by a radioactive tracer technique. A plot of the calculated distribution coefficients *vs.* the pH gave the curve shown in Fig. 2.28.

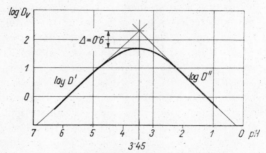

Fig. 2.28. Extraction of vanadic acid with methyl isobutyl ketone according to the results of Dyrssen and Sekine [100]. (See Example 36)

Calculate the equilibrium constants K_1 and K_2 of the reactions

$$VO_3^- + H^+ \rightleftharpoons HVO_3$$

$$HVO_3 + H^+ \rightleftharpoons VO_2^+ + H_2O$$

by using the data of the experimental curve.

The calculation is done by the two-parameter method suggested by Dyrssen and Sillén. The slopes of the asymptotes drawn to the two linear portions of the curve are $+1$ and -1, respectively, corresponding to $s = 1$, and $-(N - s) = -(2 - 1) = -1$.

From the pH at the point of intersection of the two asymptotes the overall stability constant $\beta_2 = K_1 K_2$ can be calculated by using Eq. (2.104),

$$\log K_1 + \log K_2 = 2\text{pH} = 2 \times 3.45 = 6.9 \tag{i}$$

The value of P can be calculated from the difference of the two coordinate values taken from the graph, corresponding to pH 3·45:

$$\Delta = 2\cdot 3 - 1\cdot 7 = 0\cdot 6$$

According to Eq. (2.106):

$$0\cdot 6 = \log(1 + P + 1)$$
$$4 = 2 + P$$
$$P = 2$$

According to Eq. (2.105):

$$\log K_1 - \log K_2 = 2\log 2 = 0\cdot 6 \qquad (ii)$$

From (i) and (ii)

$$\log K_1 = 3\cdot 75; \quad \log K_2 = 3\cdot 15.$$

The partition coefficient of HVO_3 is calculated from the intercept of the asymptote $\log D'_V$, Eq. (2.102) being taken into consideration:

$$\log d_{HVO_3} + \log K_1 = 5\cdot 7$$
$$\log d_{HVO_3} = 5\cdot 7 - 3\cdot 75 = 1\cdot 95$$
$$d_{HVO_3} = \frac{(HVO_3)}{[HVO_3]} = 89\cdot 0.$$

More accurate results can be obtained by using the curve-fitting method.

2.2.5. Ion-exchange methods

The principle of ion-exchange methods is similar to that of extraction methods. From the position of the distribution equilibrium of the metal ion between the ion-exchanger and solution phase a conclusion may be drawn as to the degree of complex formation with a ligand in the solution. Stability constants and overall stability constants may be calculated on the basis of the distribution equilibria measured.

The experiments can be made with either a cation- or an anion-exchange resin (or a liquid ion-exchanger). The methods described here can be used when the ligand is charged and the complex formation involves a change in the charge.

A cation-exchange method is generally used when the ligand is a multivalent anion larger in size than the metal ion. In such cases only the free metal ion (aquo-complex) is bound by the resin.

The basic equation [53] describing the ion-exchange process, if the univalent sodium ion is exchanged for a multivalent metal ion, is

$$M^{z+} + z\text{RNa} \rightleftharpoons z\text{Na}^+ + R_zM,$$

where R stands for the equivalent amount of the ion-exchange resin. The equilibrium concentration constant is:

$$K_x = \frac{(M)[\text{Na}]^z}{[M](\text{Na})^z} \qquad (2.112)$$

The symbols in round brackets denote the concentration in the ion-exchange resin, those in square brackets that in the solution.

The distribution coefficient of the metal ion M can be expressed from Eq. (2.112) as follows:

$$D_M = \frac{(M)}{[M]} = K_x \left[\frac{(\text{Na})}{[\text{Na}]}\right]^z \qquad (2.113)$$

If the concentration of the metal ion is much smaller than that of the sodium ion, then (Na) can be considered as constant and the value of D_M depends only on the concentration of sodium ion in the solution. If this is constant, then D_M is also constant.

In the presence of a negatively charged complexing ligand L, because only the free metal ion is transferred into the resin phase, the distribution coefficient of the metal ion will be:

$$D_{M(L)} = \frac{(M)}{C_M} = \frac{(M)}{[M] + [ML] + [ML_2] + \ldots}$$

$$= \frac{(M)}{[M] + \beta_1[M][L] + \beta_2[M][L]^2 \ldots} \qquad (2.114)$$

Assuming that the concentration of sodium ion in the solution is the same as in the absence of the complexing agent (i.e. the value of D_M is the same in both cases), Eq. (2.113) can be divided by Eq. (2.114), to give the A function:

$$\frac{D_M}{D_{M(L)}} = 1 + [L]\beta_1 + [L]^2\beta_2 + \ldots = A_{M(L)} \qquad (2.115)$$

Sec. 2.2]   Stability constants of mononuclear complexes

Equation (2.115) enables the $A_{M(L)}$ values for different ligand concentrations to be calculated from measurements of the distribution coefficient of the metal ion in question in the absence and presence of the complexing agent, in solutions of the same ionic strength. From the related $A_{M(L)}$ and [L] values the overall complex stability constants can be computed (see p. 114).

If the metal ion is in an oxidation state numerically greater than the negative charge on the ligand, positively charged complexes are also formed, which are absorbed by the cation-exchanger. In such cases the method developed by Fronaeus can be used for calculating overall complex stability constants [53, 101]. The error of the cation-exchange method increases with increasing ligand number. It can be used mainly in cases where $N < 4$ (see Example 37).

The use of anion-exchangers is advantageous in cases where the ligands are of small size, that is, the negatively charged complexes formed can be absorbed by the resin without any significant difficulty caused by slow diffusion, and where the formation of complexes of higher ligand numbers is being studied.

Assume that the ligand L bears one negative charge (e.g. Cl^-, Br^-, OH^-, CN^- etc.) and also that of all the species formed with a metal ion M^{z+} only the mononuclear complex species bearing p negative charges is bound on the resin. The equilibrium constant of the ion-exchange reaction will then be

$$K_x = \frac{(ML_n)[L]^p}{[ML_n](L)^p} \quad (2.116)$$

$$p = n - z \quad (2.117)$$

If the amount of metal ion is very small compared with the capacity of the resin, i.e. $(ML_n) \ll (L)$, and the ionic strength is considered as constant, then to a first approximation, K_x is independent of the concentrations.

By multiplying both the numerator and denominator of the fraction by C_M we get

$$K_x = \frac{(ML_n)}{C_M} \frac{C_M}{[ML_n]} \frac{[L]^p}{(L)^p}$$

The first factor on the right-hand side of the equation is the distribution coefficient of the metal ion M, and the second is the reciprocal value of the mole fraction of the complex species with ligand number n. That is

$$K_x = D_M \frac{1}{\Phi_n} \frac{[L]^p}{(L)^p}$$

or

$$D_M = K_x \Phi_n (L)^p [L]^{-p} \qquad (2.118)$$

If K_x and (L) are constant, the distribution coefficient is a function of the ligand concentration and of the mole-fraction of the complex species bound.

Taking logarithms and differentiating with respect to log [L] [assuming that K_x and (L) are constants under the given conditions] we obtain the following equation [102]:

$$\frac{d \log D_M}{d \log [L]} = \frac{d \log \Phi_n}{d \log [L]} - p$$

The right-hand side of the equation gives the average charge number. That is

$$\frac{d \log D_M}{d \log [L]} = \bar{p} \qquad (2.119)$$

The following relationship exists between the average charge number and ligand number if the sign is also considered:

$$\bar{p} = \sum_0^N p_i \Phi_i = z - \bar{n} \qquad (2.120)$$

If the distribution coefficient of a metal ion on an anion-exchange resin is determined for solutions in which the ligand concentration is different, and the results are plotted in a log D vs. log [L] graph, the slope of the curve at any point gives the average charge number, from which the average ligand number can be obtained from Eq. (2.120). The overall stability constants can be calculated from related $\bar{n}$ and [L] values by any of the methods described earlier in this book (see p. 119). The log D vs. log [L] curve has a maximum and a part of negative slope if a negatively charged complex ion is formed. At the maximum the slope is zero, and $\bar{p} = 0$, i.e., the uncharged species predominates. For a univalent ligand, the ligand number is equal to the oxidation state of the metal ion, $z = n$ (see Fig. 2.29) at the maximum.

In studies on complexes of low stability the experiments should be made at relatively high ligand concentrations. In such cases, however,

Sec. 2.2] Stability constants of mononuclear complexes

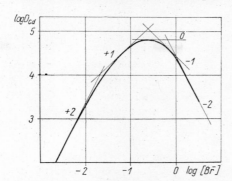

Fig. 2.29. Investigation of the cadmium–bromide system by an anion-exchange method. Distribution curve of cadmium according to measurements by Fronaeus [103]

the ionic strength cannot be maintained constant. Perchlorate, which is generally inert in complex formation reactions, is strongly bound by an anion-exchange resin and therefore cannot be used for adjusting constant ionic strength in solutions with lower ligand concentrations. If the displacement of ligand ions by perchlorate occurs, Eqs. (2.116) and (2.118) will no longer be valid. Above a certain ligand concentration limit, (L) cannot be considered as a constant, and on differentiaton of the logarithmic form of Eq. (2.118) another term remains in addition to the average charge number. Marcus and Coriell [104] calculated the dependence of (L) on [L] by taking the Donnan equilibrium into account and corrected the distribution coefficient accordingly.

The absorption of negatively charged ligands (anions of an acid) on an anion-exchange resin can be influenced by complex-forming metal ions. The distribution equilibria can be calculated by taking into account the ion-exchange equilibrium constants of the complexes formed (see p. 283). Obviously, the relationships can also be used for calculating stability constants, if the experimental relationship between the distribution coefficient and the concentration of the complex-forming metal ion is known.

Some important practical details in connection with the application of ion-exchangers are as follows. For the laboratory use of ion-exchangers see [53].

It has been assumed in deducing the equations above, that the ion-exchanger is a strong electrolyte, consequently, strongly acidic cation-exchangers or strongly basic anion-exchangers should be used in the experiments. When applying the cation-exchange method described, since the absorption of metal ions of the usual size is involved, a medium cross-linked resin (Dowex 50×8) is suitable. Similarly, a medium cross-linked anion-exchanger (Dowex 1×8) is appropriate for the investigation of small halide or cyanide complexes, etc. A macroporous resin should be chosen, however, if the ions to be absorbed are large (for example, absorption of larger ligands on an anion-exchanger). The particle size of the resin should be 50–100 mesh.

When the distribution measurements are made for a liquid ion-exchanger instead of an ion-exchange resin, the pH should generally be kept constant throughout the series of experiments, for some liquid ion-exchangers are weakly acidic or weakly basic, and their capacity depends on the pH of the solution in contact with them.

In preparing the solutions, care should be taken to use as low a metal ion concentration as possible, provided it can still be determined with sufficient accuracy. In using the cation-exchange method it is very important to use the same counter-ion in the same concentration. This ion, which does not take part in the complex formation, is responsible for maintaining the position of the ion-exchange equilibrium in each experiment. The constancy of the pH is also very important if the ligand taking part in complex formation tends to protonate (see Example 37).

It is advisable to use a static batch method in determining distribution coefficients. For the detailed description of the method see [53]. It should be borne in mind that the time required to attain equilibrium is determined by the rate of diffusion, thus the two phases should be stirred for an appropriate length of time (several hours).

After equilibrium has been reached, the concentration of the distributed component can be determined in the two separated phases by a radioactive tracer technique [105, 106]. Other instrumental methods can also be applied (such as spectrophotometry). The ions absorbed on the ion-exchanger can be determined after elution with a solution of appropriate composition.

2.2.5.1. *Worked example*

37. The distribution coefficient of nickel(II) has been determined at 80° C, on a cation-exchange resin for a solution not containing sulphosalicylate, and for solutions containing different concentrations of sulphosalicylate at pH 5 and ionic strength 0·5. The results of the measurements are as follows [107].

Sec. 2.2]   Stability constants of mononuclear complexes   171

Table 2.10

Calculation of the overall stability constants of [sulphosalicylate-nickel(II)] complexes (Example 37)

$\dfrac{c_{SS}}{\alpha_{SS}} = [SS]$	$\dfrac{D_{Ni}}{D_{Ni(SS)}} = A_{Ni(SS)}$	$\dfrac{A-1}{[SS]}$	$\dfrac{\dfrac{A-1}{[SS]} - \beta_1}{[SS]}$
$\dfrac{0\cdot 25}{2\cdot 95\times 10^5} = 8\cdot 47\times 10^{-7}$	$\dfrac{59\cdot 9}{25\cdot 2} = 2\cdot 38$	$\dfrac{1\cdot 38}{8\cdot 47\times 10^{-7}} = 1\cdot 63\times 10^6$	$\dfrac{(1\cdot 63 - 1\cdot 30)10^6}{8\cdot 47\times 10^{-7}} = 3\cdot 9\times 10^{11}$
$\dfrac{0\cdot 15}{2\cdot 95\times 10^5} = 5\cdot 08\times 10^{-7}$	$\dfrac{59\cdot 9}{34\cdot 2} = 1\cdot 75$	$\dfrac{0\cdot 75}{5\cdot 08\times 10^{-7}} = 1\cdot 48\times 10^6$	$\dfrac{(1\cdot 48 - 1\cdot 30)10^6}{5\cdot 08\times 10^{-7}} = 3\cdot 5\times 10^{11}$
$\dfrac{0\cdot 10}{2\cdot 95\times 10^5} = 3\cdot 39\times 10^{-7}$	$\dfrac{59\cdot 9}{40\cdot 3} = 1\cdot 49$	$\dfrac{0\cdot 49}{3\cdot 39\times 10^{-7}} = 1\cdot 44\times 10^6$	$\dfrac{(1\cdot 44 - 1\cdot 30)10^6}{3\cdot 39\times 10^{-7}} = 4\cdot 10\times 10^{11}$
$\dfrac{0\cdot 05}{2\cdot 95\times 10^5} = 1\cdot 69\times 10^{-7}$	$\dfrac{59\cdot 9}{49\cdot 0} = 1\cdot 23$	$\dfrac{0\cdot 23}{1\cdot 69\times 10^{-7}} = 1\cdot 36\times 10^6$	$\dfrac{(1\cdot 36 - 1\cdot 30)10^6}{1\cdot 69\times 10^{-7}} = 3\cdot 5\times 10^{11}$

C_{SS}, M	D_{Ni}
0·25	25·2
0·15	34·2
0·10	40·3
0·05	49·0
0·00	59·9

The metal ion, nickel(II), was present in trace concentration.

Calculate the overall stability constants of the complex species formed, taking into account the protonation constants of sulphosalicylate:

$\log K_1 = 10\cdot 47$; $\log K_2 = 2\cdot 30$; $\log K_3 = 1\cdot 52$ (80° C; $I = 0\cdot 5$)

The calculation is done by using Eq. (2.115) and Leden's method.

The value of the function $\alpha_{SS(H)}$ is calculated for pH 5, then the free ligand concentrations by using Eq. (1.77).

$$\alpha_{SS(H)} = 1 + 10^{-5} \times 10^{10\cdot 47} + 10^{-10} \times 10^{12\cdot 77} + 10^{-15} \times 10^{14\cdot 29} = 2\cdot 95 \times 10^5$$

The free ligand concentrations, and the calculated related values of A, $(A - 1)/[SS]$, etc. are given in Table 2.10, and the diagrams used for estimating the overall stability constants in Figs. 2.30 and 2.31.

The intercept of the line in Fig. 2.30 gives β_1 and that of the line parallel to the abscissa in Fig. 2.31 gives β_2 (at 80° C and $I = 0\cdot 5$).

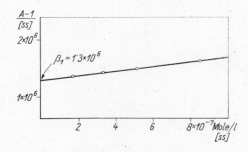

Fig. 2.30. Determination of the overall stability constants of [sulphosalicylate–nickel(II)] complexes by Leden's graphical method. Determination of β_1. (See Example 37)

Sec. 2.2] Stability constants of mononuclear complexes

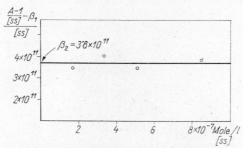

Fig. 2.31. Determination of the overall stability constant of [sulphosalicylate-nickel(II)] complexes by Leden's graphical method. Determination of β_2

As only two overall stability constants are to be calculated, let us also use the method of least squares (see p. 44).

$$\frac{A-1}{[SS]} = \beta_1 + \beta_2 [SS]$$

Related $y = (A - 1)/[SS]$ and $x = [SS]$ values are taken from Table 2.10, then x^2, xy, Σx, Σy, Σx^2 and Σxy are calculated. These values are given in Table 2.11. The number of pairs of values is $n = 4$.

Table 2.11

Calculation of the overall stability constant of [sulphosalicylate–nickel(II)] complexes by using the method of least squares (Example 37)

x	y	x^2	xy
$8\cdot 47 \times 10^{-7}$	$1\cdot 63 \times 10^6$	$71\cdot 741 \times 10^{-14}$	$13\cdot 806 \times 10^{-1}$
$5\cdot 08 \times 10^{-7}$	$1\cdot 48 \times 10^6$	$25\cdot 806 \times 10^{-14}$	$7\cdot 518 \times 10^{-1}$
$3\cdot 39 \times 10^{-7}$	$1\cdot 44 \times 10^6$	$11\cdot 492 \times 10^{-14}$	$4\cdot 882 \times 10^{-1}$
$1\cdot 69 \times 10^{-7}$	$1\cdot 36 \times 10^6$	$2\cdot 856 \times 10^{-14}$	$2\cdot 298 \times 10^{-1}$
Σx $18\cdot 630 \times 10^{-7}$	Σy $5\cdot 910 \times 10^6$	Σx^2 $111\cdot 895 \times 10^{-14}$	Σxy $28\cdot 504 \times 10^{-1}$

The values of β_1 and β_2 are calculated by solving the system of linear equations, Eqs. (1.67) and (1.68),

$$\beta_1 = \frac{\Sigma x \Sigma xy - \Sigma y \Sigma x^2}{(\Sigma x)^2 - n\Sigma x^2} = \frac{18\cdot63 \times 28\cdot504 \times 10^{-8} - 5\cdot910 \times 111\cdot895 \times 10^{-8}}{(18\cdot63)^2 \times 10^{-14} - 4 \times 111\cdot895 \times 10^{-14}}$$

$$= \frac{130\cdot27}{100\cdot503} \times 10^{-8} \times 10^{14} = 1\cdot30 \times 10^6$$

$$\beta_2 = \frac{\Sigma x \Sigma y - n\Sigma xy}{(\Sigma x)^2 - n\Sigma x^2} = \frac{18\cdot63 \times 5\cdot91 \times 10^{-1} - 4 \times 28\cdot504 \times 10^{-1}}{(18\cdot63)^2 \times 10^{-14} - 4 \times 111\cdot895 \times 10^{-14}}$$

$$= \frac{391\cdot3 \times 10^{-3}}{100\cdot503} = 3\cdot89 \times 10^{11}$$

The accuracy of the results can be increased by obtaining more experimental results.

2.2.6. Determination of the Stability Constant of Mixed-Ligand Complexes

With certain modifications the potentiometric, spectrophotometric, polarographic and distribution methods described can also be used for studying the formation of mixed-ligand complexes.

The general equation for the formation of mononuclear mixed-ligand complexes is as follows:

$$M + nL + mY \rightleftarrows ML_nY_m \qquad (2.121)$$

The overall stability constant is given by

$$\beta_{n,m} = \frac{[ML_nY_m]}{[M][L]^n[Y]^m} \qquad (2.122)$$

The complex is formed by a stepwise mechanism, and n and m may vary between zero and the maximum ligand number.

The total concentration of the metal ion in a solution containing the ligands L and Y can be given as follows:

$$C_M = [M] + [ML] + [ML_2] + \ldots + [MY] + [MY_2] + \ldots$$
$$+ [MLY] + [ML_2Y] + \ldots = [M] + [M]\sum_{i=1}^{i=N_L}\sum_{j=1}^{j=N_Y}[L]^i[Y]^j\beta_{ij} \quad (2.123)$$

On division of both sides by [M], the function $A_{M(L,Y)}$, is obtained:

$$A_{M(L,Y)} = \frac{C_M}{[M]} = 1 + \sum_{i=1}^{i=N_L}\sum_{j=1}^{j=N_Y}[L]^i[Y]^j\beta_{ij} \qquad (2.124)$$

Sec. 2.2] Stability constants of mononuclear complexes

If a set of solutions is prepared in which C_M is known and constant, the concentrations of L and Y are varied, and the free metal ion concentration is measured by direct potentiometry [108] or, in another method, the shift of the $E_{1/2}$ value is determined by polarographic measurements (see p. 147) and the value of $A_{M(L,Y)}$ calculated; the overall complex stability constants can be calculated from the related $A_{M(L,Y)}$, [L] and [Y] values by using Eq. (2.124) in the same way as has been described for simple complexes containing only one sort of ligand (see p. 114).

It is advisable to keep the concentration of the metal ion and of one of the ligands constant and vary that of the other ligand. Obviously, as the number of overall stability constants to be determined increases, the number of measurements necessary for the calculations (which become increasingly complicated) and the errors in the calculation of values of the constants also increase.

Similarly, when pH titration methods are used, the calculations become complicated if a large number of complexes are simultaneously present. However, by choosing the experimental conditions properly, it is possible to obtain only a few predominating complex species in the solution and thus the calculations can be simplified. The simplified method due to Watters [109] is only suitable for determining the overall stability constant of complexes with the maximum number of ligands. In this method the experiments are made on solutions in which the sum of average ligand numbers calculated for L and Y,

$$\bar{n}_L = \frac{C_L - [L]}{C_M}, \quad \bar{n}_Y = \frac{C_Y - [Y]}{C_M}$$

is equal to the maximum ligand number. That is,

$$\bar{n}_L + \bar{n}_Y = N$$

The formation of 'saturated' complexes is ensured by a high ligand concentration, and only the concentration ratio of the ligands is varied. In such cases the two complex formation functions are interdependent, and the calculation of overall stability constants from $\bar{n}_L$, $\bar{n}_Y$, [L] and [Y] values by means of the functions becomes easier.

Similarly, the calculation of equilibrium constants becomes simpler if one of the ligands of the mixed-ligand complex forms a very stable chelate with the metal ion, and the binding of the other ligands may proceed only through the exchange of solvent molecules (or possibly through the displacement of one of the donor atoms of the chelating ligand) according to the reaction equation

$$ML + nY \rightleftharpoons MLY_n$$

where L stands for the chelating agent. In such cases the stability constant

$$K = \frac{[MLY_n]}{[ML][Y]^n}$$

can be determined by a pH titration or spectrophotometric method, in the same way as described for simple mononuclear complexes. In the series of experiments $C_M \simeq C_{ML}$ is kept constant and the concentration of the ligand Y is varied. The determination of the stability constants of [hydroxo–EDTA–iron(III)] [110], and [halide–dimethylglyoxime–cobalt(II)] [111] mixed-ligand complexes are examples.

The formation of mixed chelate complexes is important in choosing the optimum solution composition in complexometric titrations (see p. 224).

From the analytical point of view mixed-ligand complexes are important if the properties of the mixed-ligand complexes differ significantly from those of the simple parent complexes. This usually leads to an increase in the selectivity of a particular determination [112]. For example, the [pyridine–salicylate–copper(II)] complex can be extracted with chloroform, whereas neither [pyridine–copper(II)], nor [salicylate copper(II)] complexes are extractable.

If the physical properties of the mixed-ligand complex differ markedly from those of the parent complex, its mole fraction can usually be determined by a spectrophotometric or distribution method in solutions containing different but known amounts of the ligands, and the overall complex stability constants can be calculated from related Φ, [L] and [Y] values.

In the simplest case, if the formation of the mixed-ligand complex can be described by the equation

$$ML_2 + MY_2 \rightleftharpoons 2\,MLY,$$

and if the stabilities of the parent complexes are high enough, that is, practically only the complex species given in the reaction equation are present in the solutions studied and neither of the ligands is present in excess, and if the sum of the total concentrations of the parent complexes

Sec. 2.2] Stability constants of mononuclear complexes

is constant, and their ratio is varied, then the stability constant

$$K = \frac{[MLY]^2}{[ML_2][MY_2]}$$

can be calculated if the mole fraction Φ_{MLY} and the total concentrations of the parent complexes are known, according to Spiro and Hume [113], as follows:

$$K = \frac{4\Phi^2}{(1-\Phi)^2 - \left(\dfrac{1-R}{1+R}\right)^2}$$

where

$$R = \frac{C_{MY_2}}{C_{ML_2}}$$

It is to be noted that for the calculation of Φ_{MLY} a knowledge of the molar absorptivity of the mixed-ligand complex is necessary if spectrophotometry is to be used, and of the distribution constant if the distribution method is to be used [114].

With regard to the choice of optimum concentrations and pH, the role of side-reactions, and the adjustment of constant ionic strength, the instructions given in the previous sections should be used.

In the separations by extraction used in analytical chemistry, mixed-ligand complexes that are soluble in the organic phase are of importance. Mixed-ligand complexes of this type are formed by a reaction of a metal ion with two chelating agents or by combination of a metal chelate with a donor molecule (for more details see Sections 3.8 and 3.9.4.2).

2.2.6.1. *Worked example*

38. From the data of the titration curve shown in Fig. 2.32, calculate the stability constant of EDTA–(OH)–mercury(II) and EDTA–(NH$_3$)–mercury(II).

When the protonated HgLH$_2$ complex is titrated with standard alkali, the following acid–base reactions take place.

$$\text{HgLH}_2 + \text{OH}^- \rightleftharpoons \text{HgLH}^- + \text{H}_2\text{O} \qquad \text{(i)}$$

$$\text{HgLH}^- + \text{OH}^- \rightleftharpoons \text{HgL}^{2-} + \text{H}_2\text{O} \qquad \text{(ii)}$$

$$\text{HgL}^{2-} + \text{OH}^- \rightleftharpoons \text{HgL(OH)}^{3-} + \text{H}_2\text{O} \qquad \text{(iii)}$$

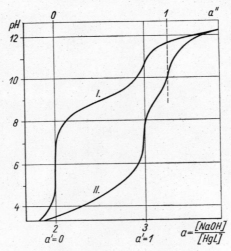

Fig. 2.32. I: Titration curve of the stoichiometric EDTA–mercury(II) complex according to measurements made by Sadek and Reilley (reprinted with permission from *Analytical Chemistry 31*, 494, 1959; copyright by the American Chemical Society) [115]. $C_{\text{Hg-EDTA}} = 0 \cdot 02M$. II: The same in the presence of $0 \cdot 025M$ ammonium nitrate. For a' and a'' see text. (See Example 38)

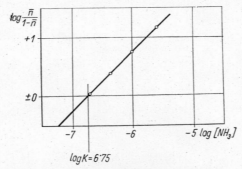

Fig. 2.33. Determination of the stability constant of the EDTA–ammine–mercury(II) mixed-ligand complex by a graphical method. (See Example 38)

Reaction (iii) proceeds in the range $2 < a < 3$. The equilibrium constant of this reaction — as estimated from the titration curve — differs markedly from those of reactions (i) and (ii). The equilibrium constant for formation of the hydroxo-complex can be calculated from values from the third section of the titration curve (see pp. 92 and 127).

The equilibrium constant of reaction (iii) is:

$$K_{\text{HgL(OH)}} = \frac{[\text{HgL(OH)}]}{[\text{HgL}][\text{OH}^-]}$$

Taking logarithms:

$$\log K_{\text{HgL(OH)}} = \log \frac{[\text{HgL(OH)}]}{[\text{HgL}]} - \log [\text{OH}^-]$$

Considering Eq. (2.17) we get:

$$\log K_{\text{HgL(OH)}} = \log \frac{a' C_{\text{HgL}} + [\text{H}^+] - [\text{OH}^-]}{(1-a') C_{\text{HgL}} - [\text{H}^+] + [\text{OH}^-]} - \log [\text{OH}^-]$$

As the concentrations $[\text{H}^+]$ and $[\text{OH}^-]$ can be neglected compared with $C_{\text{HgL}} = 2 \times 10^{-2} M$ at a degree of neutralization $a = 2\cdot 5$, where $a' = 0\cdot 5$, the first term on the right-hand side of the equation equals zero, and the negative logarithm of the hydroxide concentration in the half-titrated solution gives $\log K_{\text{HgL(OH)}}$.

$$-\log [\text{OH}^-] = 14 - \text{pH} = 14 - 8\cdot 9 = 5\cdot 1;$$

$$\log K_{\text{HgEDTA(OH)}} = 5\cdot 1$$

The stability constant of the mixed-ligand complex containing ammonia can be calculated in two different ways. First, the equilibrium constant of the acid–base reaction is calculated, taking into account the reaction

$$\text{HgL}^{2-} + \text{NH}_4^+ + \text{OH}^- \rightleftharpoons \text{HgL(NH}_3)^{2-} + \text{H}_2\text{O} \qquad \text{(iv)}$$

proceeding in the third section of the titration in the same way as described above, and then the stability constant

$$K_{\text{HgL(NH}_3)} = \frac{[\text{HgL(NH}_3)]}{[\text{HgL}][\text{NH}_3]}$$

is determined.

Alternatively, the calculation is done by taking the complex-formation reaction

$$HgL^{2-} + NH_4^+ \rightleftharpoons HgL(NH_3)^{2-} + H^+$$

into consideration, by using Calvin and Melchior's method (see pp. 119 and 127). In the latter case, however, the limits of the neutralization section in the titration diagram are determined by the amount of the protonated ligand, NH_4^+ (a'' in the figure). Using the second method, let us calculate a few related $\bar{n}$ and $[NH_3]$ data from the coordinate values of some points of the titration curve by using Eqs. (2.45) and (2.46). Then, since the complex formation reaction involves the binding of only one ligand, Eq. (2.47) may be used for the graphical evaluation of log K.

The partial results calculated by using the equations

$$\frac{(a'' - a_0'')C_{NH_4}}{C_{HgL}} = \bar{n}$$

and

$$\log [NH_3] = \log \left[C_{NH_4}(1 - [a'' - a_0'']) \right] - \log \alpha_{NH_3(H)}$$

are given in Table 2.12, and the diagram for the evaluation in Fig. 2.33.

At the intersection of the line with the abscissa, $-\log [NH_3] = \log K$ according to Eq. (2.47), i.e. log $K_{HgEDTA(NH_3)} = 6.75$.

Table 2.12

Calculation of the stability constant of the [EDTA–ammine–mercury(II)] mixed-ligand complex, with data taken from the titration curve (Example 38)

pH	$a'' - a_0''$	$\bar{n}$	log $\alpha_{NH_3(H)}$	log $[NH_3]$	log $\dfrac{\bar{n}}{1-\bar{n}}$
4·5	0·42	0·525	4·9	−6·74	+0·04
5·0	0·57	0·71	4·4	−6·37	+0·39
5·5	0·68	0·85	3·9	−6·00	+0·75
6·0	0·745	0·93	3·4	−5·60	+1·12

2.3. Determination of the Stability Constants of Polynuclear Complexes

If the degree of complex formation, $\bar{n}$, or a figure representing related properties (e.g. in the case of ion-exchange the distribution coefficient) not only depends on the concentration of the ligand but also on that of the metal ion, the formation of polynuclear complexes is encountered.

Complexes formed with a multidentate chelating ligand may be polynuclear if the number of coordination sites is greater than the coordination number of the metal ion (such as the calcium complex of triethylenetetraminehexa-acetate for example).

In practice, however, the polynuclear hydroxo-complexes already mentioned are more important. These are formed by the hydrolysis of metal salts (see p. 60).

Potentiometric, ion-exchange and partition methods are mainly used for studying polynuclear complexes, and, less frequently, spectrophotometric methods.

Methods for studying 'core and links' type complexes have been developed by Sillén. The basis for these methods is that a series of solutions is prepared with different metal ion concentrations, and the values of $\bar{n}$ or $A_{M(OH)}$ for different pH values and free ligand concentrations are calculated. The composition of the complexes and the stability constants are determined by means of $\bar{n}$ vs. pH or $A_{M(OH)}$ vs. pH plots. For more details see [6, 60, 74, 78, 116].

Chapter 3

ANALYTICAL APPLICATIONS

Complex-forming reactions can be used either directly in chemical analysis (e.g. precipitation in the form of insoluble chelates, complexometric titration, etc.) or they can be used to increase the selectivity of separations and determinations.

If the stability constants and conditional equilibrium constants are known, the optimum concentration conditions, pH, etc. to be used in an analysis can be calculated.

3.1. Gravimetric Analysis

The principle of gravimetric analysis is that an insoluble compound is prepared from the component to be determined, is separated from the liquid phase and weighed in the elemental form or in the form of an easily weighed compound of constant composition. By separation of the solid and liquid phases other constituents of the original sample are removed. (For more details see [117].)

Precipitation can be described by the general equation

$$m\text{M}^{z+} + z\text{A}^{m-} \rightleftharpoons \text{M}_m\text{A}_z \qquad (3.1)$$

where M denotes a metal ion (or possibly the hydrogen ion) and A the ion which reacts with it to form a sparingly soluble compound.

As the reaction product forms a separate homogeneous phase consisting of a single substance, the equilibrium constant of the precipitation reaction can be given as

$$K_s = \frac{1}{[\text{M}]^m[\text{A}]^z} \qquad (p, T, I = \text{const.}) \qquad (3.2)$$

i.e.

$$-\log K_s = m \log \text{M} + z \log \text{A}$$

Thus, K_s is the reciprocal of the solubility product S_0 at the given ionic strength.

$$S_0 = [\text{M}]^m[\text{A}]^z \qquad (p, T, I = \text{const.}) \qquad (3.3)$$

Sec. 3.1] Gravimetric analysis

If the precipitate formed is a strong electrolyte, the position of the equilibrium, and the amount of the component in the precipitate and in the solution under the conditions of precipitation can be calculated, provided that K_s or S_0 is known.

If either the metal ion or the anion forms a soluble compound with other components in the solution, the analytical concentrations of the ions remaining in solution after precipitation will not be equal to the concentrations of the free ions required by the equilibrium constant of the precipitation but will exceed them.

$$[M'] \geq [M] \quad \text{and} \quad [A'] \geq [A]$$

The conditional equilibrium constant

$$K'_s = \frac{1}{[M']^m[A']^z} \tag{3.4a}$$

is related to the real constant by

$$K'_s = \frac{K_S}{\alpha_M^m \alpha_A^z} \tag{3.4b}$$

i.e.

$$\log K'_s = \log K_s - m \log \alpha_M - z \log \alpha_A$$

Similarly, the conditional and real solubility products are related by

$$S'_0 = S_0 \alpha_M^m \alpha_A^z \tag{3.5}$$

i.e.

$$\log S'_0 = \log S_0 + m \log \alpha_M + z \log \alpha_A$$

where $\alpha_M = [M']/[M]$, and $\alpha_A = [A']/[A]$. Their values can be calculated if the equilibrium constants of the side-reactions and the concentration of the interfering constituent are known (see p. 52).

The analytical concentration of the metal ion remaining in solution can be calculated from Eq. (3.4a) if the precipitant A is added in excess:

$$[M'] = \sqrt[m]{\frac{1}{K'_s [A']^z}} \tag{3.4c}$$

If the concentration of the metal ion is C_M^0 in the original solution and $M']$ in the solution after the precipitation, and the change in volume is negligible, the error is

$$\Delta = -\frac{100[\text{M}']}{C_\text{M}^0}\%$$

The side-reaction of the metal ion may be hydroxo-complex formation in the given pH range. However, more important is the case when another complex-forming agent is also present in the solution and forms a soluble complex with the metal ion. In such cases the solubility may increase considerably. In analysis complexing agents are sometimes used to prevent the precipitation of a component of the solution. This is known as 'masking' the given component.

Equations (3.4b) and (3.4c) can be used to calculate the concentration of the complexing agent and the pH necessary for masking, provided that the equilibrium constant is known.

If the concentration of the component to be precipitated is $10^{-2}M$, the precipitation can be considered as quantitative, if the concentration after precipitation is equal to or lower than $10^{-5}M$ (i.e. $\leq 0.1\%$ of the substance originally present remains unprecipitated). The condition for quantitative precipitation, if a 1:1 precipitate is formed and the concentration of excess of precipitant is $10^{-2}M$, is given by

$$K_\text{s}' > \frac{1}{10^{-5} \times 10^{-2}}$$

that is

$$\log K_\text{s}' > 7 \tag{3.6}$$

In general, for precipitation not to occur it is necessary that

$$\log K_\text{s}' < 3 \tag{3.7}$$

Some precipitates are salts of weak acids. In such cases the protonation of the anion is the side-reaction which should be taken into consideration when calculating the conditional stability constants. The solubility of the precipitate depends on the pH of the solution. If the protonation constant (or constants) is (are) known, the conditional constant and the solubility of the component in question can be calculated by using the $\alpha_{\text{A(H)}}$ function.

When organic precipitants are used and the precipitate is an uncharged chelate (for example nickel dimethylglyoximate, aluminium oxinate etc.), and thus not a strong electrolyte, at equilibrium not only the disso-

Sec. 3.1]   Gravimetric analysis   185

ciated species but also the non-dissociated complex will be present in the solution in contact with the precipitate.

If a complexing agent L is used as precipitant, the composition of the precipitate is ML_2, and taking up of a further ligand (or release of one) gives a soluble complex, the following equilibrium reactions are to be taken into consideration for the solution in contact with the precipitate:

$$M + 2L \xrightleftharpoons{K_{s0}} ML_{2\,(prec)}$$

$$ML + L \xrightleftharpoons{K_{s1}} ML_{2\,(prec)}$$

$$ML_2 \xrightleftharpoons{K_{s2}} ML_{2\,(prec)}$$

$$ML_3 \xrightleftharpoons{K_{s3}} ML_{2\,(prec)} + L$$

$$ML_n \xrightleftharpoons{K_{sn}} ML_{2\,(prec)} + (n-2)L$$

where

$$K_{s0} = \frac{1}{[M][L]^2} = \frac{1}{S_0} \tag{3.8}$$

$$K_{s1} = \frac{1}{[ML][L]} \tag{3.9}$$

$$K_{s2} = \frac{1}{[ML_2]} \tag{3.10}$$

$$K_{s3} = \frac{[L]}{[ML_3]} \tag{3.11}$$

$$K_{sn} = \frac{[L]^{n-2}}{[ML_n]} \tag{3.12}$$

In general the following relationship holds between K_{sn}, K_{s0} and the overall complex stability constant β_n:

$$\beta_n = \frac{K_{s0}}{K_{sn}} \tag{3.13}$$

Provided that the overall stability constants and K_{s0} are known, $K_{s1}, K_{s2} \ldots$ can be calculated.

The optimum conditions of precipitation (pH, concentration, etc.) can be found by equilibrium calculations. It can be predicted whether quantitative precipitation or separation can or cannot be obtained under given conditions. However, the degree of co-precipitation and of contamination by adsorption, which decrease the purity of the product, and the precision of determination cannot be predicted in this way.

Another gravimetric method is electro-deposition. The metal ion to be determined is transformed into the metallic form by electrolytic reduction (or to the slightly soluble oxide by oxidation) and the product deposited on the electrode is weighed.

In the case of the reduction of a metal ion to the metal, if the electrode process is reversible, i.e. there is no overvoltage, the deposition potential will be equal to the electrode potential and can be calculated from the Nernst equation:

$$E = E_0 + \frac{0.059}{z} \log [M] \qquad (3.14)$$

In the absence of complexing agents (the value of the activity coefficient is disregarded) the free metal ion concentration is equal to the total concentration, i.e. $[M] \simeq C_M$. In the presence of a complexing agent, since

$$[M] = \frac{C_M}{\alpha_{M(L)}} \qquad (3.15)$$

the equation is modified to

$$E = E_0 - \frac{0.059}{z} \log \alpha_{M(L)} + \frac{0.059}{z} \log C_M = E_0' + \frac{0.059}{z} \log C_M \quad (3.16)$$

At a given concentration of the complexing agent, the conditional standard potential is E_0' instead of E_0, the former being always more negative than the latter.

For quantitative deposition the concentration of a metal ion should decrease by at least three orders of magnitude. According to Eq. (3.14), for the deposition of a bivalent metal ion the decrease in E is

$$\Delta E = \frac{0.059}{2} \log 10^{-3} = -0.0885 \simeq -0.1 \text{ V}.$$

In general, a quantitative separation of two metal ions can be achieved if the difference in deposition potentials is at least $2 \times 0.1 = 0.2$ V. If the potentials are closer to one another, the difference can be increased by a suitable complexing agent. The concentration of the complexing agent and the new potential can be calculated by using Eq. (3.16).

3.1.1. Worked examples

39. Calculate the solubility of calcium fluoride in water and in $0.01M$ calcium chloride solution.

$$K_s = 3 \times 10^{10}$$

Let x denote the concentration of dissolved calcium fluoride. In the first case

$$[Ca^{2+}] = x \quad \text{and} \quad [F^-] = 2x$$

$$K_s = \frac{1}{[Ca^{2+}][F^-]^2} = \frac{1}{x(2x)^2} = 3 \times 10^{10}$$

From this

$$x^3 = 8.33 \times 10^{-12}$$

$$x = 2.03 \times 10^{-4} M$$

If $0.01M$ calcium chloride is also present, compared with which the calcium ion concentration due to the solubility of calcium fluoride can be neglected, $[Ca^{2+}] \simeq 10^{-2}M$. The solubility of calcium fluoride can be calculated through the fluoride concentration in the solution.

$$\frac{1}{[Ca^{2+}][F^-]^2} = \frac{1}{10^{-2}(2x)^2} = 3 \times 10^{10}$$

from which

$$x = 2.88 \times 10^{-5} M$$

40. Calculate the total concentration of iron(III) in a solution of pH 2, in equilibrium with iron(III) hydroxide precipitate, and also the fraction of iron(III) present in the form of the binuclear hydroxo-complex. The complex formation with other ions present in the solution should be neglected ($\log K_s = 38.6$). For the overall stability constants of iron(III) hydroxo-complexes see pp. 60 and 323. The following equilibrium condition is valid for the solution in equilibrium with the precipitate [see Eq. (3.2)]:

$$\log [Fe^{3+}] + 3 \log [OH^-] = -\log K_{s0} = -38.6.$$

At pH 2

$$\log [OH^-] = -12$$

hence

$$\log [Fe^{3+}] = -38.6 + 3 \times 12 = -2.6.$$

The total concentration of iron(III) in solution will be [see Eq. (1.14)]

$$C_{Fe} = [Fe^{3+}]+[Fe^{3+}][OH^-]\beta_1 + [Fe^{3+}][OH^-]^2\beta_2 + 2[Fe^{3+}]^2[OH]^2\beta_{22}$$
$$= [Fe^{3+}](1 + [OH^-]\beta_1 + [OH^-]^2\beta_2 + 2[Fe^{3+}][OH^-]^2\beta_{22})$$
$$= 10^{-2.6}(1 + 10^{-12} \times 10^{11} + 10^{-24} \times 10^{21.7} + 2 \times 10^{-2.6} \times 10^{-24} 10^{25.1})$$
$$= 0.00251(1 + 0.100 + 0.005 + 0.063) \simeq 2.9 \times 10^{-3} M$$

The mole-fraction of the binuclear species [compare with Eqs. (1.20) and (1.25)] is given by

$$\Phi_{22} = \frac{2[Fe_2(OH)_2]}{C_{Fe}}$$
$$= \frac{2[Fe^{3+}][OH^-]^2\beta_{22}}{1 + [OH^-]\beta_1 + [OH^-]^2\beta_2 + 2[Fe^{3+}][OH^-]^2\beta_{22}}$$

(See above for the calculation of the numerator and denominator.)

$$\Phi_{22} = \frac{0.063}{1 + 0.100 + 0.005 + 0.063} = 0.054.$$

So 5.4% of iron(III) in solution is present as the binuclear complex.

41. Calculate the solubility of calcium oxalate in $10^{-2}M$ oxalate solution at pH 3. The precipitate formation constant is $\log K_s = 8.64$ ($I = 0$).

The calculation is done by using Eq. (3.4b):

$$\log K'_s = \log K_s - \log \alpha_{Ox(H)}$$

The value of $\log \alpha_{Ox(H)}$ can be found from the diagram in Fig. 1.13. At pH 3 $\log \alpha_{Ox(H)} = 0.8$.

$$\log K'_s = 8.64 - 0.8 = 7.84.$$

$$K'_s = \frac{1}{[Ox'][Ca^{2+}]}$$

The solubility of calcium oxalate is equal to the concentration of calcium ions. Therefore if $[Ca^{2+}] = x$

$$10^{-7.84} = 10^{-2} x$$

From this $x = 1.45 \times 10^{-6} M$.

42. Calculate the pH required to achieve practically quantitative precipitation of zinc anthranilate from a $10^{-2}M$ solution. Log $K_s = 9\cdot75$ and the dissociation constant of anthranilic acid is $\log k = -4\cdot9$.

To determine zinc from $10^{-2}M$ solution with an accuracy of better than $0\cdot1\%$, the concentration of zinc ions remaining in solution after precipitation must not exceed $10^{-5}M$. Assuming that dissolved zinc anthranilate is completely dissociated, the metal ion enters no side-reaction, and the analytical concentration of the excess of precipitant is $10^{-2}M$, then from Eq. (3.4a)

$$-\log K'_s = \log [Zn^{2+}] + 2\log [An'] < -5 - 2\times 2 = -9.$$

Taking into account Eq. (3.4b)

$$-\log K'_s = -\log K_s + 2\log \alpha_{An(H)} < -9$$

Therefore

$$-9\cdot75 + 2\log \alpha_{An(H)} < -9$$

and hence

$$\log \alpha_{An(H)} < \frac{9\cdot75 - 9\cdot0}{2} = 0\cdot375$$

Now

$$\alpha_{An(H)} = 1 + [H^+]10^{4\cdot9}$$

so for the following inequality to be valid

$$10^{4\cdot9} \times [H^+] < 2\cdot37 - 1$$

it follows that

$$[H^+] < \frac{1\cdot37}{10^{4\cdot9}} = 10^{-4\cdot76}$$

i.e.

$$pH > 4\cdot76$$

The pH given in the book by Erdey [117] is close to the value calculated here.

43. Calculate the concentration of nickel(II) remaining in solution when nickel(II) dimethylglyoximate is precipitated at pH 5, from 5% tartaric acid solution. Log $K_{s0} = 23\cdot66$, log $K_{s1} = 11\cdot9$ and log $K_{s2} = 5\cdot68$ [118]. The stability constant of the nickel(II) tartrate complex is log $\gamma_1 \simeq 3\cdot0$. The protonation constants of tartrate ion are log $K_1 = 4\cdot1$ and log $K_2 = 2\cdot9$. The protonation constant of dimethylglyoxime is log $K_1 = 10\cdot6$ [118].

Let [Ni'] denote the concentration of nickel(II) remaining in solution owing to the solubility of the dimethylglyoxime complexes and to complexation by tartrate

$$[Ni'] = [Ni^{2+}] + [Ni(DMG)] + [Ni(DMG)_2] + [NiT]$$

From the overall complex stability constants and the protonation of the ligand, it follows that

$$\frac{[Ni']}{[Ni^{2+}]} = \left[1 + \frac{[DMG']}{\alpha_{DMG(H)}}\beta_1 + \left(\frac{[DMG']}{\alpha_{DMG(H)}}\right)^2 \beta_2 + \frac{[T']}{\alpha_{T(H)}}\gamma_1\right] = \alpha_{Ni}$$

Assuming that the concentration of excess of precipitant $[DMG'] = 10^{-2}M$, the concentration of nickel(II) is determined by K'_{s0}. [See Eq. (3.4c).]

$$\log[Ni'] = -\log K'_{s0} - 2\log[DMG']$$

The values of $\alpha_{DMG(H)}$, $\alpha_{T(H)}$ and α_{Ni} calculated for pH 5 by using Eqs. (1.70) and (1.72) are

$$\alpha_{DMG(H)} = 1 + 10^{-5} \times 10^{10.6} = 10^{5.6}$$

$$\alpha_{T(H)} = 1 + 10^5 \times 10^{4.1} + 10^{-10} \times 10^7 = 10^{0.05}$$

$$\alpha_{Ni} = 1 + 10^{-2} \times 10^{-5.6} \times 10^{11.76} + 10^{-4} \times 10^{-11.2} \times 10^{17.98}$$

$$+ 10^{-0.48} \times 10^{0.05} \times 10^3 = 10^{4.16}$$

The value of K'_{s0} is calculated using Eq. (3.4b), those of β_1 and β_2 by using Eq. (3.13).

$$\log K'_{s0} = \log K_{s0} - 2\log \alpha_{DMG(H)} - \log \alpha_{Ni} = 23.66 - 2 \times 5.6 - 4.16 = 8.30$$

$$\log \beta_1 = \log K_{s0} - \log K_{s1} = 23.66 - 11.9 = 11.76$$

$$\log \beta_2 = \log K_{s0} - \log K_{s2} = 23.66 - 5.68 = 17.98$$

The total concentration of nickel(II) in the solution is

$$\log[Ni'] = -8.30 + 2 \times 2 = -4.30$$

$$[Ni'] = 5 \times 10^{-5} M$$

44. The formation constant of lead oxalate precipitate is log $K_s = 10\cdot5$, that of calcium oxalate is log $K_s = 8\cdot64$. Calculate whether calcium ions can be selectively precipitated with oxalate from a solution that is $10^{-2}M$ with respect to both calcium and lead, if EDTA is used as masking agent. Log $K_{\text{Pb EDTA}} = 18\cdot0$; log $K_{\text{Ca EDTA}} = 10\cdot7$.

The selective precipitation of calcium will be quantitative when the concentration of excess of the precipitant is $10^{-2}M$ if

$$\log K'_{s(\text{Ca})} = -\log [\text{Ca}'] - \log [\text{Ox}'] \geq 5 + 2 = 7 \qquad \text{(i)}$$

and at the same time

$$\log K'_{s(\text{Pb})} = -\log [\text{Pb}'] - \log [\text{Ox}'] \leq 2 + 2 = 4 \qquad \text{(ii)}$$

From Eq. (3.4b), and the thermodynamic precipitate formation constant, and assuming that no side-reaction of calcium occurs, condition (i) can be written in the form

$$8\cdot64 - \log \alpha_{\text{Ox(H)}} > 7$$

i.e.

$$\log \alpha_{\text{Ox(H)}} < 1\cdot64$$

As shown by Fig. 1.13, this last condition is fulfilled at pH $> 2\cdot5$.

The masking agent will also slightly affect precipitation of the calcium (i.e. $\alpha_{\text{Ca(EDTA)}} > 1$), therefore, a slightly higher pH is chosen. At pH $= 4$ log $\alpha_{\text{Ox(H)}} = 0\cdot1$.

From Eq. (3.4b) and taking into consideration the real value of the constant, condition (ii) is modified to

$$\log K_{s(\text{Pb})} - \log \alpha_{\text{Pb(EDTA)}} - \log_{\text{Ox(H)}} < 4$$

$$10\cdot5 - \log \alpha_{\text{Pb(EDTA)}} - 0\cdot1 < 4$$

From this

$$\log \alpha_{\text{Pb(EDTA)}} > 6\cdot4$$

As, according to the definition of α

$$\alpha_{\text{Pb(EDTA)}} = 1 + \left[\frac{[\text{EDTA}']}{\alpha_{\text{EDTA(H)}}}\right] K,$$

and as the value of $\log \alpha_{EDTA(H)}$ is 8·6 at pH 4 from Fig. 1.14, and 1 can be neglected compared with the second term on the right-hand side of the equation, then

$$\log \alpha_{Pb(EDTA)} = \log [EDTA'] - 8\cdot 6 + 18$$

and

$$6\cdot 4 < \log [EDTA'] - 8\cdot 6 + 18.$$

From this

$$\log [EDTA'] > -3.$$

Thus the excess EDTA has to exceed $10^{-3}M$ in order to prevent the precipitation of lead(II).

The maximum permissible concentration of EDTA is calculated by considering condition (i)

$$\log K_{s(Ca)} - \log \alpha_{Ca(EDTA)} - 0\cdot 1 > 7$$

$$8\cdot 64 - \log \alpha_{Ca(EDTA)} - 0\cdot 1 > 7$$

$$\log \alpha_{Ca(EDTA)} < 1\cdot 54$$

At pH 4

$$\alpha_{Ca(EDTA)} = 1 + \left(\frac{[EDTA']}{10^{8\cdot 6}}\right) 10^{10\cdot 7}$$

and the maximum permissible concentration of EDTA is given by

$$[EDTA'] < 0\cdot 26M.$$

Thus to obtain safe separation, EDTA is added to the solution at pH 4 in an amount such that after complexing the lead(II) it is present in a concentration of $2 \times 10^{-2}M$. In this case the conditional constants are as follows:

$$\log K'_{s(Pb)} = 10\cdot 5 - 7\cdot 7 - 0\cdot 1 = 2\cdot 7$$

$$\log K'_{s(Ca)} = 8\cdot 64 - 0\cdot 54 - 0\cdot 1 = 8\cdot 0$$

According to measurements made by the author [119] the determination can be carried out with an error of $\pm 0\cdot 5\%$.

The solubilities of calcium and lead have been calculated and are shown diagrammatically for different pH values, in the absence and in the presence of masking agent, by means of Eqs. (3.4b) and (3.4c), for [Ox']

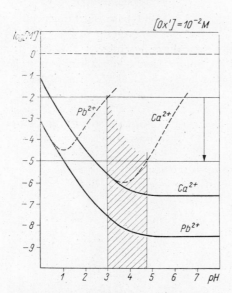

Fig. 3.1. The solubility of calcium(II) and lead(II) as a function of pH at a constant $10^{-2}M$ precipitant (oxalate) concentration; solid lines: without complexant; broken lines: in the presence of EDTA; $[EDTA'] = 10^{-2}M$. In the shaded pH range calcium can be selectively precipitated [119]

$= 10^{-2}M$ in Fig. 3.1. The pH range applicable in the selective precipitation of calcium in the presence of the masking agent can be estimated from the figure.

45. Calculate the deposition potential of silver from a $10^{-2}M$ silver nitrate solution on a silver electrode if the solution is $10^{-1}M$ in potassium cyanide, $E_0 = +0.80$ V and the overall stability constants of cyanosilver complexes are $\log \beta_2 = 21\cdot1$; $\log \beta_3 = 21\cdot8$; and $\log \beta_4 = 20\cdot7$ [120].

$\alpha_{Ag(CN)}$ is calculated first:

$$\alpha_{Ag(CN)} = 1 + 10^{-2} \times 10^{21\cdot1} + 10^{-3} \times 10^{21\cdot8} + 10^{-4} \times 10^{20\cdot7} = 10^{19\cdot27}$$

then the deposition potential from Eq. (3.16)

$$E = 0\cdot80 - 0\cdot059 \times 19\cdot27 - 0\cdot059 \times 2 = -0\cdot46 \text{ V}.$$

3.2. Acid-base Titrations

The theory and practical applications of acid–base titrations are given in detail in other textbooks. Here the author only wishes to draw attention to some points which make calculations easier.

3.2.1. Calculation of the hydrogen ion concentration

If only one strong monobasic acid is present in a solution, it follows from the principle of electroneutrality that

$$[H^+] = [A^-] + [OH^-]$$

For a strong acid $[A^-]$ is equal to the analytical concentration, C_{HA}. In aqueous solution, if $C_{HA} > 10^{-6}M$, the second term on the right-hand side can be neglected, and

$$[H^+] = C_{HA} \quad (C_{HA} > 10^{-6}M) \tag{3.17}$$

For solutions of strong bases the hydroxide concentration can be similarly calculated:

$$[OH^-] = C_B \quad (C_B > 10^{-6}M) \tag{3.18}$$

If only one weak monobasic acid, HA, is present in the solution, it follows from the principle of electroneutrality that

$$[H^+] = [A^-] + [OH^-]$$

Taking into consideration the definitive equation of the protonation constant and the ionic product of water, the square of the hydrogen ion concentration is given by

$$[H^+]^2 = \frac{[HA]}{K} + k_w \tag{3.19}$$

where K is the protonation constant of the conjugate base, and $k_w = 10^{-14}$, the ionic product of water. As

$$[HA] = C_{HA} - [H^+] + [OH^-]$$

(where the last term can be neglected) Eq. (3.19) becomes a quadratic from which the hydrogen ion concentration can be calculated for given values of C_{HA} and K.

If, however, $C_{HA} > 20 [H^+]$ and $[H^+] > 10^{-6}$, then Eq. (3.19) can be reduced to

$$[H^+]^2 \simeq \frac{C_{HA}}{K} \qquad (0.05\, C_{HA} > [H^+] > 10^{-6}) \qquad (3.20)$$

When more than one weak acid is present, the square of the hydrogen ion concentration can be calculated from similar considerations. As

$$[H^+] = [A_1^-] + [A_2^-] + \ldots + [OH^-],$$

introducing the protonation constants gives

$$[H^+]^2 = [HA_1]\frac{1}{K_1} + [HA_2]\frac{1}{K_2} + \ldots + k_w \qquad (3.21)$$

where $K_1, K_2 \ldots$ denote the protonation constants of the conjugate bases of the acids HA_1, HA_2, etc.

A comparison of Eq. (3.21) with Eqs. (3.19) and (3.20) shows that in the square of the hydrogen ion concentration the contribution of the acids is the same whether they are present together or alone in the solution. Generally one or two terms predominate in Eq. (3.21) and the others, including k_w, can be neglected.

The following equations can be similarly deduced for solutions of one weak base and of several weak bases.

$$[OH^-]^2 = [B]k_w K + k_w \qquad (3.22)$$

$$[OH^-]^2 \simeq C_B k_w K \qquad (0.05\, C_B > [OH^-] > 10^{-6}) \qquad (3.23)$$

$$[OH^-]^2 = [B_1]k_w K_1 + [B_2]k_w K_2 + \ldots + k_w \qquad (3.24)$$

where K_1 is the protonation constant of the base B_1, K_2 that of the base B_2, etc.

If the conjugate base is also present in the solution of a weak acid (acetic acid and sodium acetate for example), then the hydrogen ion concentration can be calculated, provided that the analytical concentrations and protonation constants are known, by using the following equation [compare with Eqs. (2.16) and [2.17)]

$$[\text{H}^+] = \frac{C_{\text{HA}} - [\text{H}^+] + [\text{OH}^-]}{C_{\text{A}} + [\text{H}^+] - [\text{OH}^-]} \times \frac{1}{K} \qquad (3.25)$$

If [H$^+$] and [OH$^-$] on the right-hand side can be neglected, the equation reduces to

$$[\text{H}^+] \simeq \frac{C_{\text{HA}}}{C_{\text{A}} K} \qquad (0.05\, C_{\text{HA}} > [\text{H}^+] > 10^{-6}) \qquad (3.26)$$

Equations (3.25) and (3.26) can also be used for calculating the hydrogen ion concentration if a strong base, or a base stronger than the conjugate base is added to the solution of the weak acid (e.g. partial neutralization of acetic acid with ammonia). Thus, if $K_\text{B} > 20\, K_\text{A}$, the neutralization reaction

$$\text{HA} + \text{B} \rightleftharpoons \text{A}^- + \text{BH}^+$$

takes place and the hydrogen ion concentration will be determined by the ratio of the concentration of the residual stronger acid HA to that of the conjugate base, A, formed. That is

$$[\text{H}^+] = \frac{[\text{HA}]}{[\text{A}^-] K_\text{A}} \simeq \frac{C_{\text{HA}} - C_\text{B}}{C_\text{B} K_\text{A}} \qquad (3.27)$$

provided that $0.05(C_{\text{HA}} - C_\text{B}) > [\text{H}^+] > 10^{-6}$ and $C_{\text{HA}} > C_\text{B}$, where C_{HA} and C_B stand for total concentrations, and K_A and K_B for the protonation constants of the conjugate base and the second base, respectively.

If a base weaker than the conjugate base is added to the solution of the weak acid, that is

$$K_\text{A} > 20\, K_\text{B}$$

the base B is practically unprotonated in the solution of the weak acid, that is $[\text{B}] \simeq C_\text{B}$, and the modified hydrogen ion concentration can be calculated according to Ringbom [15] by using the $\alpha_{\text{H(B)}}$ function. The reaction of the base with hydrogen ions can be considered as an interfering side reaction, namely,

$$[\text{H}']^2 \simeq \frac{C_{\text{AH}}}{K'_\text{A}} = \frac{C_{\text{AH}} \alpha_{\text{H(B)}}}{K_\text{A}} \qquad (3.28)$$

Sec. 3.2] Acid-base titrations

As $[H^+] = [H']/\alpha_{H(B)}$, the square of the true hydrogen ion concentration can be given as

$$[H^+]^2 = \frac{C_{AH}}{K_A \alpha_{H(B)}}, \qquad (3.29)$$

where

$$\alpha_{H(B)} = 1 + [B]K_B$$

(ammonium chloride plus a small amount of sodium acetate, for example). The small neutralization effect of the base can be considered as a side-reaction of hydrogen ions.

It follows from Eqs. (3.28) and (3.29) that if C_{HA} is equal to C_B, and 1 can be neglected in comparison with $C_B K_B$, the square of the hydrogen on concentration can be approximated by:

$$[H^+]^2 \simeq \frac{1}{K_A K_B} \qquad (3.30)$$

Equation (3.30) is suitable for calculating the pH in a solution containing the salt of a weak acid and a weak base, or in that of an ampholyte (acidic salt).

It can be similarly deduced that when an acid stronger than the conjugate acid is added to a solution of a weak base, the hydroxide ion concentration in the solution depends on the ratio of the concentration o- the non-protonated base to that of the protonated base after the neutralfization (for example ammonia and a small amount of acetic acid).

$$[OH^-] = \frac{[B]}{[BH^+]} K_B k_w \simeq \frac{[C_B - C_{HA}}{C_{HA}} K_B k_w \qquad (3.31)$$

(if $K_B > 20 K_A$; $C_B > C_{HA}$)

If the acid added is weaker than the conjugate acid (e.g. sodium acetate plus a small amount of ammonium chloride), then the following equation can be used for an approximate calculation [see Eq. (3.23)]

$$[OH^-]^2 \simeq C_B k_w K_B \frac{1}{\alpha_{OH(A)}} \quad (20 K_B < K_A) \qquad (3.32)$$

where

$$\alpha_{OH(A)} = 1 + [HA]K_A \qquad (3.33)$$

To calculate the hydrogen ion concentration of solutions of polybasic acids the following equation may be written on the basis of the electroneutrality principle (for simplicity, the charges on the ions are omitted):

$$[H^+] = [H_{n-1}A] + 2[H_{n-2}A] + \ldots + n[A] + [OH^-]$$

Introducing the protonation constants and the ionic product of water gives

$$[H^+]^2 = \frac{[H_nA]}{K_n} + \frac{2[H_{n-1}A]}{K_{n-1}} + \ldots + \frac{n[HA]}{K_1} + k_w \quad (3.34)$$

When the relationship

$$C_{H_nA} = [H_nA] + [H_{n-1}A] + \ldots + [A] = [H^+]^n[A]K_1 \ldots K_n$$
$$+ [H^+]^{n-1}[A]K_1 \ldots K_{n-1} \ldots + [A] \quad (3.35)$$

is also introduced, Eq. (3.34) becomes an equation of order $(n+2)$, the solution of which is complicated. It can be seen, however, from Eq. (3.34) that the square of the hydrogen ion concentration is mainly determined by the value of the first term, as the subsequent terms rapidly decrease with the increase in the protonation constant values and the decrease in the concentrations of the products of dissociation.

Accordingly, the hydrogen ion concentration can be approximated in the same way as for solutions of monobasic acids:

$$[H^+]^2 = \frac{C_{H_nA} - [H^+]}{r} \quad (3.36)$$

$$[H^+]^2 \simeq C_{H_nA} k_1 \quad (3.37)$$

where K_n is the nth protonation constant and k_1 is the first dissociation constant (see p. 36).

The validity of this approximation can be tested by calculating the mole-fractions of the various species of the acid, from the protonation constants, approximate $[H^+]$, and C_{H_nA} concentration values.

The hydrogen ion concentration of solutions of polyacidic bases can be calculated by similar considerations. (See Example 47.)

Sec. 3.2] **Acid-base titrations** 199

3.2.2. Determination of Acids and Bases

When a strong base is titrated with a strong acid or vice versa, the following reaction takes place

$$H_3O^+ + OH^- \to 2H_2O$$

As the concentration of water is practically constant, and its pH is 7·0 the hydrogen ion concentration at the equivalence point of the titration is

$$[H^+]_{eq} = [OH^-]_{eq} = 10^{-7.0} \qquad (3.38)$$

If a weak acid is titrated with a strong base, a new base is formed in addition to water in the neutralization reaction.

$$HA + OH^- \rightleftharpoons H_2O + A^-$$

Although the new base is weaker than the primary base, its concentration and protonation constant determine the pH at the equivalence point. If the hydroxide ion concentration originating from the dissociation of water is disregarded [see Eq. (3.22)] then

$$[OH^-]^2_{eq} = [A^-]k_w K_A \qquad (3.39)$$

$$\log [OH^-]_{eq} = \frac{1}{2} \log [A^-] - 7 + \frac{1}{2} \log K_A$$

Similarly, if a weak base is titrated with a strong acid, the hydrogen ion concentration at the equivalence point will depend on the concentration and protonation constant of the weak acid (protonated base) formed [see Eq. (3.19)]:

$$[H^+]^2_{eq} = \frac{[HB^+]}{K_B} \qquad (3.40)$$

$$\log [H]_{eq} = \frac{1}{2} \log [HB] - 1/2 \log K_B.$$

In the course of the titration of acids and bases the change in pH of the solution gives a characteristic S-shaped curve, as shown in Figs. 2.1 and 2.2 (for more details see [52]).

Polybasic acids can be titrated partially in successive steps if the ratio of the successive protonation constants exceeds 10^4. At the partial equivalence point, the pH is independent of the concentration (in a certain range) of the titrated acid, since the square of the hydrogen ion concentration in the solution of the product of the titration, as calculated according to Eq. (3.30) is:

$$[H^+]^2_{eq} \simeq \frac{1}{K_{n-1}K_n} \qquad (3.41)$$

$$\log [H^+]_{eq} \simeq -\frac{1}{2}(\log K_{n-1} + \log K_n).$$

Acid–base indicators are acids or bases which change colour on taking up or releasing a proton. For example proton uptake by Methyl Orange (the sodium salt of dimethylaminoazobenzene sulphonic acid) can be described as follows:

$$I^- + H^+ \rightarrow {}^-IH^+$$

where I^- is the indicator base. The protonation constant of Methyl Orange is $K_1 = 10^4$; the base is yellow, whereas the protonated species is red. The indicator is of transition colour if the concentration of the yellow form is just equal to that of the red one. From the equation for the protonation constant, if

$$[I] = [IH],$$

then

$$K_I = \frac{1}{[H^+]}$$

i.e. the pH when the colour change occurs (transition point) is given by

$$pH = \log K_I$$

Plots of Φ vs. pH for Methyl Orange and phenolphthalein are given in Fig. 3.2.

With phenolphthalein, only one form of the indicator being coloured, the pH at which the colour appears or disappears (transition point) depends on the concentration of the indicator and on the limit of perception of the colour of the indicator. (See Example 51.) For more details concerning chemical indicators see [121]. If the transition point does not coincide with the equivalence point, an indicator error arises.

A titration error arises from the fact that the end-point of the titration does not always coincide with the equivalence point. The error is usually

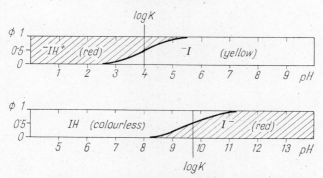

Fig. 3.2. The distribution of the dissociation products of the acid–base indicators Methyl Orange (log $K = 4$) and phenolphthalein (log $K = 9\cdot 75$), as a function of pH

due to two reasons, one being that the end-point as shown by an indicator occurs before or after the equivalence point. This error is called the indicator error. The other reason is the uncertainty involved in the determination of the end-point (which may cause either positive or negative errors) but this can be evaluated statistically.

In a determination the important factor for the analytical chemist is the relative error. If C_d is the total concentration of the titrand and C_t that of the titrant in the solution titrated, the mole-fraction of the lack or excess of the titrant can be expressed generally as

$$\frac{C_t - C_d}{C_d} \tag{3.42}$$

The error is positive for overtitration ($C_t > C_d$), negative for undertitration ($C_d > C_t$), and zero if $C_d = C_t$.

Multiplying the mole-fraction by 100, and choosing the correct sign gives the percentual relative error of the titration.

Consider the error in the titration of a monobasic weak acid with a strong base, for example sodium hydroxide.

The analytical concentration of the titrant at the end-point can be found from the charge-balance equation:

$$C_{\text{Na(end)}} + [\text{H}^+] = [\text{A}^-] + [\text{OH}^-] \tag{3.43}$$

The total concentration of the acid is

$$C_{\text{HA}} = [\text{A}^-] + [\text{HA}] \tag{3.44}$$

Substitution of (3.43) and (3.44) into (3.42) gives

$$\frac{C_{Na} - C_{HA}}{C_{HA}} = \frac{[A^-] + [OH^-] - [H^+] - [A^-] - [HA]}{C_{HA}}$$

$$= \frac{[OH^-] - [H^-]}{C_{HA}} - \frac{[HA]}{C_{HA}} \qquad (3.45)$$

The last term on the right-hand side is the mole-fraction of the protonated species, i.e.

$$\frac{[HA]}{C_{HA}} = \Phi_{HA} = \frac{[H^+]K}{1 + [H^+]K}$$

The relative error for the titration of an acid with a strong base will be:

$$\Delta = 100\left[\frac{[OH^-] - [H^+]}{C_{HA}} - \Phi_{HA}\right]\% \qquad (3.46)$$

The error can be calculated from Eq. (3.46) provided that the concentration and dissociation constant of the titrated acid and the pH at the end-point of the titration are known [122].

For the titration of weak acids, if $K > 10^2$, Eq. (3.46) can be simplified (since near the equivalence point $[H^+]K \ll 1$ and $[OH^-] \gg [H^+]$) to give

$$\Delta \simeq 100\left[\frac{k_w}{[H^+]C_{HA}} - [H^+]K\right]\% \qquad (3.47)$$

The following conclusions can be drawn from Eq. (3.46).

1. For the titration of strong acids, $\Phi = 0$ even in strongly acidic media, since $K < 1$, and only one term remains on the right-hand side in Eq. (3.46). The value of the error, which changes sign at pH 7, will be determined by $[H^+]$ before the equivalence point and $[OH^-]$ after it. The error increases with decreasing concentration of the acid.

2. For the titration of weak acids, the error is determined by the value of Φ, i.e. of $K[H^+]$ (which is independent of the concentration) before the equivalence point. Accordingly, if two acids of different strengths are undertitrated to the same extent, the error will be greater for the weaker acid (since, if K is large, Φ is also large). In the vicinity of the equivalence point and after it the first term on the right-hand side of (3.46) behaves as in the case of the titration of a strong acid.

In Figs. 3.3 and 3.4 the logarithm of the error is plotted *vs.* the pH of the end-point of the titration, calculated from Eq. (3.47) for a strong

Acid-base titrations

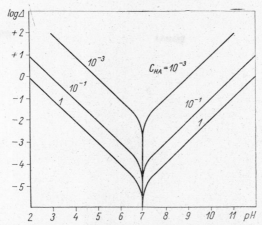

Fig. 3.3. Relationship between the logarithm of the titration error and the pH of the end-point in the titration of a strong acid of various concentrations [122]

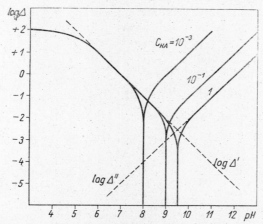

Fig. 3.4. Relationship between the logarithm of the titration error and the pH of the end-point in the titration of a weak acid of various concentrations ($\log K = 5$) [122]

and a weak acid ($K = 10^5$) of various concentrations. The equivalence point is at the point where the titration error becomes zero ($\log \Delta = -\infty$). It is clearly shown in Fig. 3.4 that the position of the equivalence point changes with the concentration of the titrated weak acid [see Eq. (3.39)].

It can be seen from the curves in Fig. 3.4 that if the colour change of the indicator used occurs just at the equivalence point (i.e. the indicator error is zero), but the end-point can be observed with an uncertainty of ± 0.5 pH, the value of $\log \Delta$ is -0.5 ($\Delta \simeq \pm 3.2\%$) for the titration of $10^{-3} M$ acid, and $\log \Delta = -2$ ($\Delta \simeq \pm 0.01\%$) for that of $1 M$ acid.

The equations of the asymptotes drawn in broken lines in Fig. 3.4 are as follows:

$$\log \Delta' = (2 + \log K) - \text{pH};$$

$$\log \Delta'' = \text{pH} - (14 - 2 + \log C)$$

[see Eq. (3.47)].

The pH at the point of intersection of the two lines gives the pH of the equivalence point.

$$(\text{pH})_{eq} = 7 + 1/2 \log K + \frac{1}{2} \log C_{HA} \qquad (3.48)$$

[See Eq. (3.39).]

For the stepwise titration of a polybasic acid the error depends on the values of two consecutive protonation constants. For example, if a dibasic acid is titrated to the first equivalence point

$$C_{Na} = [OH^-] - [H^+] + 2[A^{2-}] + [AH^-]$$

$$\frac{C_{Na}}{C_{H_2A}} - 1 = \frac{[OH^-] - [H^+]}{C_{H_2A}} + \frac{2 + [H^+]K_1}{1 + [H^+]K_1 + [H^+]^2 K_1 K_2} - 1$$

If $C_{H_2A} > 20[H^+], 20[OH^-]$, the first term can be neglected, and

$$\Delta \simeq 100 \left[\frac{1 - [H^+]^2 K_1 K_2}{1 + [H^+]K_1 + [H^+]^2 K_1 K_2} \right] = 100 \left[\Phi_A - \Phi_{H_2A} \right] \%$$

In the titration of strong and weak bases the equation derived for the error is

$$\Delta = 100 \left[\frac{[H^+] - [OH^-]}{C_B} - \Phi_B \right] \% \qquad (3.49a)$$

Sec. 3.2] Acid-base titrations

where Φ_B is the mole-fraction of the non-protonated base:

$$\Phi_B = \frac{1}{1 + [H^+]K} \qquad (3.50)$$

The error in the titration of weak bases, since near the equivalence point $[H^+]K \gg 1$, can be approximated by the equation:

$$\Delta \simeq 100 \left[\frac{[H^+]}{C_B} - \frac{1}{[H^+]K} \right] \% \qquad (3.49b)$$

Very weak acids ($K_A > 10^9$) and very weak bases ($K_B < 10^5$) can either not be titrated at all in aqueous solution, or only with a high error. The pH jump around the equivalence point becomes smaller in the titration of acids as the protonation constant increases, and in the titration of bases, as it decreases. Consequently, the determination of the end-point becomes more and more uncertain.

The sensitivity of end-point indication can be increased by using instrumental methods (the ±0·5 pH uncertainty of the visual end-point can be reduced to ±0·1 pH) or by converting the titration curve into a linear form. According to Gran [123], if the volume of standard solution added during the titration is multiplied by the related hydrogen ion concentration and by the protonation constant, and the result is plotted against the volume of standard solution added, a straight line is obtained which intersects the abscissa at the volume of standard solution consumed at the equivalence point.

For the titration of a monobasic weak acid with a strong base the ratio of the concentration of the protonated species to that of the non-protonated species can be expressed in terms of the degree of neutralization (see pp. 91 and 93).

$$\frac{[AH]}{[A]} \sim \frac{1-a}{a} \qquad (3.51)$$

Let the volume of standard solution added at the equivalence point be denoted by V_{eq}, and smaller volumes by V. Then

$$\frac{1-a}{a} = \frac{V_{eq} - V}{V} \qquad (3.52)$$

since $aV_{eq} = V$. Substituting this in the equation for the protonation constant [see Eq. (2.2)] yields

$$K = \frac{V_{eq} - V}{V[H^+]}$$

i.e.
$$VK[H^+] = V_{eq} - V \qquad (3.53)$$

which is a linear equation.

It should, however, be noted that the approximation given by Eq. (3.51) is justified only if the concentration of the acid exceeds $10^{-2}M$ and the pH values (including that of the equivalence point) measured during the titration fall within the pH range 5–9 (see p. 93). Therefore the points calculated for the beginning of the titration (pH $<$ 5) deviate from the straight line. If a very weak acid is titrated (log $K > 9$), or the concentration of the titrated acid is low ($<10^{-2}M$), only the complete equation deduced by Ingman and Still [124] (which does not neglect any terms) may be used; or also, the recent method of Midgeley and McCallum [124a].

In the titration of a very weak acid, if the base formed can be complexed by a suitable reagent (see pp. 117 and 199), i.e. its basic strength reduced, the equilibrium of the neutralization reaction can be shifted in the desired direction. By complexation of the base the value of the conditional protonation constant is reduced to an extent which enables the titration to be carried out satisfactorily (e.g. by using mannitol in the case of titration of boric acid).

The pH at the equivalence point of the titration in the presence of a complexing agent can also be calculated by using Eq. (3.39) according to Ringbom [15], if the conditional protonation constant K' is inserted in the place of the real value K:

$$\log K' = \log K - \log \alpha_{A(M)} \qquad (3.54)$$

The base (ligand) formed during the titration can also be complexed by multivalent metal ions (see Example 53).

Very weak bases can be similarly titrated if the weak acid formed is bound or extracted into another phase.

Obviously, the strength of the base or acid formed during the titration can be reduced by using an appropriate solvent instead of water. This is one of the advantages of titrations in non-aqueous solvents.

If the consecutive protonation constants do not differ much, it cannot be decided on the basis of the pH titration curve whether the unknown is a mono- or polybasic acid. This problem can be solved according to Sturrock [125] on the basis of the difference between the pH values at one quarter and three quarters of the complete neutralization curve.

Neglecting the concentrations [H^+] and [OH^-] in Eq. (2.17), and

Sec. 3.2] Acid-base titrations

expressing the pH for 25% and 75% neutralization ($a = 0.25$ and $a = 0.75$) gives

$$\text{pH} = \log K - \log 3 = \log K - 0.477$$

and

$$\text{pH} = \log K + \log 3 = \log K + 0.477.$$

The difference of the two pH values is given by

$$\Delta \text{pH} = 0.954 \tag{3.55}$$

For the titration of any monobasic acid (if the titration curve falls within the pH range 5–9, and $C > 10^{-2}M$), ΔpH has the same value. If ΔpH > 0.954 then the degree of neutralization was determined incorrectly, and a polybasic acid was involved. A more extensive set of criteria has been developed by Meites et al. [125a, b].

3.2.3. Indirect determination of complex-forming ions by acid–base titration

The reaction of the protonated ligand with the metal ion

$$n\text{HL} + \text{M} \rightleftharpoons \text{ML}_n + n\text{H}^+$$

can be used for the indirect determination of the metal ion M, if $C_{\text{HL}} > nC_{\text{M}}$, and the reaction proceeds quantitatively to the right. According to Schwarzenbach and Biederman [126], cadmium ions for example can be determined by adding the disodium salt of EDTA to the neutral solution, then titrating the hydrogen ions liberated in the reaction

$$\text{H}_2\text{L}^{2-} + \text{Cd}^{2+} \rightleftharpoons 2\text{H}^+ + \text{CdL}^{2-}$$

For this type of determination to be feasible it is necessary that the equilibrium constant of the reaction should be ≥ 1:

$$K = \frac{K_{\text{CdL}}}{K_1 K_2} \geq 1$$

(where K_1 and K_2 are the protonation constants).

Not only proton–metal ion exchange but also ligand exchange can be used for the indirect determination of some ions or compounds. For example, according to Beck and Szabó [127], aluminium can be determined indirectly by titrating the hydroxide ions released in the displacement reaction

$$Al(OH)_6^{3-} + 6F^- \rightleftharpoons AlF_6^{3-} + 6OH^-$$

The condition that displacement reactions are applicable in quantitative determinations is that the ratio of the overall formation constants of the complexes involved is sufficiently high (see pp. 226, 229, 248 and 253).

3.2.4. WORKED EXAMPLES

46a. Calculate the pH of a $0.01M$ hydrogen peroxide solution Log $K = 11.75$. First try Eq. (3.20):

$$[H^+]^2 \simeq \frac{10^{-2}}{10^{11.75}} = 10^{-13.75}$$

$$[H^+] \simeq 10^{-6.87}, \quad \text{i.e.} \quad pH \simeq 6.87$$

As $C_{H_2O_2} \gg 20\,[H^+]$, $C_{H_2O_2} - [H^+] + [OH^-]$ may be approximated by $C_{H_2O_2}$. However, because $[H^+] < 10^{-6}$, the ionic product of water has also to be taken into account. That is

$$[H^+]^2 = \frac{10^{-2}}{10^{11.75}} + 10^{-14} = 10^{-13.56}$$

$$[H^+] = 10^{-6.78}, \quad \text{i.e.} \quad pH = 6.78.$$

46b. Calculate the pH of the hydrogen peroxide solution in example 46a in the presence of $10^{-3}M$ formic acid. (The protonation constant of formate is log $K = 3.77$.)

The calculation is done by using Eq. (3.21). As a first approximation, the total concentration of the acid is taken instead of [HA].

$$[H^+]^2 \simeq \frac{10^{-3}}{10^{3.77}} + \frac{10^{-2}}{10^{11.75}} + 10^{-14} \simeq 10^{-6.77};$$

$$[H^+] \simeq 10^{-3.38}, \quad \text{i.e.} \quad pH \simeq 3.38$$

As shown by the calculation, the hydrogen ion concentration of the solution is determined practically solely by the formic acid. However, since C_{HCOOH} only slightly exceeds $[H^+]$, the latter must also be taken into consideration. Thus

$$[H^+]^2 = \frac{10^{-3} - [H^+]}{10^{3\cdot77}}$$

$$10^{3\cdot77}[H^+]^2 + [H^+] - 10^{-3} = 0$$

From this

$$[H^+] = 3\cdot36 \times 10^{-4}, \quad \text{i.e.} \quad pH = 3\cdot48.$$

47. Calculate the pH of $0\cdot1 M$ sodium carbonate. The protonation constants of carbonate are $\log K_1 = 10\cdot1$; $\log K_2 = 6\cdot3$.

As a first approximation, the hydroxide ion concentration can be calculated from Eq. (3.23).

$$[OH^-]^2 \simeq 10^{-1} \times 10^{-14} \times 10^{10\cdot1} = 10^{-4\cdot9}$$

$$[OH^-] \simeq 10^{-2\cdot45}$$

$$pH = 14 - 2\cdot45 = 11\cdot55$$

As pH > 11, the concentration of bicarbonate is very low, and its effect can be neglected. The pH calculated can be accepted as quite correct, since the condition $C_{CO_3} > 20\,[OH^-]$ is also fulfilled.

48. Calculate the pH of a $0\cdot1 M$ sodium carbonate solution to which oxalic acid has been added to give a total oxalate concentration of $0\cdot02 M$.

$$\log K_{CO_3} = 10\cdot1 \quad \text{and} \quad \log K_{Ox} = 3\cdot8$$

As $K_{CO_3} \gg K_{Ox}$, the reaction

$$2CO_3^{2-} + H_2Ox \rightarrow 2HCO_3^- + Ox^{2-}$$

proceeds practically quantitatively.

The concentration of carbonate will be $0\cdot1 - 2 \times 0\cdot02 = 0\cdot06 M$, that of bicarbonate $2 \times 0\cdot02 = 0\cdot04 M$. Oxalate is a very weak base compared with carbonate, therefore its effect on the hydroxide concentration can be neglected. According to Eq. (3.31)

$$[OH^-] = \frac{0\cdot06}{0\cdot04} \times 10^{10\cdot1} \times 10^{-14} = 10^{-3\cdot72}$$

$$pH = 14 - 3\cdot72 = 10\cdot28$$

49. Calculate the pH of a solution which is $0.1M$ in ammonium chloride and $10^{-3}M$ in sodium benzoate. Log $K_{NH_3} = 9.35$ and log $K_B = 4.12$ ($I = 0.1$).

As $K_{NH_3} > 20\ K_B$, in the weakly acid solution containing ammonium ions, benzoate is presumably only slightly protonated. The hydrogen ion concentration is calculated by using Eq. (3.29):

$$[H^+]^2 = \frac{10^{-1}}{10^{9.35}} \times \frac{1}{1 + 10^{-3} \times 10^{4.12}} = 10^{-11.50}$$

$$[H^+] = 10^{-5.75}; \quad pH = 5.75$$

50. Calculate the pH of a $0.1M$ NaH_2PO_4 solution. Log $K_{HPO_4} = 6.9$ and log $K_{H_2PO_4} = 2.0$ ($I = 0.1$).

The calculation is done using Eq. (3.30):

$$[H^+]^2 = \frac{1}{10^2 \times 10^{6.9}} = 10^{-8.9}; \quad pH = 4.45$$

51. Calculate the pH at which the colour change of phenolphthalein occurs in the titration of an acid with a strong base. Log $K_I = 9.6$ ($I = 0.1$) [128]. Three drops of 0.1% ($3 \times 10^{-3}M$) phenolphthalein solution are added to 50 ml of the solution to be titrated. For the pink colour to be observable it is necessary that the concentration of the red conjugate base is about $5 \times 10^{-7}M$.

The total concentration of the indicator is calculated first. The volume of one drop is about 0.05 ml, so

$$C_I = 3 \times 10^{-3} \times \frac{3 \times 0.05}{50} = 9 \times 10^{-6}M.$$

The calculation is continued by using the equation for the protonation constant

$$[H^+] = \frac{[IH]}{[I]K_I}$$

$$[H^+] = \frac{9 \times 10^{-6} - 5 \times 10^{-7}}{5 \times 10^{-7}} \times 10^{-9.6} \simeq 10^{-8.4}; \quad pH \simeq 8.4$$

52. Malonic acid is to be titrated with $0.1M$ sodium hydroxide solution. Calculate the pH of the solution at the equivalence point, if the

Sec. 3.2]   Acid-base titrations   211

concentration of malonic acid, C_{H_2A}, is $10^{-2}M$. Log $K_1 = 5 \cdot 66$ and log $K_2 = 2 \cdot 85$. Estimate the indicator error if phenolphthalein is used as indicator.

Malonic acid is a dibasic acid. The difference in the logarithms of the protonation constants is less than 4, thus this acid cannot be titrated with adequate accuracy to the first equivalence point. The pH at the second equivalence point is determined by the concentration and first protonation constant of the bivalent malonate ion. According to Eq. (3.39)

$$\log [OH^-]_{eq} = 1/2 (-2) - 7 + 1/2 \times 5 \cdot 66 = -5 \cdot 17$$

$$(pH)_{eq} = 14 - 5 \cdot 17 = 8 \cdot 83$$

The colour change of phenolphthalein can be seen at pH 8·4 (Example 51).

The indicator error is calculated by an equation similar to Eq. (3.47) deduced for the titration of dibasic acids.

$$\Delta \simeq 100 \left[\frac{k_w}{2[H^+]C_{H_2A}} - 1/2 \; [H^+]K_1 \right]\%$$

$$= 100 \left[\frac{10^{-14}}{2 \times 10^{-8 \cdot 4} \times 10^{-2}} - 1/2 \times 10^{-8 \cdot 4} \times 10^{5 \cdot 66} \right]$$

$$= -0 \cdot 078\%.$$

Thus when phenolphthalein is used as indicator the solution is undertitrated, but the error is less than 0·1%, and the accuracy is acceptable.

53. Can acetylacetone, being a very weak acid, be titrated with 0·1N standard base solution if nickel(II) is used to promote the titration? Calculate the pH of the equivalence point as well as the titration error in the absence and presence of the nickel, provided that the endpoint of the titration can be determined with an uncertainty of $\pm 0 \cdot 1$ and $\pm 0 \cdot 5$ pH unit, respectively. The protonation constant is log $K = 8 \cdot 9$ ($I = 0 \cdot 1$). The overall stability constants of acetylacetone–nickel(II) complexes are log $\beta_1 = 5 \cdot 6$, log $\beta_2 = 10 \cdot 0$ and log $\beta_3 = 12 \cdot 4$ ($I = 0 \cdot 1$).

Acetylacetone is poorly soluble in water but readily soluble in alcohol. The solution to be titrated contains 10% alcohol, is $5 \times 10^{-3}M$ in acetylacetone, and 0·1M in nickel(II).

14*

The pH at the equivalence point in the absence of nickel can be calculated by using Eq. (3.39):

$$\log [OH^-]_{eq} = 1/2\,(-2\cdot3) - 7 + 1/2 \times 8\cdot9 = -3\cdot7;$$

$$(pH)_{eq} = 10\cdot3$$

If the titration were carried out by a potentiometric method, without an indicator, and the uncertainty of the determination of the equivalence point were $\pm 0\cdot1$ pH, the error would be, according to Eq. (3.47):

$$\Delta \simeq 100 \left[\frac{10^{-14}}{10^{-10\cdot2} \times 10^{-2\cdot3}} - 10^{-10\cdot2} \times 10^{8\cdot9} \right] = -1\cdot85\%$$

or

$$\Delta \simeq 100 \left[\frac{10^{-14}}{10^{-10\cdot4} \times 10^{-2\cdot3}} - 10^{-10\cdot4} \times 10^{8\cdot9} \right] = +1\cdot85\%$$

The error of the titration would be $\pm 1\cdot85\%$ even though the end-point were determined by an instrumental method instead of with an indicator.

In the presence of nickel(II) the conditional protonation constant is calculated from Eq. (3.54). Because nickel(II) tends to form hydroxo-complexes, the conditional stability constant is calculated for a pH 8·5 solution which is $0\cdot1 M$ in nickel(II). As nickel(II) is present in excess over acetylacetone the formation of only the 1 : 1 complex is considered.

At pH 8·5 $\log \alpha_{Ni(OH)} = 0\cdot1$ according to Fig. 1.10, so

$$\alpha_{A(Ni)} = 1 + \left[\frac{[Ni']}{\alpha_{Ni(OH)}} \right] \beta_1 = 1 + \left[\frac{10^{-1}}{10^{0\cdot1}} \right] \times 10^{5\cdot6} = 10^{4\cdot5}$$

The conditional protonation constant is given by

$$\log K' = 8\cdot9 - 4\cdot5 = 4\cdot4.$$

The pH at the equivalence point is given by

$$\log [OH]_{eq} = 1/2 \times (-2\cdot3) - 7 + 1/2 \times 4\cdot4 = -5\cdot95$$

$$(pH)_{eq} = 14 - 5\cdot95 = 8\cdot05$$

The indicator error when Neutral Red ($pK_I = 7\cdot 4$) is used is given by

$$\Delta \simeq 100 \left[\frac{10^{-14}}{10^{-7\cdot 4} \times 10^{-2\cdot 3}} - 10^{-7\cdot 4} \times 10^{4\cdot 4} \right] = -0\cdot 095\%$$

If the end-point can be determined with an uncertainty of $\pm 0\cdot 5$ pH unit, the error of the determination is

$$\Delta \simeq 100 \left[\frac{10^{-14}}{10^{-6\cdot 9} \times 10^{-2\cdot 3}} - 10^{-6\cdot 9} \times 10^{4\cdot 4} \right] = -0\cdot 31\%$$

or

$$\Delta \sim 100 \left[\frac{10^{-14}}{10^{-7\cdot 9} \times 10^{-2\cdot 3}} - 10^{-7\cdot 9} \times 10^{4\cdot 4} \right] = -0\cdot 015\%$$

The determination can thus be carried out with a maximum error of $-0\cdot 31\%$.

3.3. Precipitation Titrations

The basis of precipitation titrations is the formation of a precipitate in the reaction of titrand with titrant. If the determinand is a metal ion M which forms a 1:1 compound (which is slightly soluble in water) with the titrant ion A [e.g. titration of silver(I) with chloride], the equilibrium of the precipitation reaction,

$$M + A \rightleftharpoons MA_{(prec)} \tag{3.56}$$

is determined by the value of the equilibrium constant

$$K_s = \frac{1}{[M][A]} \tag{3.57}$$

(see p. 182). On plotting the negative logarithm of the metal ion concentration, pM, calculated by Eq. (3.57) vs. the volume of standard solution added, an S-shaped titration curve is obtained. The part of the curve around the equivalence point is dependent on K_s (or on its reciprocal, the solubility product; see Fig. 3.5). At the equivalence point

$$[M]_{eq} = [A]_{eq}$$

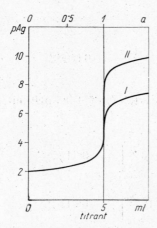

Fig. 3.5. Titration of 50 ml of $10^{-2}M$ silver nitrate solution. I: With $0.1M$ sodium chloride standard solution. II: With $0.1M$ potassium bromide standard solution

thus from Eq. (3.57)

$$\frac{1}{[M]_{eq}^2} = K_s \qquad (3.58)$$

or

$$-\log [M]_{eq} = (pM)_{eq} = \frac{1}{2} \log K_s$$

If the change in the metal ion concentration during the titration can be recorded, the equivalence point can be determined from the titration curve. Otherwise, the end-point of the titration may be detected by a suitable indicator. For more details see [52].

The precipitation reaction can also be used for determining the ion A by titration with a standard solution containing M (e.g. determination of bromide by argentometry).

If the composition of the titrated solution is such that the ions taking part in the precipitation reaction are in equilibrium with other components, the conditional constant of precipitate formation can be calculated in the usual way. The steepness of the titration curve decreases with decreasing conditional precipitation formation constant, and accordingly, the uncertainty of the end-point detection and determination increases.

The error of a precipitation titration, if a metal ion is titrated and a 1:1 precipitate is formed, can be calculated from the following considerations. The total amounts of the titrand and of the titrant (in moles) at the end-point of the titration are given by

$$N_M = [M]V + N_{(prec)}$$

$$N_A = [A]V + N_{(prec)}$$

where V is the final volume, and N_{prec} the number of moles of the precipitate.

Sec. 3.3] Precipitation titrations

The titration error (see p. 201) is

$$\frac{N_A - N_M}{N_M} = \frac{([A] - [M])V}{N_M}$$

From Eq. (3.57) and $N_M/V = C_M$:

$$\frac{[A] - [M]}{C_M} = \frac{\dfrac{1}{[M]K_s} - [M]}{C_M}$$

and the relative error is given by

$$\varDelta = 100 \left[\frac{1}{[M]K_s C_M} - \frac{[M]}{C_M} \right] \% \qquad (3.59a)$$

where C_M is the theoretical final total concentration of the metal ion, calculated from the total amount of the metal ion and final volume. If the change in volume of the solution during titration can be neglected, then C_M will be equal to the initial concentration. If $[M] = [A]$, that is, the end-point coincides with the equivalence point, then $\varDelta$ calculated by using Eq. (3.59) is zero.

The form of Eq. (3.59) when A is the determinand is

$$\varDelta = 100 \left[\frac{1}{[A]K_s C_A} - \frac{[A]}{C_A} \right] \% \qquad (3.59b)$$

For a precipitate of the composition M_2A the error involved in the determination of the metal ion can be calculated from the equation

$$\varDelta = 100 \left[\frac{2}{[M]^2 K_s C_M} - \frac{[M]}{C_M} \right] \% \qquad (3.60)$$

Two ions of similar behaviour can be determined simultaneously with the same titrant if the ratio of the precipitate formation constants exceeds 10^4 (for example, the simultaneous determination of chloride and iodide with silver nitrate standard solution) but errors may arise if the more soluble species is coprecipitated by occlusion so that the exchange reaction $MX + Y \rightarrow MY + X$ is mechanically hindered. If the constants K_s are closer, then the interfering component may be masked by an appropriate complexing agent used at a suitable pH, i.e. the conditional precipitate formation constant may be reduced to the extent that the other component becomes selectively titrated.

For an ion to be determined from a $10^{-2}M$ solution with an error $\leq 0.1\%$, if a 1 : 1 precipitate is formed and the uncertainty in the determination of the equivalence point is ΔpM or ΔpA $= \pm 0.2$, it is necessary, according to Eq. (3.59), that

$$K_s > 10^{10}.$$

3.3.1. Electrodes of the second kind and membrane electrodes

By means of these electrodes, the activity of one of the ions of the precipitate in the solution in contact with the electrode can be measured by a potentiometric method. Common electrodes of the second kind are silver/silver halide electrodes which are suitable for measuring the activity of halide ions.

The potential of a silver electrode is directly determined by the activity, or if the activity coefficient is disregarded, by the concentration of free silver(I) ions:

$$E = E^0_{Ag} + 0.059 \log [Ag^+] \qquad (3.61)$$

In the presence of halides the concentration of silver(I) will be determined by the value of the precipitate formation constant:

$$K_s = \frac{1}{[Ag^+][A^-]} \qquad (3.62)$$

where A^- is a halide. From Eq. (3.62) it follows that

$$E = E^0_{Ag} - 0.059 \log K_s - 0.059 \log [A^-]$$
$$= E^0_A - 0.059 \log [A^-] \qquad (3.63)$$

where E^0_A is the standard potential of the halide electrode, which is constant at constant temperature and ionic strength.

It can similarly be deduced that a silver electrode can also be used for measuring divalent sulphide, oxalate etc. as an electrode of the second kind:

$$E = E^0_X - \frac{0.059}{2} \log [X^{2-}]$$

where

$$E^0_X = E^0_{Ag} - \frac{0.059}{2} \log K_{s(Ag_2X)}$$

Sec. 3.3] Precipitation titrations 217

If the ion X is involved in a side-reaction (e.g. protonation), then the potential of the electrode is given by

$$E = E_X^0 + \frac{0 \cdot 059}{2} \log \alpha_{X(H)} - \frac{0 \cdot 059}{2} \log C_X \qquad (3.64)$$

where C_X is the analytical concentration.

Membrane electrodes are suitable for the direct measurement of the activity or concentration in the solution of one (or both) of the ions of precipitate. The membrane is composed of a slightly soluble metal salt, or metal complex, sometimes embedded in an inert carrier.

The half-cell

| Ag | AgBr; | KBr internal solution | AgBr membrane | Br⁻ external solution |

is suitable for measuring the activity of bromide in the external solution in contact with the membrane. If certain conditions are fulfilled (if the concentration of the solutions in contact with the membrane exceeds that due to the solubility of the precipitate), the potential difference between the two sides of the membrane depends only on the bromide activity, or, if the activity coefficients are constant, on the concentrations in the two solutions in contact with the membrane [129] and is given by

$$\Delta E_m = -0 \cdot 059 \log \frac{[Br]}{[Br]_{in}} \qquad (3.65)$$

where $[Br]_{in}$ refers to the internal solution. As $[Br]_{in}$ is constant, the activity of bromide in the external solution can be determined by measuring the membrane potential:

$$\Delta E_m = E_{Br} - 0 \cdot 059 \log [Br^-] \qquad (3.66)$$

ΔE_m is zero if the bromide activity in the internal solution is equal to that in the external one.

In practice the potential is measured with respect to a reference electrode of constant potential (e.g. calomel).

By use of appropriate precipitates to produce the membrane, electrodes can be made for the selective determination of a number of ions (such as halides, cyanide, thiocyanate, sulphide, etc.).

Pungor and Tóth have interpreted the mechanism and also the selectivity of precipitate-based membrane electrodes [130].

If a silver bromide membrane electrode comes into contact with another ion capable of forming a precipitate with silver, then an exchange reaction may occur according to the equation:

$$AgBr_{(prec)} + A^- \rightleftharpoons AgA_{(prec)} + Br^- \qquad (3.67)$$

where A^- is the precipitate-forming ion. The bromide selectivity of the electrode is determined by the formation constants of the precipitates AgBr and AgA.

The selectivity constant, K_d, is defined as

$$K_{d(Br/A)} = \frac{K_{s(AgBr)}}{K_{s(AgA)}} \qquad (3.68)$$

If the ion A is also present in the solution of bromide to be measured by the silver bromide membrane electrode, Eq. (3.65) is modified to

$$\Delta E_m = -0.059 \log \frac{[Br^-] + [A^-]/K_d}{[Br]_{in}} \qquad (3.69)$$

The ion A does not interfere with the determination of bromide as long as $[A^-]/K_d \ll [Br^-]$.

Small amounts of chloride do not interfere with the determination of bromide since $K_{s(AgBr)} > K_{s(AgCl)}$. In the presence of iodide, however, the electrode becomes selective for iodide as $K_{s(AgBr)} < K_{s(AgI)}$.

Ions which form stable complexes with silver will also interfere with the determination. The selectivity constants in the presence of complexing ions can be calculated in essentially the same way as in the case of precipitate-forming ions.

If the precipitate-forming anions tend to protonate, the conditional formation constants should be inserted in Eq. (3.68) in the place of the thermodynamic precipitation constants. The selectivity for ions which tend to protonate decreases with decreasing pH.

3.3.2. Worked examples

54. Calculate the indicator error for the titration of $0.01M$ silver nitrate with $0.1M$ ammonium thiocyanate, in the presence of iron(III) ammonium sulphate as indicator. The titration is carried out in nitric acid solution. It is necessary that the thiocyanate concentration is

$10^{-5}M$ for the red colour of the thiocyanato-iron(III) complex to be just perceptible. Log $K_s = 11.7$ ($I = 0.1$).

The calculation is done by using Eqs. (3.4c) and (3.59).

$$[Ag^+]_{end} = 10^{-11.7} \times 10^5 = 10^{-6.7}$$

$$\Delta = 100 \left[\frac{1}{10^{-6.7} \times 10^{11.7} \times 10^{-2}} - \frac{10^{-6.7}}{10^{-2}} \right] = +0.1\%$$

55. Calculate the maximum bromide concentration which does not interfere with the determination of $10^{-4}M$ iodide when a silver iodide membrane electrode is used. Log $K_{s(AgI)} = 16.08$; log $K_{s(AgBr)} = 12.3$.

According to Eq. (3.68) the selectivity of an iodide electrode for bromide ions is

$$K_{d(I/Br)} = \frac{K_{s(AgI)}}{K_{s(AgBr)}} = \frac{10^{16.08}}{10^{12.3}} = 10^{3.78}$$

From Eq. (3.69) for practical purposes bromide does not affect the determination of iodide if

$$\frac{[I^-]}{[Br^-]/K_d} \geq 20$$

From this

$$[Br^-] \leq \frac{[I^-]K_d}{20} = \frac{10^{-4} \times 10^{3.78}}{20} = 3 \times 10^{-2} M$$

The maximum allowable bromide concentration determined experimentally [129] agrees well with the calculated value.

3.4. Complexometric Titrations

Complexometric titrations are the most widely used analytical application of complex-formation reactions. Their importance has increased remarkably since the EDTA-type chelating agents have become widely used as titrants. (For more details see [24, 131].)

3.4.1. Theory of complexometric titrations using chelating agent as titrant

The principle of complexometric titrations is that a metal ion is titrated with a standard solution containing a chelating agent, the end-point of the complex-formation reaction being indicated either chemically or instrumentally. The equation for the formation of a 1:1 complex is

$$M + L \rightleftharpoons ML \tag{3.70}$$

The titration curve can be constructed in the same way as for acid–base or precipitation titrations, the coordinates being pM and the volume of standard solution added. If pM′, the negative logarithm of the conditional metal ion concentration, is plotted instead of pM against the degree of titration, then very useful data can be obtained from the titration curve.

Because chelate-forming ligands usually have a tendency to protonate, the conditional constant should be used in practice.

The titration curve before the equivalence point can be constructed simply by using the relationship

$$[M'] = C_M(1 - a) \tag{3.71}$$

if the stability of the complex is high enough ($K' > 10^7$), where a, the degree of titration, is similar to the degree of neutralization, being the fraction of the stoichiometric amount of complexing agent added to the titrand. At the equivalence point

$$[M']_{eq} = [L']_{eq}$$

or

$$[M']_{eq}^2 = \frac{[ML]}{K'} \simeq \frac{C_M}{K'}$$

$$(pM')_{eq} \simeq \frac{1}{2}(\log K' - \log C_M) \tag{3.72}$$

The titration curve after the equivalence point can be calculated from

$$[M'] = \frac{1}{(a - 1)K'} \tag{3.73}$$

Sec. 3.4] Complexometric titrations

The complexometric titration curve for calcium titrated with the disodium salt of EDTA is shown in Fig. 3.6.

If the conditional stability constant of the complex formed is low ($K' < 10^7$) then the calculation cannot be done by using the simplified equation given above. The related [M'] and C_M values can be calculated by solving the simultaneous equations

$$[ML] = C_M - [M'] \quad (3.74)$$

$$[ML] = C_L - [L'] \quad (3.75)$$

$$[ML] = [M'][L']K' \quad (3.76)$$

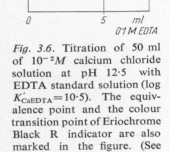

Fig. 3.6. Titration of 50 ml of $10^{-2}M$ calcium chloride solution at pH 12·5 with EDTA standard solution (log $K'_{CaEDTA}=10·5$). The equivalence point and the colour transition point of Eriochrome Black R indicator are also marked in the figure. (See Example 56)

If the end-point of the titration does not coincide with the equivalence point, then a titration error occurs, which can be calculated similarly to that for acid–base and precipitation titrations, on the basis of the following considerations [122].

If the titrant is a solution of the complexing ligand L, and a complex with composition ML is formed, the total concentrations of the titrant and titrand at the equivalence point are

$$C_L = [L'] + [ML] \quad (3.77)$$

$$C_M = [M'] + [ML] \quad (3.78)$$

respectively.

From these equations the relative error is

$$\frac{C_L - C_M}{C_M} = \frac{[L'] - [M']}{C_M} \quad (3.79)$$

Introducing the conditional stability constant into Eq. (3.78) and rearranging gives

$$C_M = [M'] + [M'][L']K'$$

$$[L'] = \frac{C_M - [M']}{[M']K'}$$

which on substitution into Eq. (3.79) yields

$$\Delta = 100 \left[\frac{C_M - [M']}{C_M [M'] K'} - \frac{[M']}{C_M} \right] \% \qquad (3.80)$$

As $C_M \gg [M']$, in the vicinity of the equivalence point, $[M']$ in the numerator of the first term on the right-hand side can be ignored and Eq. (3.80) takes the simpler form

$$\Delta \simeq 100 \left[\frac{1}{[M'] K'} - \frac{[M']}{C_M} \right] \% \qquad (3.81)$$

Compare Eq. (3.81) with Eq. (3.49b).

Hence, if the free metal ion concentration at the end-point, the total concentration of the metal ion and the stability constant of the complex are known, the titration error can be calculated by using Eq. (3.81). If the free metal ion concentration at the end-point is equal to the concentration at the equivalence point, Δ is zero. The smaller the value of K and C_M, the greater is the titration error.

The error curves of the titrations based on the formation of complexes with different stabilities are shown in Figs. 3.7 and 3.8. The equations of the asymptotes drawn to the curve, which intersect at the equivalence point, are

$$\log \Delta' = 2 - \log C_M - \mathrm{pM} \qquad (3.82)$$

$$\log \Delta'' = \mathrm{pM} - \log K + 2 \qquad (3.83)$$

At the point of intersection

$$\mathrm{pM} = \frac{1}{2} (\log K - \log C_M) \qquad (3.84)$$

which is the same as Eq. (3.72).

Before the equivalence point, the second term on the right-hand side of Eq. (3.81) (the value of Φ_M) predominates, and the error is negative, whereas after the equivalence point the first term predominates and the error is positive.

It can be seen from Fig. 3.7 that if $K' = 10^{12}$, and $C_M = 10^{-3}$, and the error should not exceed $0\cdot1\%$, then pM at the end-point may differ from the value at the equivalence point by a maximum of $1\cdot5$ units.

Complexometric titrations

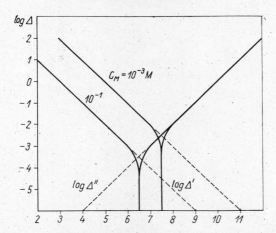

Fig. 3.7. Relation between the logarithm of the titration error and pM value at the end-point in complexometric titrations (log K'_{ML} = 12·0) [122]

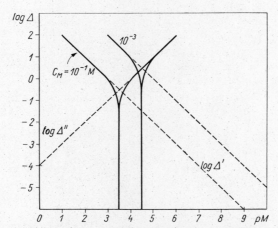

Fig. 3.8. Relation between the logarithm of the titration error and the pM value at the end-point in complexometric titrations (log K'_{ML} = 6·0) [122]

If, however $K' = 10^6$ and $C_M = 10^{-3}$ (see Fig. 3.8), the titration will have an error range of $\pm 9\%$ even if an indicator is available which changes its colour just at the equivalence point but there is an uncertainty of ± 0.5 pM unit in the determination of the end-point.

It follows from Eq. (3.81) that if $C_M = 10^{-2}M$ and the uncertainty in the determination of the end-point is ± 0.5 pM unit, the condition for the titration to be feasible is that

$$\log K' > 7. \tag{3.85}$$

In this case the titration error is less than $\pm 1\%$.

The conditional stability of a chelate is increased by the formation of a mixed-ligand complex as can be seen from Eq. (1.87), and this can be taken into consideration when predicting the optimum solution composition, and auxiliary complexing agents may be used. For example, the conditional stability constant of EDTA-titanium(IV) is increased by hydrogen peroxide through the formation of a mixed-ligand complex.

3.4.2. End-point detection

The end-point of complexometric titrations can be detected by means of metal or redox indicators, by visual, photometric, amperometric, conductometric, potentiometric methods etc. For more details see [15]. Only the most common visual method using metal indicators, and the potentiometric method, will be dealt with here.

In one group of metal indicators are acid–base type indicators which also change colour during complex formation (e.g. Eriochrome Black T), and in the other group are indicators which are either colourless or slightly coloured, but give coloured metal complexes, e.g. tiron (pyrocatechol disulphonic acid) and sulphosalicylic acid.

For colour-changing indicators, if the pH-dependent protonation of the indicator and the possible side reactions of the metal ion are considered, the conditional stability constant of the 1:1 indicator complex is as follows

$$K'_{MI} = \frac{[MI]}{[M'][I']} = \frac{K_{MI}}{\alpha_M \alpha_I} \tag{3.86}$$

where I stands for the indicator ligand. The condition for the operation of the indicator is that the stability of the metal complex with the titrant is higher than that of the indicator complex. That is,

$$K'_{ML} > K'_{MI} \tag{3.87}$$

The transition point occurs when the concentrations of the two forms of different colour are the same, that is $[I'] = [MI]$. Accordingly, the metal ion concentration at the transition point, is given by

$$[M']_t = \frac{1}{K'_{MI}}$$

i.e.

$$(pM)_t = \log K'_{MI} \tag{3.88}$$

If the conditional stability constant of the indicator complex is known, pM' at the transition point can be calculated. If $(pM')_t \neq (pM')_{eq}$, then an indicator error occurs (see Example 56).

For visual observation the uncertainty in the determination of the transition point is ± 0.5 pM. For photometric indication this can be reduced to ± 0.1 pM, by use of two colour filters [132].

When a one-colour indicator is used a greater indicator concentration is allowed, as the non-complexed indicator does not affect the observation of the end-point. As the end-point is poorer with this type of indicator, the concentration of the indicator should be about $10^{-4}M$. The indicator complex concentration which just shows no noticeable colour is usually about $10^{-5}M$. (This corresponds to the transition point.)

If the metal ion titrated does not form a sufficiently stable complex with any indicator, and there is no appropriate colour transition, indirect indication may be used. A small amount of another metal ion (N), which forms a more stable complex with the indicator and a less stable complex with the titrant than does the metal ion (M) to be titrated, is added to the titrated solution in the form of its complex with the titrant. Naturally, NI must be less stable than NL or the indicator would be 'blocked'. The conditions are

$$K'_{ML} > K'_{NL} > K'_{NI} \tag{3.89}$$

and

$$K'_{MI} < K'_{NI} \tag{3.90}$$

For example calcium can be titrated with EDTA, with Eriochrome Black T as indicator, in the presence of MgEDTA.

During the titration of M, N is only replaced from its complex NI at the end-point. As the ion N is added to the solution in the form of NL, no overconsumption of the titrant occurs. The pM at the end-point of the titration can be calculated according to Ringbom [15] as follows.

From the complex formation equilibria for the two metal ions we can write

$$\frac{[ML]}{[NL]} = \frac{[M'][L']K'_{ML}}{[N'][L']K'_{NL}}$$

and pM can be expressed from the logarithmic form of this equation as follows

$$pM' = pN' + \log\frac{K'_{ML}}{K'_{NL}} + \log\frac{[NL]}{[ML]} \qquad (3.91)$$

The value of pN' at the transition point can be calculated by using Eq. (3.88). The last term on the right-hand side of Eq. (3.91) can be calculated from the amounts of the titrated ion and the 'indicator metal' ion. The latter is chosen so that pM calculated by means of the equation is as close as possible to the pM for the equivalence point.

It should be noted that if the indicator is very sensitive to N, i.e. the value of pN is very high, this indirect end-point detection can be used even if the values of the further terms in Eq. (3.91) are negative. For example PAN–copper(II) can be used in the titration of zinc with EDTA, even though $K_{Zn\,EDTA} < K_{Cu\,EDTA}$.

Direct potentiometric end-point detection can only rarely be used in complexometric titrations, as there are only a few ions for which the activity can be measured directly. According to measurements by Reilley and Schmid [133] a mercury/mercury(II) electrode can be used in a certain pH interval as an electrode of the second kind for monitoring the free ligand, or, as an electrode, of the third kind for monitoring a second metal ion.

The potential of a mercury electrode, as given by the Nernst equation, if the activity coefficient is disregarded, is

$$E = E^0_{Hg} + \frac{0.059}{2} \log [Hg^{2+}] \qquad (3.92)$$

In the presence of a chelating agent, from Eq. (3.16):

$$E = E^0_{Hg} + \frac{0.059}{2} \log C_{Hg} - \frac{0.059}{2} \log \{1 + [L]K_{HgL}\}$$

Sec. 3.4]   **Complexometric titrations**   227

If $C_{Hg} < C_L$ and $[L]K_{HgL} \gg 1$, then the simplified form of this equation will be:

$$E = E^0_{Hg} + \frac{0.059}{2} \log \frac{C_{Hg}}{K_{HgL}} - \frac{0.059}{2} \log [L]$$

$$= E^{0\prime}_L - \frac{0.059}{2} \log [L] \qquad (3.93)$$

In a given pH range, if the total concentration of mercury, C_{Hg}, is constant, the electrode potential depends only on the logarithm of the free ligand concentration (compare with the comments on second kind electrodes, p. 216).

If a metal ion M, which forms a complex ML with the chelating ligand, is also present in the solution, the electrode potential will be

$$E = E^0_{Hg} + \frac{0.059}{2} \log \frac{[HgL]K_{ML}}{K_{HgL}} + \frac{0.059}{2} \log \frac{[M]}{[ML]}$$

$$= E^{0\prime}_M + \frac{0.059}{2} \log \frac{[M]}{[ML]} \qquad (3.94)$$

Hence, by measuring the electrode potential, the change in the [M]/[ML] concentration ratio can be followed during the titration.

3.4.3. Analysis of mixtures of metals

It frequently occurs that as well as the metal ion to be determined another metal ion is present which also forms a complex with the titrant. If the ratio of the conditional stability constants of the complexes formed with the titrant ligand is high enough, then the metal ion which forms the more stable complex can be determined in the presence of the other. If the conditional stability constant of the complex of the second metal ion exceeds 10^7, then it can also be determined by successive titration. Obviously, a suitable method of end-point detection is also necessary. Let the metal ion forming the more stable complex be denoted by M. The interference from the metal ion N is considered as a side-reaction of the ligand [15, 134].

$$\alpha_{L(N)} = 1 + [N']K'_{NL} \qquad (3.95)$$

15*

If the complexation of N by the ligand is the only side-reaction of the ligand, then the conditional constant of the main reaction will be

$$K'_{ML} = \frac{K_{ML}}{\alpha_{M(Y)}\alpha_{L(N)}} \tag{3.96}$$

If the metal ion M enters into no side-reactions and $[N']K'_{NL} > 1$, then the conditional constant can be approximated by

$$K'_{ML} \simeq \frac{K_{ML}}{K'_{NL}C_N} \tag{3.97}$$

or

$$\log K'_{ML} \simeq \log K_{ML} - \log K'_{NL} - \log C_N$$

The condition for the feasibility of the titration (p. 224) is that

$$\log K'_{ML} > 7$$

If the value of $\alpha_{L(H)}$ due to the protonation of the ligand is similar to $\alpha_{L(N)}$, then the former also has to be taken into consideration when calculating the conditional stability constant. That is

$$\alpha_L = \alpha_{L(H)} + \alpha_{L(N)} - 1 \tag{3.98}$$

[see Eq. (1.75)]. It follows from Eq. (3.98) that N interferes with the titration of M only if $\alpha_{L(N)}$ is greater than or similar in value to $\alpha_{L(H)}$. Otherwise the latter will determine the gross α_L value. In this case, N will not interfere if the acidity is adjusted so that $\log K'_{ML} > 7$ and $\log K_{NL} < 3$ (i.e. if $K'_{ML}/K'_{NL} > 10^4$).

The interference from N can be reduced significantly according to Eqs. (3.95) and (3.97) if the conditional stability constant of the complex NL is reduced by adding a suitable auxiliary complexant, that is, by masking the interfering ion. It is important that the masking agent does not react at all or only to a small extent with the metal ion M to be determined under the given conditions (see Example 57).

Back-titration may be used if a metal ion cannot be titrated directly (e.g. there is no suitable indicator or the complex formation reaction is slow). The standard solution of complexant is added in excess, and the excess is back-titrated with a standard solution containing N. The equilibrium condition for the feasibility of the titration is

$$\log K'_{ML} > \log K'_{NL} > 7 \tag{3.99}$$

It should be noted that if the complex ML is highly inert, the titration can be carried out even if $K'_{ML} < K'_{NL}$ [for example, the back-titration of EDTA with zinc(II) in the determination of aluminium].

Displacement reactions can be used for determining some metal ions. The principle of this indirect determination is that a complex NL is added to the solution of M in excess, then the amount of N liberated (equivalent to M) is titrated.

$$M + NY \rightleftharpoons MY + N \tag{3.100}$$

The condition for the displacement titration method is that the equilibrium constant of reaction (3.100) should be at least 10^3. A typical example is displacement of nickel from its tetracyano-complex by silver: $2Ag^+ + Ni(CN)_4^{2-} \rightarrow 2Ag(CN)_2^- + Ni^{2+}$.

3.4.4. Worked Examples

56. Calculate the indicator error in the titration of $10^{-2}M$ calcium with EDTA in the presence of Eriochrome Black R as indicator at pH 12·5. The complex stability constants are $\log K_{CaEDTA} = 10·7$ and $\log K_{CaI} = 5·3$. The protonation constants of the indicator are $\log K_1 = 13·5$ and $\log K_2 = 7·0$. The stability constant of the hydroxo-complex of calcium is $\log \beta_1 = 1·3$.

The indicator is red in acid solution (pH < 7), blue above pH 7, and orange above pH 13·5. The calcium complex is red. This indicator can therefore be used for calcium titration only in the pH range of 8–12·5 if a good blue transition is to be obtained (see Fig. 3.9). To calculate the transition point of the indicator, the α_{Ca} and α_I functions are calculated first:

$$\alpha_{Ca(OH)} = 1 + [OH^-]\beta_1 = 1 + 10^{-1·5} \times 10^{1·3} = 1·6$$

$$\log \alpha_{Ca(OH)} = 0·2$$

$$\alpha_{I(H)} = 1 + 10^{-12·5} \times 10^{13·5} + 10^{-25} \times 10^{20·5} = 11$$

$$\log \alpha_{I(H)} = 1·04$$

The conditional stability constant of the calcium complex of the indicator is given by

$$\log K'_{CaI} = 5·3 - 0·2 - 1·04 = 4·06$$

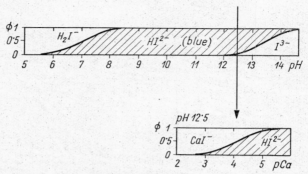

Fig. 3.9. Mole-fraction distribution of the protonation products and calcium complex of the indicator Eriochrome Black R as a function of pH and pCa, respectively. (See Example 56)

This, according to Eq. (3.88), is equal to pCa'_t. Thus at the colour transition indicating the end-point of the titration $pCa'_t = 4.06$. The value of pCa' at the equivalence point can be calculated by using Eq. (3.72):

$$\log K'_{CaEDTA} = \log K - \log \alpha_{Ca(OH)} - \log \alpha_{EDTA(H)} = 10.7 - 0.2 - 0 = 10.5$$

$$pCa'_{eq} = \frac{1}{2}(10.5 + 2) = 6.25$$

The transition point of the indicator does not coincide with the equivalence point (see Fig. 3.6); the colour change occurs before the equivalence point. The error, according to Eq. (3.81), is

$$\Delta = \left[\frac{1}{10^{-4.06} \times 10^{10.5}} - \frac{10^{-4.06}}{10^{-2}}\right] 100 = -0.87\%$$

The pH cannot be increased in order to reduce the indicator error, without affecting the sharpness of the red to blue colour change.

The stability constant of the calcium complex of Murexide indicator is somewhat lower than that of the Eriochrome Black R complex, but the conditional stability constant is higher at pH 12.5; $\log K'_{CaI} = 4.8$. With Murexide the error is smaller:

$$\Delta = \left[\frac{1}{10^{-4.8} \times 10^{10.5}} - \frac{10^{-4.80}}{10^{-2}}\right] 100 = -0.16\%.$$

However, the uncertainty of the visual observation of the colour transition is greater when Murexide is used instead of Eriochrome Black R.

57. Is it possible to determine $10^{-2}M$ magnesium with EDTA in the presence of $10^{-2}M$ zinc when potassium cyanide is present as a masking agent and Eriochrome Black T as indicator? The stability constants are $\log K_{\text{MgEDTA}} = 8\cdot6$ and $\log K_{\text{ZnEDTA}} = 16\cdot5$. The overall stability constant of the tetracyanozinc complex is $\log \beta_4 = 16\cdot7$.

The titration is carried out at pH 10, in the presence of an ammonia–ammonium chloride buffer.

Magnesium can be determined if

$$\log K'_{\text{MgEDTA}} > 7$$

or

$$\log K_{\text{MgEDTA}} - \log \alpha_{\text{Mg}} - \log \alpha_{\text{EDTA}} > 7.$$

Magnesium does not form complexes with cyanide. Complexation of magnesium by ammonia is small even in $0\cdot1M$ ammonia (see Fig. 1.11).

$$\log \alpha_{\text{Mg(NH}_3)} \simeq 0\cdot1$$

From the stability constant of the EDTA–magnesium complex, the required condition is

$$8\cdot6 - 0\cdot1 - \log \alpha_{\text{EDTA}} > 7,$$

from which

$$\log \alpha_{\text{EDTA}} < 1\cdot5$$

Log $\alpha_{\text{EDTA(H)}}$ at pH 10 is $0\cdot5$ according to Fig. 1.14. As the interference from zinc is more important, the effect of protonation can be neglected. Thus

$$\log \alpha_{\text{EDTA(Zn)}} < 1\cdot5$$

$$\alpha_{\text{EDTA(Zn)}} < 10^{1\cdot5}$$

According to Eq. (3.95):

$$\alpha_{\text{EDTA(Zn)}} = 1 + [\text{Zn}']K'_{\text{ZnEDTA}}$$

from which

$$1 + 10^{-2}K'_{\text{ZnEDTA}} < 10^{1\cdot5}$$

$$K'_{\text{ZnEDTA}} < 10^{3\cdot5}$$

Thus sufficient cyanide is to be added to reduce the conditional stability constant of the zinc–EDTA complex to below $10^{3.5}$.

The conditional stability constant of the zinc–EDTA complex is

$$\log K'_{ZnEDTA} = \log K_{ZnEDTA} - \log \alpha_{Zn} - \log \alpha_{EDTA},$$

where $\log \alpha_{EDTA} = 0.5$. The possible side-reactions of zinc at pH 10 with ammonia or hydroxide are not sufficient to mask zinc (see Figs. 1.10 and 1.11). To meet the condition

$$16.5 - \log \alpha_{Zn} - 0.5 < 3.5$$

the value of $\log \alpha_{Zn(CN)}$ must be > 12.5.

The concentration of cyanide necessary to mask zinc is calculated from this inequality.

$$\alpha_{Zn(CN)} = 1 + [CN^-]^4 \beta_4$$

$$1 + [CN^-]^4 \times 10^{16.7} > 10^{12.5}$$

As $10^{12.5} \gg 1$,

$$[CN^-]^4 > 10^{-4.2}$$

$$[CN^-] > 0.089 M$$

If potassium cyanide is added in an amount sufficient for complexation of zinc and to ensure a free cyanide concentration of about $0.1M$, magnesium can be determined satisfactorily.

Since about 10% of the cyanide is protonated at pH 10 ($\log K = 9.2$) the required total concentration of cyanide is

$$C_{CN} \simeq 0.04 + 0.1 + 0.01 = 0.15 M$$

3.5. Redox Titrations

The principle of redox titrations is that the solution of a reducing agent is titrated with a solution of an oxidizing agent (or vice versa) and the end-point of the reaction

$$y Ox_t + z Red_d \rightleftharpoons y Red_t + z Ox_d \qquad (3.101)$$

is indicated by a visual redox indicator or by a potentiometric method (t indicates the titrant, d the titrand, z and y are the stoichiometric coefficients). Plotting the negative logarithm of the electron activity (or the redox potential proportional to it) *vs.* the degree of titration gives a curve similar to those obtained in the case of acid–base, complexometric or precipitation titrations (Fig. 3.10). The shape of the curve depends on the equilibrium constant of the reaction and on the number of electrons transferred during the reaction (see later).

In what follows we shall consider the titrant to be the oxidant. At the equivalence point

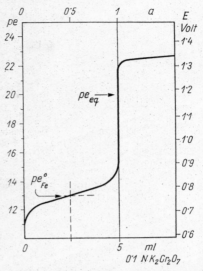

Fig. 3.10. Titration of 50 ml of $10^{-2}M$ iron(II) sulphate solution (pH=0) with $0·1N$ potassium dichromate standard solution

$$y[Ox_t] + y[Red_t] = z[Ox_d] + z[Red_d] \tag{3.102}$$

Taking into account that the following equation is valid for the entire course of the titration (if all the substance to be determined was present in the reduced form at the beginning of the titration)

$$y[Red_t] = z[Ox_d] \tag{3.103}$$

and that, according to Eq. (1.49)

$$[Ox_t] = \frac{[Red_t]}{K_{r_t}[e]^z} \tag{3.104}$$

$$[Red_d] = K_{r_d}[Ox_d][e]^y \tag{3.105}$$

Eq. (3.102) can be written as

$$\frac{1}{K_{r_t}[e]^z_{eq}} = K_{r_d}[e]^y_{eq}$$

From this, pe at the equivalence point [cf. 135] is given by

$$(pe)_{eq} = -\log [e]_{eq} = \frac{1}{z+y} (\log K_{rd} + \log K_{rt}) \qquad (3.106)$$

where z stands for the number of electrons involved in the titrant system and y for that in the titrand system, i.e.

$$Ox_t + ze \rightleftharpoons Red_t$$

$$Ox_d + ye \rightleftharpoons Red_d$$

If pe at the end-point of the titration is not equal to that at the equivalence point $[(pe)_{end} \gtrless (pe)_{eq}]$ a titration error occurs.

For calculating the relative titration error, the total concentrations of the titrand and titrant in the titrated solution at the end-point should be written:

$$C_t = [Ox_t]_{end} + [Red_t]_{end}$$

$$C_d = [Ox_d]_{end} + [Red_d]_{end}$$

As Eq. (3.103) is also valid at the end-point, the relative error when the titrant is the oxidant (see p. 201) can be given in the following form:

$$\frac{yC_t - zC_d}{zC_d} = \frac{y[Ox_t]_{end} - z[Red_d]_{end}}{zC_d}$$

From Eqs. (1.49) and (3.103)–(3.105):

$$\frac{\dfrac{z[Ox_d]}{K_{rt}[e]^z} - zK_{rd}[Ox_d][e]^y}{z[Ox_d] + z[Ox_d]K_{rd}[e]^y} = \frac{\dfrac{1}{K_{rt}[e]^z} - K_{rd}[e]^y}{1 + K_{rd}[e]^y}$$

The relative error [cf. 122, 135] is:

$$\Delta = 100 \left[\frac{\dfrac{1}{K_{rt}[e]^z} - K_{rd}[e]^y}{1 + K_{rd}[e]^y} \right] \% \qquad (3.107)$$

Near the equivalence point, $K_{rd}[e]^y \ll 1$, consequently

$$\Delta \simeq 100 \left[\frac{1}{K_t^r[e]^z} - K_d^r[e]^y \right] \qquad (3.108a)$$

Sec. 3.5] Redox titrations

It is clearly shown by Eqs. (3.107) and (3.108a) that the titration error in redox titrations depends only on the redox equilibrium constants and on the electron activity at the end-point. To a first approximation, the error is independent of the concentration. In fact, the titration error does also depend on the concentration, as the values of the equilibrium constants are affected by the ionic strength. When the titrant is the reductant the error is

$$\Delta \simeq 100 \left[K_{r_t}[e]^y - \frac{1}{K_{rd}[e]^z} \right] \% \qquad (3.108b)$$

where y stands for the number of electrons involved in the titrant reaction.

The error curves for the titration of thallium(III) with ascorbic acid and the titration of iron(II) with dichromate, calculated by using Eq. (3.107), are shown in Fig. 3.11.

The equations of the tangents drawn to the descending and ascending parts of the curves are

$$\log \Delta' = \log K_{r_1} + 2 - y\text{pe}$$

$$\log \Delta'' = z\text{pe} + 2 - \log K_{r_2}$$

The abscissa at the point of intersection

$$\frac{1}{z+y} (\log K_{r_1} + \log K_{r_2})$$

gives the pe of the equivalence point of the titration [see Eqs. (3.106) and (3.108a, b)].

If the oxidized or reduced product present in the solution containing the redox system takes part in a side-reaction, and the equilibrium position of this reaction can be kept constant, by maintaining suitable experimental conditions, the conditional redox equilibrium constant can be used in the calculations.

The conditional redox equilibrium constant is, in general, given by

$$K'_r = \frac{[\text{Red}']}{[\text{Ox}'][e]^z} \qquad (3.109)$$

or

$$K'_r = K_r \frac{\alpha_{\text{Red}(A)}}{\alpha_{\text{Ox}(B)}} \qquad (3.110)$$

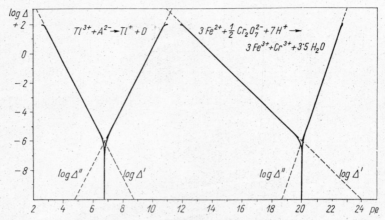

Fig. 3.11. Relation between the logarithm of the titration error and pe at the end-point in the titration of thallium(III) with ascorbic acid (pH = 4) and in that of iron(II) with potassium dichromate (pH = 0)

where A and B denote the substances reacting with the reduced and oxidized substance, respectively.

In practice, the most important side-reactions are complex formation and protonation. The oxidized and reduced form of a metal ion may form complexes of different stabilities with the complexing ligand L. As the value of pe^0 is related to the redox equilibrium constant [see Eq. (1.52)] the new $pe^{0'}$ of the redox system in the presence of the complexing agent, taking Eq. (3.110) into account will be

$$pe^{0'} = \frac{1}{z} \log K'_r = \frac{1}{z} \log K_r + \frac{1}{z} \log \frac{\alpha_{Red(L)}}{\alpha_{Ox(L)}} \qquad (3.111)$$

The pe of a solution containing various amounts of the oxidized and reduced form of a species and a complexing ligand can be calculated from the equation

$$pe = pe^0 + \frac{1}{z} \log \frac{\alpha_{Red(L)}}{\alpha_{Ox(L)}} + \frac{1}{z} \log \frac{[Ox']}{[Red']}$$

$$= pe^{0'} + \frac{1}{z} \log \frac{[Ox']}{[Red']} \qquad (3.112)$$

where [Ox'] and [Red'] are the total analytical concentrations.

Sec. 3.5] **Redox titrations**

The value of pe⁰ changes in the presence of the complexing agent, depending on the relative stability of the complexes with different oxidation states of the metal.

Complexes of higher oxidation states are generally more stable (see p. 70) and accordingly, pe⁰ is usually decreased. [The situation is reversed in the case of the 1,10-phenanthroline complexes of iron(II) and iron(III)].

According to Eq. (1.54) pe is proportional to the redox potential of solutions containing oxidizing or reducing species. From Eqs. (3.111) and (3.112), the potential of a redox system of a metal ion in the presence of the complexing ligand L can be given as

$$E = E^0 + \frac{0.059}{z} \log \frac{\alpha_{Red(L)}}{\alpha_{Ox(L)}} + \frac{0.059}{z} \log \frac{[Ox']}{[Red']}$$

$$= E^{0'} + \frac{0.059}{z} \log \frac{[Ox']}{[Red']} \qquad (3.113)$$

where $E^{0'}$ is the conditional standard redox potential observed in the presence of the complexing agent.

The shift of standard potential is

$$\Delta E^0 = \frac{0.059}{z} \log \frac{\alpha_{Red(L)}}{\alpha_{Ox(L)}} \qquad (3.114)$$

[compare with Eq. (3.16)].

When the complexing agent forms 1 : 1 complexes of high stability with the metal ion in both the oxidized and reduced form, the conditional standard redox potential can be calculated from the simplified equation

$$E^{0'} \simeq E^0 + \frac{0.059}{z} \log \frac{K_{RedL}}{K_{OxL}}$$

where K_{RedL} and K_{OxL} are the stability constants of the complexes. Thus the standard redox potential is independent of the pH and the concentration of the ligand over a certain range. [For example, the iron(III)–iron(II) system in the presence of EDTA, see Example 58.]

The standard pe, or the standard redox potential of redox systems with acidic or basic character (or involving protons or hydroxide ions in the half-cell) can be altered by varying the pH or by complexation with a metal ion. Equations (3.111) and (3.114) can also be used in this case if $\alpha_{Ox(H)}$ or $\alpha_{Ox(M)}$ is written in place of $\alpha_{Ox(L)}$ and $\alpha_{Red(H)}$ or $\alpha_{Red(M)}$ in place of $\alpha_{\text{Red(L)}}$; E^0 is the standard potential of the redox system containing the non-protonated (not complexed) species (see Examples 60 and 62).

It can be seen from Eqs. (3.111)–(3.114) that by appropriate choice of pH and complexing agent, the value of pe^0 or of the standard redox potential can be altered as required. This shift is of importance in practice from two points of view. First, for the oxidation (or reduction) to be quantitative the amount of determinand remaining untitrated at the equivalence point should not exceed 0·1% of that originally present, i.e.

$$\log \frac{[\text{Ox}']_d}{[\text{Red}']_d} \geq 3$$

and

$$(pe)_{eq} > (pe^0)_d + \frac{3}{y} \tag{3.115}$$

From Eqs. (3.106) and (1.52), and from the condition given by Eq. (3.114) it follows [cf. 135] that:

$$y \log K_{r_t} - z \log K_{r_d} > 3(z + y) \tag{3.116}$$

or

$$E_t^0 - E_d^0 > 0 \cdot 18 \left(\frac{z + y}{zy} \right) \tag{3.117}$$

If the equilibrium constants of the titrand and titrant systems do not satisfy Eq. (3.116), the conditions can often be altered to meet this requirement, by using a complexing agent (or changing the pH) to change the conditional redox equilibrium constants in the appropriate direction. For example, according to Vydra and Přibil [136], cobalt(II) ions can be titrated with iron(III) ions if 1,10-phenanthroline is added to the solution, and the pH adjusted to 3 even though $K_{rCo} \gg K_{rFe}$. Secondly, if as well as the determinand another component present in the solution has similar oxidizing or reducing properties, then this interfering species can be masked, so that the conditional redox constant of the interfering system is changed to such an extent that it no longer interferes with the main reaction. For example, chromium(VI) can be determined by iodometry in the presence of iron(III) if the latter is masked with EDTA [137].

The end-point of redox titrations can be detected by colour-changing redox indicators, by potentiometric methods or by other instrumental methods (e.g. amperometry).

Redox indicators, similarly to acid–base or complexometric indicators,

Sec. 3.5]   **Redox titrations**   239

may be two-colour (e.g. ferroin) or one-colour (diphenylamine). (However, at the concentrations normally used, the oxidized form of ferroin is so pale in colour that the indicator acts as a one-colour indicator.) For a two-colour indicator the pe of the colour transition coincides with the pe$^{0'}$ of the indicator: for a one-colour indicator it also depends on the concentration of the indicator (see Example 59).

In the case of potentiometric titrations the titration curve can be transformed into linear form to enable a more precise determination to be made of the amount of standard solution consumed. Gran's method [123] can be used also for redox titrations [138].

Consider the case where a reducing species is titrated with a standard oxidizing solution. Then

$$\frac{[\text{Red}]_d}{[\text{Ox}]_d} = \frac{V_{eq} - V}{V} \tag{3.118}$$

can be written for the course of the titration, where V is the volume of titrant added, and V_{eq} the volume of titrant added at the equivalence point. Substituting this into the basic Eq. (1.49) we obtain

$$K_r = \frac{V_{eq} - V}{V[e]^z} \tag{3.119}$$

Since, according to Eq. (1.52)

$$K_r = \frac{1}{[e]_0^z}$$

where $[e]_0$ is the standard electron activity, Eq. (3.119) can be written as

$$V[e]^z = V_{eq}[e]_0^z - V[e]_0^z \tag{3.120}$$

Plotting $V[e]^z$ against the volume V of titrant added gives a straight line which intercepts the abscissa at V_{eq}. The slope of the line is $-[e]_0^z$ [cf. Eq. (3.53)].

If the redox potential E is measured during the titration, the related $[e]$ value can be calculated by using the relationship

$$[e] = 10^{-\frac{E}{0.059}} \tag{3.121}$$

See Eqs. (1.51), (1.54), (1.55) and Example 63.

3.5.1. Worked examples

58. Calculate the conditional standard redox potential of the iron(III)–iron(II) system in the presence of a $10^{-2}M$ excess of EDTA and determine the pH above which it can be considered as constant.

$\log K_{Fe(III)EDTA} = 25 \cdot 1$, $\log K_{Fe(II)EDTA} = 14 \cdot 3$ and $E_{Fe}^0 = +0 \cdot 77$ V

The calculation is done by using Eq. (3.113).

$$\alpha_{Fe(II)(EDTA)} = 1 + \left[\frac{10^{-2}}{\alpha_{EDTA(H)}}\right] 10^{14 \cdot 3} \quad (i)$$

$$\alpha_{Fe(III)(EDTA)} = 1 + \left[\frac{10^{-2}}{\alpha_{EDTA(H)}}\right] 10^{25 \cdot 1} \quad (ii)$$

If pH > 3, then, according to Fig. 1.14, $\alpha_{EDTA(H)} < 10^{11}$, and in both equations $\left[\dfrac{10^{-2}}{\alpha_{EDTA(H)}}\right] K \gg 1$, thus 1 can be neglected.

It follows that

$$E^{0'} = +0 \cdot 77 + 0 \cdot 059(14 \cdot 3 - 25 \cdot 1) = +0 \cdot 13 \text{ V} \quad (\text{if pH} > 3)$$

59. Calculate the pe value at which the transition of the redox indicator Variamine Blue occurs at pH 4, if the concentration of the indicator is $10^{-4}M$. The concentration of the oxidized form, $[V'_{Ox}]$, should be $10^{-5}M$ for the blue colour to be observable.

Both the reduced and oxidized forms of Variamine Blue undergo protonation in acidic solution, the protonation constants being $\log K_{Red} = 5 \cdot 9$ and $\log K_{Ox} = 6 \cdot 6$ [139]. The reduced form is colourless, whether protonated or not, whereas the non-protonated form of the oxidized compound is yellow, and the protonated form blue in solution. Hence, the indicator can only be used in solutions of pH lower than 6 (see Fig. 3.12). The redox equation and equilibrium constant of the indicator redox system are

$$V_{Ox} + 2e + 2H^+ \rightleftharpoons V_{Red}$$

$$\log K_r = 24 \cdot 8$$

From Eq. (3.112)

$$pe = pe^0 + \frac{1}{2} \log \frac{\alpha_{V_{Red}(H)}}{\alpha_{V_{Ox}(H)}} + \frac{1}{2} \log \frac{[V'_{Ox}]}{[V'_{Red}]} - pH$$

Sec. 3.5] Redox titrations 241

At pH 4

$$\alpha_{V_{Red(H)}} = 1 + 10^{-4} \times 10^{5\cdot 9} = 10^{1\cdot 9}$$

$$\alpha_{V_{Ox(H)}} = 1 + 10^{-4} \times 10^{6\cdot 6} = 10^{2\cdot 6}$$

$$pe = 12\cdot 4 + \frac{1}{2}(1\cdot 9 - 2\cdot 6) + \frac{1}{2}\log\frac{1\times 10^{-5}}{9\times 10^{-5}} - 4$$

$$pe = 7\cdot 58$$

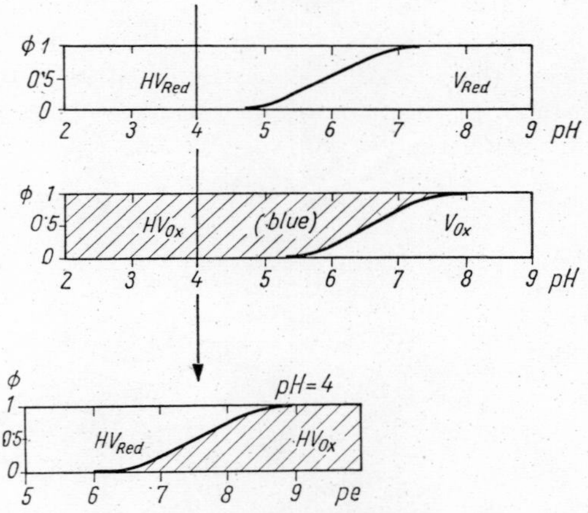

Fig. 3.12. Mole-fraction distribution of the protonated and non-protonated forms of the reduced and oxidized products of the indicator Variamine Blue as a function of pH. Mole-fraction distribution of the reduced and oxidized forms at pH 4 as a function of pe

60. Calculate the indicator error in the titration of thallium(III) with ascorbic acid, if Variamine Blue is used to detect the end-point. According to the procedure given by Erdey *et al.* [140], the thallium(I) is first oxidized to thallium(III) by adding bromine water, then the excess of bromine is reduced to bromide and the pH adjusted to 4 with a formate buffer. The concentration of bromide is about $10^{-1}M$ in the solution before and during titration.

The redox couples and equilibrium constants are as follows:

$$Tl^{3+} + 2e \rightleftharpoons Tl^+ \qquad \log K_{r_{Tl}} = 42\cdot 6$$

$$D + 2e \rightleftharpoons A^{2-} \qquad \log K_{r_A} = -2\cdot 5$$

where D and A^{2-} represent dehydroascorbic acid and ascorbinate ion, respectively.

The overall stability constants of the bromo-complexes of thallium(III) are $\log \beta_1 = 8\cdot3$, $\log \beta_2 = 14\cdot6$, $\log \beta_3 = 19\cdot2$, $\log \beta_4 = 22\cdot3$, $\log \beta_5 = 24\cdot8$ and $\log \beta_6 = 26\cdot5$; those of the hydroxo-complexes are $\log \beta_1 = 12\cdot9$ and $\log \beta_2 = 25\cdot4$; those of the bromo-complexes of thallium(I) are $\log \beta_1 = 0\cdot92$, $\log \beta_2 = 0\cdot92$ and $\log \beta_3 = 0\cdot40$. The protonation constants of the ascorbinate ion are $\log K_1 = 11\cdot56$ and $\log K_2 = 4\cdot17$. The transition pe value of Variamine Blue at pH 4 is 7·58 (see Example 59).

First the values of the α-functions are calculated, then the conditional equilibrium constants of the titrand and titrant systems and finally the indicator error by using Eq. (3.108b)

$$\alpha_{Tl(III)(Br)} = 1 + 10^{-1} \times 10^{8\cdot3} + 10^{-2} \times 10^{14\cdot6} + 10^{-3} \times 10^{19\cdot2}$$
$$+ 10^{-4} \times 10^{22\cdot3} + 10^{-5} \times 10^{24\cdot8} + 10^{-6} \times 10^{26\cdot5} = 10^{20\cdot6}$$

$$\alpha_{Tl(III)(OH)} = 1 + 10^{-10} \times 10^{12\cdot9} + 10^{-20} \times 10^{25\cdot4} = 10^{5\cdot4}$$

$$\alpha_{Tl(III)} = 10^{20\cdot6} + 10^{5\cdot4} - 1 \simeq 10^{20\cdot6}$$

$$\alpha_{Tl(I)(Br)} = 1 + 10^{-1} \times 10^{0\cdot92} + 10^{-2} \times 10^{0\cdot92} + 10^{-3} \times 10^{0\cdot4} = 10^{0\cdot28}$$

The conditional equilibrium constant of the thallium(III)–thallium(I) redox system is

$$\log K'_{rTl} = 42\cdot6 + \log \alpha_{Tl(I)} - \log \alpha_{Tl(III)} = 42\cdot6 + 0\cdot28 - 20\cdot6 = 22\cdot28$$

The α-function of the ascorbinate ion at pH 4 is

$$\alpha_{A(H)} = 1 + 10^{-4} \times 10^{11\cdot56} + 10^{-8} \times 10^{15\cdot73} = 10^{7\cdot95}$$

The conditional equilibrium constant is

$$\log K'_{rA} = -2\cdot5 + 7\cdot95 = 5\cdot45$$

The titration error is given by

$$\Delta = 100 \left[10^{5\cdot45} \times (10^{-7\cdot58})^2 - \frac{1}{10^{22\cdot28} \times (10^{-7\cdot58})^2} \right] = -7\cdot6 \times 10^{-6} \%$$

Hence, the indicator selected is very good. According to Eq. (3.106), pe at the equivalence point is

$$(pe)_{eq} = \frac{1}{4}(22\cdot28 + 5\cdot45) = 6\cdot93$$

61. The oxidizing power of the copper(II)–copper(I) redox system is greatly increased by iodide, owing to the formation of copper(I) iodide which is only slightly soluble in water. This is the basis of the iodometric determination of copper(II). In weakly acidic solutions copper(II) oxidizes iodide to iodine which can be titrated with standard sodium thiosulphate solution. The interference from copper(II) in the iodometric determination of other substances can be eliminated by using masking agents [141].

Calculate the amount of sodium tartrate which should be added to a solution of $2\times10^{-2}M$ in copper(II) to prevent iodine formation, if the concentration of iodide is $10^{-1}M$ and the pH 6·1.

The equations and standard redox potentials of the iodine–iodide and copper(II)–copper(I) complex are

$$I_3^- + 2e \rightleftharpoons 3I^- \qquad +0\cdot536 \text{ V}$$

$$Cu^{2+} + e \rightleftharpoons Cu^+ \qquad +0\cdot16 \text{ V}$$

The formation constant of the copper(I) iodide precipitate is $\log K_s = 11\cdot7$ ($I = 0\cdot1$). The protonation constants of tartrate are $\log K_1 = 4\cdot1$ and $\log K_2 = 3\cdot9$. The overall stability constants of the tartrate–copper(II) complexes are $\log \beta_1 = 3\cdot2$; $\log \beta_2 = 5\cdot1$; $\log \beta_3 = 5\cdot8$ and $\log \beta_4 = 6\cdot2$. It should be stressed that in deducing the standard redox potential for the $I_3^-/3I^-$ system, the side-reaction of the oxidized form (iodine) with iodide ions ($K = 10^{2\cdot9}$) has been taken into consideration.

If the starch indicator is not to give a colour with the iodine, the concentration of I_3^- must be lower than $10^{-5}M$. The highest permissible redox potential in the solution is therefore

$$E_I = +0\cdot536 + \frac{0\cdot059}{2} \log \frac{10^{-5}}{(10^{-1})^3} = 0\cdot477 \text{ V}$$

For calculation of the potential for the copper(II)–copper(I) couple, $[Cu^{2+}] = 2\times10^{-2}M$ and $[Cu^+]$, which is dependent on the iodide concentration, can be calculated from Eq. (3.4c):

$$[Cu^+] = \frac{1}{10^{-1} \times 10^{11 \cdot 7}} = 10^{-10 \cdot 7} M$$

The redox potential, as calculated by using Eq. (3.113) is

$$E_{Cu} = +0 \cdot 16 + 0 \cdot 059 \log \frac{10^{-1 \cdot 7}}{10^{-10 \cdot 7}} - 0 \cdot 059 \log \alpha_{Cu(T)}$$

$$= +0 \cdot 69 - 0 \cdot 059 \log \alpha_{Cu(T)}.$$

To eliminate the oxidation by copper(II) it is necessary that $E_{Cu} < E_I$, that is

$$0 \cdot 69 - 0 \cdot 059 \log \alpha_{Cu(T)} < 0 \cdot 477$$

From this

$$\log \alpha_{Cu(T)} > 3 \cdot 61$$

The equation of the $\alpha_{Cu(T)}$ function, after insertion of the β values for the tartrate complexes, is

$$\alpha_{Cu(T)} = 1 + [T]10^{3 \cdot 2} + [T]^2 10^{5 \cdot 1} + [T]^3 10^{5 \cdot 8} + [T]^4 10^{6 \cdot 2}$$

The first protonation constant of tartrate is $10^{4 \cdot 1}$, i.e. there is practically no protonation at pH 6·1 and $[T] \simeq C_T$. It can be established from this equation that copper(II) will not be masked properly if the concentration of tartrate ions is $10^{-1} M$, but a $0 \cdot 4 M$ concentration will be enough:

$$\alpha_{Cu(T)} = 1 + 10^{-0 \cdot 4} \times 10^{3 \cdot 2} + 10^{-0 \cdot 8} \times 10^{5 \cdot 1} + 10^{-1 \cdot 2} \times 10^{5 \cdot 8}$$

$$+ 10^{-1 \cdot 6} \times 10^{6 \cdot 2} = 10^5$$

and

$$10^5 > 10^{3 \cdot 61}$$

Now the acidity of the solution required for the subsequent iodometric determination of copper(II) under these conditions will be found.

The iodometric determination of copper(II) is based on the reaction:

$$2Cu^{2+} + 5I^- \rightleftharpoons I_3^- + 2CuI_{prec}$$

Potassium iodide is added to the solution of copper(II) so that the concentration of iodide is approximately $0 \cdot 5 M$. For the reaction to proceed quantitatively it is necessary that the redox potential of the copper(II)–copper(I) system be equal to or higher than that of the iodine–iodide system after the reduction of 99·9% of the copper(II).

$$[Cu^{2+}]_{end} = 2 \times 10^{-5} M; \quad [Cu^+]_{end} = 10^{-11\cdot 4} M; \quad [I_3^-]_{end} = 10^{-2} M$$

$$E_{Cu} = 0\cdot 16 + 0\cdot 059 \log \frac{10^{-4\cdot 7}}{10^{-11\cdot 4}} - 0\cdot 059 \log \alpha_{Cu(T)}$$

$$E_I = 0\cdot 536 + \frac{0\cdot 059}{2} \log \frac{10^{-2}}{(10^{-0\cdot 3})^3} = +0\cdot 5$$

Hence
$$0\cdot 56 - 0\cdot 059 \log \alpha_{Cu(T)} > 0\cdot 5$$

$$\log \alpha_{Cu(T)} < 1\cdot 0$$

By acidification of the solution, the concentration of the free ligand is reduced to the necessary extent, according to the relationship

$$[T] = \frac{C_T}{\alpha_{T(H)}}$$

If the pH is 2·5, then

$$\alpha_{T(H)} = 1 + 10^{-2\cdot 5} \times 10^{4\cdot 1} + 10^{-5} \times 10^7 = 10^{2\cdot 15}$$

$$[T] = \frac{10^{-0\cdot 4}}{10^{2\cdot 15}} = 10^{-2\cdot 55}$$

$$\alpha_{Cu(T)} = 1 + 10^{-2\cdot 55} \times 10^{3\cdot 2} + 10^{-5\cdot 1} \times 10^{5\cdot 1} + 10^{-7\cdot 65} + 10^{5\cdot 8} +$$
$$+ 10^{-10\cdot 2} \times 10^{6\cdot 2} = 10^{0\cdot 81}$$

Since $0\cdot 81 < 1\cdot 0$, after adjustment of the pH to 2·5, the copper(II) can be determined.

62. Calculate the pH necessary for the accurate direct titration of potassium hexacyanoferrate(III) with ascorbic acid, given that $\log K'_{rFe(CN)_6} = 6\cdot 1$; the protonation constants of hexacyanoferrate(II) are $\log K_1 = 4\cdot 17$, $\log K_2 = 2\cdot 22$, $\log K_3 < 1$, $\log K_4 < 1$; the logarithms of all the protonation constants of hexacyanoferrate(III) are <1. For the equilibrium constant of the dehydroascorbic acid–ascorbinate redox system, and protonation constants of the ascorbinate ion, see Example 60.

The criterion for the feasibility of the titration, according to (3.116) is

$$2 \log K'_{rFe(CN)_6} - \log K'_{rA} > 3(1 + 2) = 9$$

From the protonation constants of hexacyanoferrate(II), if the pH $> 5\cdot 5$, then $\alpha_{Fe(CN)_6(H)} = 1$ and $\log K'_{rFe(CN)_6} = \log K_{rFe(CN)_6} = 6\cdot 1$.

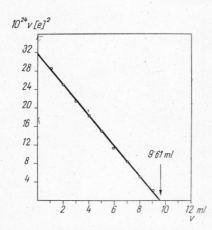

Fig. 3.13. Gran's plot for the titration of hydroxyquinone with cerium(IV) sulphate standard solution (pH $\simeq$ 0.2). (See Example 63)

Therefore $2 \times 6.1 - \log K'_{rA} > 9$, or $\log K'_{rA} < 3.2$. If pH = 6, then

$$\alpha_{A(H)} = 1 + 10^{-6} \times 10^{11.57}$$
$$+ 10^{-12} \times 10^{15.73} = 10^{5.58}$$

$$\log K'_{rA} = -2.5 + 5.6 = 3.1.$$

Thus, if the pH $\geq$ 6, the titration can be performed with adequate accuracy (see [142]).

63. Hydroquinone has been titrated in $1N$ sulphuric acid medium (pH $\simeq$ 0.2) with a standard cerium (IV) sulphate solution. The titrant has been added in 1 ml portions and the redox potentials (vs. the standard hydrogen electrode) noted (see Table 3.1). Using Gran's method, calculate the volume of standard solution added at the equivalence point.

The calculation is done by using Eqs. (3.120) and (3.121). The calculated values of $V[e]^z$ are plotted against V. For details of the calculation see Table 3.1; Gran's plot constructed from the calculated data is shown in Fig. 3.13.

Table 3.1
Calculation of data necessary for the linearization of the titration curve, for the titration of hydroquinone with standard cerium(IV) sulphate solution.
pH $\simeq$ 0.2

V, ml	E V	$[e] = 10^{-\frac{E}{0.059}}$	$V[e]^z$
1.0	0.665	$10^{-11.27}$	28.8×10^{-24}
2.0	0.676	$10^{-11.45}$	25.2×10^{-24}
3.0	0.683	$10^{-11.57}$	21.7×10^{-24}
4.0	0.689	$10^{-11.67}$	18.3×10^{-24}
5.0	0.694	$10^{-11.76}$	15.1×10^{-24}
6.0	0.700	$10^{-11.86}$	11.4×10^{-24}
7.0	0.706	$10^{-11.96}$	8.5×10^{-24}
8.0	0.713	$10^{-12.08}$	5.4×10^{-24}
9.0	0.725	$10^{-12.28}$	2.4×10^{-24}

The volume of standard solution added at the equivalence point, V_{eq}, is 9·61 ml.

The reliability of the result can be increased by using the method of least squares to calculate the slope and intercept of the line best fitting the results obtained (see p. 44).

From the intercept, the conditional $pe^{0'}$ and conditional standard redox potential of hydroquinone can be obtained

$$31·72 \times 10^{-24} = 9·61 \, [e]_0^2$$

From this

$$[e]_0 = 1·82 \times 10^{-12}$$

$$pe^{0'} = -\log [e]_0 = 11·74 \quad (pH = 0·2)$$

The conditional standard redox potential is

$$E^{0'} = 0·059 \times 11·74 = 0·69 \text{ V} \quad (pH = 0·2)$$

By use of the protonation constants of hydroquinone, the value of the α-function and the thermodynamic standard potential can be calculated.

3.6. Polarography

In polarographic analysis, complex formation reactions can be used in the following ways.

(i) Half-wave potentials can be shifted by using complexing agents. By the appropriate choice of complexant, pH etc. the polarographic waves of metal ions having half-wave potentials very close to each other can be separated and thus the selectivity of determinations can be increased.

(ii) Polarographically inactive metal ions or ions which are difficult to determine can be determined indirectly, through complex displacement reaction.

(iii) Metal ions difficult to determine by direct polarography, may be determined indirectly by use of complexing agents which have polarographic waves which are dependent on the degree of complex formation.

There are concise papers in the literature on the application of complexing agents in polarography (e.g. [143]). For details of the methods of polarographic analysis see [89].

The equation giving the shift of half-wave potential on the addition of a complexing agent is [15]:

$$\Delta E_{1/2} = -\frac{0\cdot 059}{z}\log \alpha_{M(L)} + \frac{0\cdot 059}{z}\log \frac{i_{dM}}{i_{dML}} \qquad (3.122)$$

(see p. 149). In approximate calculations the second term on the right-hand side can be neglected. The equation may be used to calculate the shift in the half-wave potential and its new value in the presence of a complexing agent provided that the concentration, pH and the original half-wave potential in the absence of the complexant are known (see Example 64).

If an appropriate selective complexing agent can be chosen which forms highly stable complexes with metal ions interfering with the determination, but does not complex the ion to be determined, then the interferences can be masked to such an extent that their polarographic waves do not appear at all in the potential range studied. For example in the determination of uranium(VI), the manganese(II), nickel(II), cobalt(II), zinc(II) etc. can be masked with EDTA or DCTA [144].

The basic equation of the displacement polarographic technique is

$$M + NL \rightleftharpoons ML + N \qquad (3.123)$$

A complex of a polarographically active metal ion NL is added in excess to the solution of the ion to be determined and the amount of N (equivalent to M) liberated in the displacement reaction is determined by polarography. In this way the actual amount of M is obtained indirectly. The equilibrium constant of the reaction described by Eq. (3.123) is the ratio of the conditional stability constants of the two complexes.

$$K = \frac{K'_{ML}}{K'_{NL}} \qquad (3.124)$$

$$\log K = \log K'_{ML} - \log K'_{NL}$$

For the displacement reaction to proceed quantitatively (i.e. the concentration of liberated N is 99·9% of that of M) it is necessary that the ratio of the two conditional stability constants is at least 10^3. At the same time the value of K'_{NL} should be at least 10^6. The α-functions can be used to select the optimum conditions of displacement determinations, and to calculate the conditional constants (see Example 65). Another

Sec. 3.6]   **Polarography**

condition for the analytical application of complex displacement reactions is rapid attainment of equilibrium.

The polarographic behaviour of a group of complex-forming azo-dyes, which have hydroxo groups on carbon atoms next to the azo group (e.g. Solochrome Violet R, Eriochrome Black T, etc.), changes on complex formation [145]. As a consequence of complex formation their polarographic wave splits and a new wave appears at a more negative potential. The height of the new wave is, under certain conditions, proportional to the concentration of the metal ion, i.e. metal complex formed. According to the measurements by the author, magnesium can be determined in a certain concentration range by means of Calmagite.

3.6.1. Worked examples

64. Thallium(I) is reduced at -0.50 V, lead(II) at -0.46 V on a dropping mercury electrode in weakly acidic solution. Determine whether the two ions can be determined simultaneously by polarography after addition of EDTA to the solution and adjustment of the pH to 5. Thallium(I) is not complexed by EDTA. The stability constant of the EDTA–lead(II) complex is log $K = 18$.

The calculation is done by using Eq. (3.122). The concentration of EDTA used is $5 \times 10^{-2} M$. The value of log $\alpha_{\text{EDTA(H)}}$ is 6.6 at pH 5 according to Fig. 1.14.

$$\log K' = 18 - 6.6 = 11.4$$

$$\alpha_{\text{Pb(EDTA)}} = 1 + 10^{-1.3} \times 10^{11.4} = 10^{10.1}$$

The shift of the half-wave potential of lead is

$$\Delta E_{1/2} = -\frac{0.059}{2} \times 10.1 = -0.30 \text{ V.}$$

Hence, in the presence of EDTA at pH 5 thallium(I) and lead(II) can be determined by polarography when present together [146]. The thallium(I) wave appears at -0.50 V, and the lead(II) wave at -0.76 V.

65. Consider whether the polarographically inactive rare earths, such as samarium, can be determined after ion-exchange separation in a solution of $1M$ ammonium lactate by a complex displacement method, if the EDTA–copper(II) complex is used as reagent. If it is possible, then what should the pH be? Log $K_{\text{CuEDTA}} = 18.8$; log $K_{\text{SmEDTA}} = 17.14$.

The overall stability constants of lactate–copper(II) complexes are $\log \beta_1 = 3\cdot 02$ and $\log \beta_2 = 4\cdot 84$. Those of the lactate–samarium(III) complexes are $\log \beta_1 = 2\cdot 56$, $\log \beta_2 = 4\cdot 58$ and $\log \beta_3 = 5\cdot 9$.

For the reaction

$$\text{CuEDTA} + \text{Sm}^{3+} \rightleftarrows \text{SmEDTA} + \text{Cu}^{2+}$$

to proceed quantitatively it is necessary that the conditional stability constant of the EDTA–copper(II) complex be reduced to about 10^7. That is, the sum $\log \alpha_{Cu} + \log \alpha_{EDTA}$ should be $11\cdot 8$. The conditional constant is to be reduced mainly by increasing α_{Cu}, since if α_{EDTA} increases, the conditional stability constant of rare earths will also decrease; α_{Cu} can be increased by increasing the concentration of ammonia. According to Fig. 1.11, when $[NH_3] = 0\cdot 5$, then $\log \alpha_{Cu(NH_3)} \simeq 11$. Since in $1M$ ammonium lactate $[NH_4^+] \simeq 1$, and the protonation constant of ammonia is $\log K = 9\cdot 3$, the pH of the solution will be 9:

$$\frac{[NH_4^+]}{[NH_3][H^+]} = \frac{1}{0\cdot 5[H^+]} = 10^{9\cdot 3}$$

$$[H^+] = 10^{-9}$$

At pH 9, $\alpha_{EDTA(H)} = 1\cdot 4$ (from Fig. 1.14).

Lactate complexes copper(II) to a much lower extent even at $1M$ concentration than does ammonia, therefore the effect of the former can be ignored ($\alpha_{Cu(NH_3)} \gg \alpha_{Cu(L)}$).

$$\log K'_{CuEDTA} = 18\cdot 8 - 11 - 1\cdot 4 = 6\cdot 4$$

As samarium is complexed by lactate but not by ammonia, only the former is to be considered when calculating the conditional stability constant.

$$\alpha_{Sm(L)} = 1 + 1 \times 10^{2\cdot 56} + 1 \times 10^{4\cdot 58} + 1 \times 10^{5\cdot 9} = 10^{5\cdot 92}$$

$$\log K'_{SmEDTA} = 17\cdot 14 - 5\cdot 92 - 1\cdot 4 = 9\cdot 82$$

The equilibrium constant of the displacement reaction is given by

$$\log K = 9\cdot 82 - 6\cdot 4 = 3\cdot 42$$

and hence the determination can be performed by polarography [92].

66. It is required to determine magnesium by indirect polarography, using Calmagite (1-hydroxy-4-methyl-2-phenylazo-2-naphthol sulphonic acid) as reagent. What pH should be used and over what concentration range can the method be used for determining magnesium? Owing to its limited solubility, the concentration of Calmagite may not exceed $10^{-3}M$. Magnesium forms a 1 : 1 complex with Calmagite and $\log K_{MgCal} = 8\cdot1$. The protonation constants of Calmagite are $\log K_1 = 12\cdot4$ and $\log K_2 = 8\cdot1$. The formation constant of magnesium hydroxide precipitate is $\log K_s = 10\cdot7$. The stability constant of the magnesium hydroxo-complex is $\log \beta_1 = 2\cdot6$.

If 99·9% of the magnesium present is complexed, then

$$\frac{[MgCal]}{[Mg']} = K'_{MgCal}[Cal'] \geq 10^3$$

To ensure a high value of the conditional stability constant a high pH is required. The upper limit of pH is, however, set by the formation of a magnesium hydroxide precipitate. At pH 12

$$\alpha_{Cal(H)} = 1 + 10^{-12} \times 10^{12\cdot4} + 10^{-24} \times 10^{20\cdot5} = 10^{0\cdot55}$$

$$\alpha_{Mg(OH)} = 1 + 10^{-2} \times 10^{2\cdot6} = 10^{0\cdot7}$$

$$\log K'_{MgCal} = 8\cdot1 - 0\cdot7 - 0\cdot55 = 6\cdot85$$

The required condition is that

$$10^{6\cdot85}[Cal'] \geq 10^3$$

whence

$$[Cal'] \geq 1\cdot4 \times 10^{-4}M$$

If the total concentration of the reagent is $C_{Cal} = 10^{-3}M$ and pH = 12, then the maximum concentration of Calmagite available to complex magnesium is $10^{-3} - 1\cdot4 \times 10^{-4} = 8\cdot6 \times 10^{-4}M$; that is, the maximum concentration of a magnesium that can be determined with adequate accuracy is $8\cdot6 \times 10^{-4}M$.

Consider now the possible formation of a magnesium hydroxide precipitate. At pH 12 the maximum concentration of free magnesium ions, according to Eq. (3.4c), if a precipitate is *not* to form, is

$$[Mg^{2+}] = \frac{1}{10^{10\cdot7} \times (10^{-2})^2} = 10^{-6\cdot7}M$$

If $C_{Mg} = 8\cdot 6 \times 10^{-4}$ and $[Cal'] = 1\cdot 4 \times 10^{-6} M$, the concentration of free magnesium ions can be obtained from the equation

$$\frac{C_{Mg}}{[Mg^{2+}]} = \alpha_{Mg} = 1 + \left[\frac{[Cal']}{\alpha_{Cal(H)}}\right] K_{MgCal} + [OH]\beta_1$$

$$= 1 + 10^{-3\cdot 85} \times 10^{-0\cdot 55} \times 10^{8\cdot 1} + 10^{-2} \times 10^{2\cdot 6} = 10^{3\cdot 7}$$

$$[Mg^{2+}] = 8\cdot 6 \times 10^{-4} \times 10^{-3\cdot 7} = 10^{-6\cdot 76} M$$

Thus, the concentration of free magnesium ions is slightly lower than the allowed value and therefore no precipitation occurs.

Experimental results prove the correctness of the calculation.

3.7. Spectrophotometry

In spectrophotometric determinations complexing agents are used mainly as spectrophotometric reagents. By measurement of the absorbance of the coloured complex the amount of one of the complex-forming constituents (usually the metal ion) is determined.

Furthermore, complex formation reactions can be used for increasing the selectivity of spectrophotometric determinations if the interference can be masked by a suitable complexing agent.

Finally, some ions or compounds can be determined by indirect methods through complex displacement reactions.

When the complexing agent is used as a spectrophotometric reagent, the accuracy of the determination is generally dependent on the quantitativeness of the complex formation reaction. (It is to be noted, however, that under certain conditions, a non-quantitative reaction can also be used for quantitative determination.) According to Ringbom [15] conditional stability constants and overall stability constants can be used to advantage for calculating the optimum concentrations and pH in spectrophotometric determinations. If the ion M is to be determined, and a light-absorbing complex ML_n is formed with the spectrophotometric reagent, the condition

$$\frac{[ML_n]}{[M']} = \beta'_n [L]^n \geq 10^2 \tag{3.125}$$

should be fulfilled for 99% of the metal ion to be complexed, i.e. for the complex formation reaction to be practically quantitative. This condition enables the optimum reagent excess and pH to be calculated.

If the spectrophotometric method is based on the measurement of the absorbance of the species ML_n only, but depending on the pH, other complex species with different composition, and possibly unwanted mixed-ligand complexes may also be formed, then the latter can be taken into account when calculating the overall conditional stability constant by means of the α-functions, provided that the corresponding stability constants are known (see pp. 61 and 140).

If another component similar in behaviour to the one to be determined is also present in the solution, and also forms a complex with the spectrophotometric reagent, the interference can be eliminated by using a selective masking agent. In the case of the determination of metal ions the masking can be considered as complete, if the complexation of the interfering ion by the spectrophotometric reagent is of such a small extent that it does not practically affect the determination. In calculating the suitable concentration of the masking agent it should be considered that the maximum allowable value of the overall conditional stability constant of the interfering complex depends on its molar absorptivity, and also that if the masking agent reacts with the ion to be determined, this modifies the conditional stability constant of its light-absorbing complex. Obviously, the molar absorptivity of the interfering complex should be taken at the wavelength of the determination.

Aluminium interferes in the spectrophotometric determination of beryllium as its sulphosalicylate. According to Přibil [147] the determination can be carried out even in the presence of aluminium if EDTA is added to the solution.

The use of a complex displacement reaction is advantageous in spectrophotometric analysis if the component in question cannot be determined directly. For example, several spectrophotometric methods for the determination of fluoride are based on reactions of the type

$$ML_m + nF^- \rightleftharpoons MF_n + mL$$

and the fluoride is determined by measuring the reduction in absorbance of a solution containing a light-absorbing complex ML_m, when fluoride is added [148, 149].

A complex displacement reaction can be combined with a spectrophotometric complex reaction, if the ion displaced from the complex is measured by means of a spectrophotometric reagent.

Small amounts of chloride can be determined by adding to the solution mercury(II) thiocyanate and iron(III) sulphate acidified with nitric acid [150]. The thiocyanate liberated in the reaction

$$2Cl^- + Hg(SCN)_2 \rightleftharpoons HgCl_2 + 2SCN^-$$

forms a red complex with iron(III) which can be determined spectrophotometrically. If the iron is present in excess practically only the FeSCN$^+$ complex is formed.

Rare earths can be determined on a similar basis, by adding to their solution EDTA–copper(II) complex and PAN indicator. The copper(II) displaced from the complex in stoichiometric amounts by rare earths reacts with the PAN indicator and the absorbance of the violet solution can be measured [151]. Since the metal ion liberated in the displacement reaction of this type forms a new complex, the latter complex formation can be considered as a side-reaction of the displaced metal ion, and should be taken into account by means of the α-function. In general, if 1 : 1 complexes are involved in both reactions, the reaction equations will be

$$M + NL \rightleftharpoons ML + N \qquad (3.126)$$

$$N + Y \rightleftharpoons NY \qquad (3.127)$$

For the displacement reaction to be quantitative the equilibrium constant should exceed 10^2

$$K_{(1)} = \frac{K'_{ML}}{K'_{NL}} \geq 10^2 \qquad (3.128)$$

[see Eq. (3.124)]. In view of the definition of the conditional constant

$$K_{(1)} = \frac{K_{ML}\alpha_N\alpha_L}{K_{NL}\alpha_M\alpha_L} \geq 10^2 \qquad (3.129)$$

Presumably, reaction (3.127) predominates among the side-reactions of N, i.e.

$$\alpha_N \simeq \alpha_{N(Y)} = 1 + [Y']K'_{NY}$$

$$\alpha_N \simeq [Y']\frac{K_{NY}}{\alpha_{Y(H)}} \qquad (3.130)$$

Inserting this into Eq. (3.129) gives

$$\frac{K_{ML}K_{NY}}{K_{NL}\alpha_M\alpha_{Y(H)}}[Y'] \geq 10^2 \qquad (3.131)$$

Sec. 3.7]   Spectrophotometry

An additional condition is that the displacement reaction

$$NL + Y \rightleftharpoons NY + L \qquad (3.132)$$

does not take place between the complex NL and Y. That is

$$K_{(2)} = \frac{K'_{NL}}{K'_{NY}} \geq 10^2$$

$$\frac{K_{NL}\alpha_{Y(H)}}{K_{NY}\alpha_{L(H)}} \geq 10^2 \qquad (3.133)$$

The third condition is the completeness of the colour reaction (3.127). According to Eq. (3.125)

$$K'_{NY}[Y'] \geq 10^2 \qquad (3.134)$$

A comparison of conditions (3.131) and (3.134) clearly shows that the spectrophotometric method combined with a displacement reaction can be used in the case of a pair of metal ions which form complexes of similar stability with a given chelating agent but only one of which reacts with the chromogenic reagent.

Indirect spectrophotometric determinations can be assessed by using (3.131), (3.133) and (3.134), and the optimum pH and reagent concentration can be calculated (see Example 68).

3.7.1. Worked examples

67. Calculate whether pH 2 or 5 should be used in the spectrophotometric determination of iron(II) with 2,2′-bipyridyl. The total concentration of the acetate buffer used to adjust the pH is $10^{-1}M$, the concentration of excess of reagent is $10^{-3}M$. The overall stability constants of the bipyridyl–iron(II) complexes are $\log \beta_1 = 4.4$, $\log \beta_2 = 8.0$ and $\log \beta_3 = 17.6$. The protonation constant of 2,2′-bipyridyl is $\log K_1 = 4.4$. The stability constant of the acetato–iron(II) complex is $\log \beta_1 = 1.4$, that of the the hydroxo-iron(II) complex is $\log \beta_1 = 4.5$.

The calculation is done using (3.125):

$$\frac{[\text{Fe bipy}_3]}{[\text{Fe}']} = \beta'_3[\text{bipy}']^3 \geq 10^2$$

At pH 2,
$$\log \alpha_{Ac(H)} = 2.8 \quad \text{(see Fig. 1.13)};$$

$$\alpha_{Fe(Ac)} = 1 + \left[\frac{10^{-1}}{10^{2.8}}\right] 10^{1.4} \simeq 1$$

$$\alpha_{bipy(H)} = 1 + 10^{-2} \times 10^{4.4} = 10^{2.4}$$

The conditional stability constant is:
$$\log \beta_3' = 17.6 - 0 - 3 \times 2.4 = 10.4$$

The ratio of the concentration of complexed to non-complexed iron is
$$\frac{[\text{Fe bipy}_3]}{[\text{Fe}']} = 10^{10.4} \times (10^{-3})^3 = 25.1$$

$25.1 < 10^2$, and only
$$\frac{25.1}{26.1} \times 10 \simeq 96\%$$

of iron(II) is complexed.

At pH 5,
$$\log \alpha_{Ac(H)} = 0.4 \quad \text{(see Fig. 1.13)}$$

$$\alpha_{Fe(Ac)} = 1 + 10^{-1.4} \times 10^{1.4} = 10^{0.3}$$

$$\alpha_{Fe(OH)} = 1 + 10^{-9} \times 10^{4.5} \simeq 1$$

$$\alpha_{bipy(H)} = 1 + 10^{-5} \times 10^{4.4} = 10^{0.1}$$

$$\beta_3' = 17.6 - 0.3 - 3 \times 0.1 = 17.0$$

$$\frac{[\text{Fe bipy}_3]}{[\text{Fe}']} = 10^{17} \times (10^{-3})^3 = 10^8 \gg 10^2$$

Thus virtually all the iron(II) is complexed and the spectrophotometric complex formation reaction is quantitative.

68. After the ion-exchange separation of rare earths the single ions are obtained in solutions containing ammonium lactate at pH 3–5. Consider whether the following complex displacement and spectrophotometric method can be used for determining erbium under the conditions given. EDTA–copper(II) complex of stoichiometric composition and PAN [1-(2-pyridylazo)-2-naphthol] indicator is added to the ammonium lactate solution (pH 4) containing a small amount of erbium

($<10^{-4}M$) and the absorbance of the reddish violet solution [colour of the PAN–copper(II) complex] measured. The concentration of lactate is $10^{-1}M$. Log $K_{\text{CuEDTA}} = 18\cdot8$, log $K_{\text{ErEDTA}} = 18\cdot98$ and log $K_{\text{CuPAN}} = 16\cdot0$. The overall stability constants of the lactate–erbium(III) complexes are log $\beta_1 = 2\cdot77$, log $\beta_2 = 5\cdot11$ and log $\beta_3 = 6\cdot7$; those of the lactate–copper(II) complexes are log $\beta_1 = 3\cdot02$ and log $\beta_2 = 4\cdot84$. The protonation constant of lactate is log $K = 3\cdot75$ ($I = 0\cdot1$), those of PAN are log $K_1 = 12\cdot2$ and log $K_2 = 1\cdot9$.

In the calculations, (3.131), (3.133) and (3.134) are used. At pH 4,

$$\alpha_{L(H)} = 1 + 10^{-4} \times 10^{3\cdot75} = 10^{0\cdot2}$$

$$\alpha_{Er(L)} = 1 + 10^{-1\cdot2} \times 10^{2\cdot77} + 10^{-2\cdot4} \times 10^{5\cdot11} + 10^{-3\cdot6} \times 10^{6\cdot7} = 10^{3\cdot25}$$

$$\alpha_{PAN(H)} = 1 + 10^{-4} \times 10^{12\cdot2} + 10^{-8} \times 10^{14\cdot1} = 10^{8\cdot2}$$

According to (3.131)

$$\log K_{\text{ErEDTA}} + \log K_{\text{CuPAN}} + \log [\text{PAN}'] - \log K_{\text{CuEDTA}} - \log \alpha_{Er(L)}$$
$$- \log \alpha_{PAN(H)} \geq 2$$
$$18\cdot98 + 16\cdot0 + \log [\text{PAN}'] - 18\cdot8 - 3\cdot25 - 8\cdot2 \geq 2$$

From this
$$\log [\text{PAN}'] \geq -2\cdot73$$

i.e. the concentration of PAN should be at least $2 \times 10^{-3}M$.

Next consider if the other two conditions can also be fulfilled under the conditions given. Log $\alpha_{\text{EDTA(H)}}$ for pH 4 is taken from Fig. 1.14.

According to (3.133)

$$\log K_{\text{CuEDTA}} + \log \alpha_{\text{PAN(H)}} - \log K_{\text{CuPAN}} - \log \alpha_{\text{EDTA(H)}} \geq 2$$
$$18\cdot8 + 8\cdot2 - 16 - 8\cdot6 = 2\cdot4 > 2$$

$$\alpha_{Cu(L)} = 1 + 10^{-1\cdot2} \times 10^{3\cdot02} + 10^{-2\cdot4} \times 10^{4\cdot84} = 10^{2\cdot53}$$

$$\log K'_{\text{CuPAN}} = 16 - 2\cdot53 - 8\cdot2 = 5\cdot27$$

According to (3.134)

$$\log K'_{\text{CuPAN}} + \log [\text{PAN}'] \geq 2$$
$$5\cdot27 - 2\cdot7 = 2\cdot57 > 2$$

As all three conditions are satisfied the complex displacement method combined with spectrophotometry is applicable to the determination of erbium (see [151, 152]).

3.8. Liquid-liquid Extraction

Extraction methods have greatly increased in importance in the past few years, since they permit extremely selective separations of trace amounts of ions by use of an appropriate complexing agent and solvent. Extraction techniques can be used not only for separation, but also if the complex present in the organic phase is coloured, for selective spectrophotometric determination and for end-point detection in complexometric titrations [153].

If the equilibrium distribution of metals between two liquids can be influenced by complex formation, chromatographic separations can also be performed. In such cases quantitative separation can be achieved even if the distribution coefficients of the components to be separated do not differ significantly (see later). In chromatographic separations the stationary phase may be the aqueous phase and the mobile phase an organic solvent containing a chelating agent or a liquid ion-exchanger, or, the reverse may be the case as in so-called reversed-phase chromatography [154].

Liquid ion exchange and chelate formation distribution equilibria have gained importance in connection with modern liquid ion-selective membrane electrodes [155, 156].

The basic equilibria and simple laboratory extraction separations will be dealt with here.

Most extraction methods are based on complex-formation reactions. According to the type of the basic chemical reaction these methods can be classified into three groups.

(i) Methods based on the formation of simple ion-association complexes [e.g. extraction of iron(III) from hydrochloric acid solution with ether, etc.].

(ii) Methods based on the formation of uncharged chelates [extraction of zinc(II) with carbon tetrachloride, in the form of dithizonate–zinc(II) complex etc.].

(iii) Extraction by the use of liquid ion exchangers. The chemical reactions taking place are similar to the formation of ion-association complexes [e.g. extraction of iron(III) from hydrochloric acid solution with a solution of a long-chain amine in xylene]. Liquid ion exchangers will be dealt with in Section 3.9.4 in more detail (see p. 289).

3.8.1. EXTRACTION EQUILIBRIUM FOR SIMPLE OR ION-ASSOCIATION COMPLEXES

The extraction equilibria are based on the Nernst distribution law (see p. 155). Certain inorganic salts or complex acids are readily soluble in certain organic solvents. This can be accounted for by the formation of ion-pairs, or ion-association complexes, which have increased covalent nature. Strong electrolytes are soluble in water, but electrically neutral complexes dissolve better in organic solvents having a lower dielectric constant than water. The nature of the solvent is of great importance in determining the solubility.

Some strongly covalent salts are readily soluble in non-polar organic solvents (e.g. benzene, carbon tetrachloride), but the solubility of complex compounds with a slightly ionic character may increase significantly if the solvent, owing to its polar character, may also form part of the co-ordination sphere. Water-immiscible ketones, ethers and esters which contain oxygen donor groups find wide application.

Similarly, the solubility in an organic solvent can be increased by the formation of mixed-ligand complexes, when uncharged organic bases enter into the coordination sphere [e.g. thiocyanatopyridinecopper(II) can be extracted into chloroform; long-chain amines having anion-exchange properties will extract cadmium hydroxide into benzene].

In general, the distribution coefficient of a metal ion M, when it is extracted in the presence of a complexing agent, is given by

$$D_M = \frac{(ML_n)}{C_M} = \frac{d_{ML_n}[L]^n \beta_n}{1 + [L]\beta_1 + [L]^2\beta_2 + \ldots [L]^N \beta_N} \quad (3.135)$$

where d_{ML_n} is the distribution constant of the extractable species ML_n, which depends on the nature of the solvent used [see Eq. (2.97)]. Equation (3.135) may be used to calculate the ligand concentration for optimum extraction (maximum D_M) provided that the composition of the extractable species and the overall stability constants are known. The equation can also be used when the solvent molecules take part in complex formation, since the constant characteristic of the complex formation equilibrium is involved in the distribution constant. In the case of the formation of real mixed-ligand complexes, when the ligand concentration can be changed, the distribution coefficient is given by

$$D_M = \frac{(ML_nY_m)}{C_M} = \frac{d_{ML_nY_m}[L]^n[Y]^m \beta_{n,m}}{\sum_0^{N_L} \sum_0^{N_Y} [L]^i [Y]^j \beta_{i,j}} \quad (3.136)$$

[see Eq. (2.123)].

If in an aqueous solution a ligand is present which forms non-extractable complexes with the metal ion, then this must be taken into account when calculating the distribution coefficient. In such cases the denominator of the fraction on the right-hand side of Eqs. (3.135) and (3.136) will also contain the terms

$$[X]\gamma_1 + [X]^2\gamma_2 + \ldots [X]^N\gamma_N = \alpha_{M(X)} - 1 \qquad (3.137)$$

where $[X]$ is the concentration of the ligand in question and γ_1, γ_2, etc. are the overall stability constants [see Eq. (2.99)].

By use of an appropriate complexant X the distribution coefficient can be reduced to such an extent that practically no extraction occurs, i.e. the metal ion is masked.

The metal ion in the organic phase can be stripped into the aqueous solution by decomposing the complex (e.g. by acidification, or by a stronger complexing agent forming anionic complexes).

3.8.2. Extraction equilibrium for chelates

Extraction methods based on the formation of extractable chelates are of great importance in analytical chemistry, since they give extremely selective separations. Uncharged metal chelates are usually more soluble in organic solvents than in water, thus the complexed metal ion can be transferred into the organic phase with great efficiency under suitable conditions (pH, concentration, solvent).

If the aim is the extraction of a metal ion present in small amount, the chelating agent is usually dissolved in an organic solvent and this solution is used as extractant.

When the two phases are in contact, the equilibrium processes taking place on extraction are as follows.

(i) Formation of the chelate

$$M + nL \rightleftharpoons ML_n$$

The position of the equilibrium is determined by the overall stability constants $(\beta_1, \beta_2 \ldots)$.

(ii) Distribution of the complex between the two phases. The equilibrium is dependent on the distribution constant (see pp. 156 ff.)

$$d_{ML_n} = \frac{(ML)_n}{[ML]_n} \qquad (3.138)$$

(iii) Distribution of the chelating agent between the two phases. If it is a monobasic acid, the distribution constant is

$$d_{HL} = \frac{(HL)}{[HL]} \quad (3.139)$$

(iv) Finally, the protonation of the ligand. The position of the equilibrium is determined by the protonation constant K of the ligand.

To simplify the calculations the following assumptions are made:
(i) the solubility of the chelating agent in water is neglected,
(ii) only one extractable species of the composition ML_n is formed,
(iii) the chelating agent is a monobasic acid, HL,
(iv) the metal ion does not take part in any side-reaction in the aqueous phase.

The extraction process is as follows:

$$M^{n+}_{(aq)} + nHL_{(org)} \rightleftharpoons ML_{n(org)} + nH^{+}_{(aq)} \quad (3.140)$$

where the subscript aq indicates the aqueous phase, and org the organic phase. The so-called extraction constant, K_e, is given by

$$K_e = \frac{(ML_n)[H^+]^n}{[M^{n+}](HL)^n} \quad (3.141)$$

Round brackets indicate concentrations in the organic phase, square brackets those in the aqueous phase. The distribution coefficient of the metal ion is given by

$$D_M = \frac{(ML_n)}{[M^{n+}]} = K_e \left(\frac{(HL)}{[H^+]} \right)^n$$

or

$$\log D_M = \log K_e + n \log (HL) + n\text{pH} \quad (3.142)$$

It can be deduced that

$$K_e = \frac{\beta_n d_{ML_n}}{(Kd_{HL})^n} \quad (3.143)$$

According to Eq. (3.143), the greater the value of β_n and d_{ML_n} and the smaller that of the protonation constant and d_{HL}, the greater is the extraction constant. (If the distribution constants are known for different solvents, K_e for one solvent can be calculated from that for another.)

To a first approximation the distribution constant of the extractable species is equal to the ratio of the solubilities. For phases in equilibrium, in contact with a precipitate of composition ML_n,

$$\beta_n d_{ML_n} = \frac{(ML_n)}{[M][L]^n} \simeq (\text{Sol.}) K_s$$

where (Sol.) is the solubility in the organic phase and K_s the precipitation constant.

From Eq. (3.142) it can be seen that the distribution coefficient of the metal ion depends mainly on the value of the extraction constant, and also on the concentration of the chelating agent and the pH in the aqueous solution. For a given solvent and reagent concentration, the logarithm of the distribution coefficient is proportional to the pH, in the ideal case. The slope of log D vs. pH plots (see Fig. 3.14) is n, and the intercept on the abscissa is at

$$(\text{pH})_{1/2} = -\frac{1}{n} \log K_e - \log (\text{HL}) \tag{3.144}$$

The pH value where log $D = 0$ or $D_M = 1$ at a given reagent concentration has been denoted by $(\text{pH})_{1/2}$ by Irving and Williams [157]. If pH > $(\text{pH})_{1/2}$, then most of the metal is in the organic phase, if pH < $(\text{pH})_{1/2}$, then most is in the aqueous phase at equilibrium.

In the ideal case Eq. (3.142) is suitable for the calculation of the distribution coefficient of the metal ion provided that the pH and the reagent concentration are known (see the extraction constants in Section 4.4).

Considering that the values of the protonation constant and the distribution constant are usually high, the concentration (HL) is practically equal to the total concentration of the reagent. If, however, pH $\geq$ log d_{HL} + log K, then (HL) is different from the total concentration (since the ionic L^- is more soluble in water than in the solvent). According to Ringbom [15], the relationship can be calculated as given below.

The total concentration of the reagent is C_{HL}. Part of this is in the organic phase (HL), the rest in the aqueous phase ([HL] + [L]).

$$\frac{C_{HL}}{(HL)} = \frac{(HL) + [HL] + [L]}{(HL)} = \frac{d_{HL}[HL] + [HL] + [L]}{d_{HL}[HL]}$$

$$= \frac{d_{HL}[H^+]K + [H^+]K + 1}{d_{HL}[H^+]K} = 1 + \frac{1}{d_{HL}} + \frac{1}{d_{HL}[H^+]K}$$

Sec. 3.8] Liquid–liquid extraction

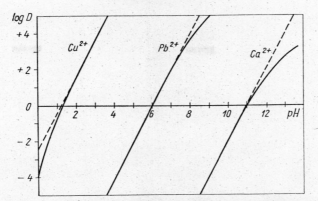

Fig. 3.14. Logarithm of the distribution coefficients of copper(II), lead(II) and calcium(II) as a function of pH in the extraction with a $10^{-2}M$ solution of 8-hydroxyquinoline in chloroform

The second term can be neglected compared with the first. If the volumes (V_{aq} and V_{org}) of the aqueous and organic phases are not the same, they must also be taken into account:

$$\frac{C_{HL}}{(HL)} = \alpha_{HL(aq)} \simeq 1 + \frac{1}{[H^+]Kd_{HL}} \cdot \frac{V_{aq}}{V_{org}} \quad (3.145)$$

(HL) can be obtained by dividing the total concentration by the value of α calculated from Eq. (3.145).

For example 8-hydroxyquinoline can take up or release a proton (see p. 38). The non-dissociated molecule HL is more soluble in organic solvents whereas the ions L^- and H_2L^+ are more soluble in water. The function $\alpha_{HL(aq)}$ deduced for 8-hydroxyquinoline is

$$\alpha_{HL(aq)} = 1 + \left([H^+]K_2 + \frac{1}{[H^+]K_1}\right)\frac{1}{d_{HL}}\frac{V_{aq}}{V_{org}} \quad (3.146)$$

If the metal ion is involved in a side-reaction in the aqueous phase, then this fact can be taken into account in the usual way by means of the $\alpha_{M(X)}$ function. Reactions which must be considered as side-reactions are not only those in which the metal ion forms a non-extractable complex with a foreign ligand but also those reactions with the ligand L which lead to non-extractable species (e.g. ML_{n-1}, ML_{n+1}). The conditional extraction constant is, with the usual notation,

$$K'_e = \frac{K_e}{\alpha_{M(X)}} \tag{3.147}$$

By choice of a suitable masking agent X, the conditional extraction constant can be reduced to such an extent that the distribution coefficient is so low that practically no extraction takes place (see Example 71).

The modified form of Eq. (3.142) is

$$\log D_M = \log K_e - \log \alpha_M + n\,[\log C_{HL} - \log \alpha_{HL(aq)}] + n\text{pH}$$

$$= \log K'_e + n\,[\log C_{HL} - \log \alpha_{HL(aq)}] + n\text{pH} \tag{3.148}$$

The distribution coefficients have been calculated by using Eq. (3.142), for the extraction of copper(II), lead(II) and calcium(II) with chloroform and $\log D$ plotted *vs.* pH to give the broken lines shown in Fig. 3.14. When the solubility of the reagent in acid and alkaline aqueous medium is also taken into account according to Eq. (3.146), and the formation of hydroxo-complexes of lead(II) and calcium(II) by using Eq. (3.147), the full lines in Fig. 3.14 are obtained. It is easily seen from the figure that copper(II) and lead(II) can be separated from each other from solutions at pH 3–4, and lead(II) and calcium(II) at pH 8–9.

3.8.3. Selectivity and degree of extraction

The degree of extraction of a metal ion with an organic solvent depends not only on the distribution coefficient, that is, on the position of the equilibrium, but also on the ratio of the volumes of the phases and on the number of successive extractions.

For a single extraction operation the fraction of the metal ion extracted can be calculated from the distribution coefficient.

Let m_0 be the total amount of metal ion, m_{org} the amount extracted into the organic phase of volume V_{org}, and $m_0 - m_{org} = m_{aq}$ the amount of metal left in the aqueous phase of volume V_{aq}. By definition, the distribution coefficient is given by

$$D = \frac{m_{org}/V_{org}}{(m_0 - m_{org})/V_{aq}} = \frac{m_{org}}{m_0 - m_{org}} \cdot \frac{V_{aq}}{V_{org}} \tag{3.149}$$

Sec. 3.8] Liquid–liquid extraction

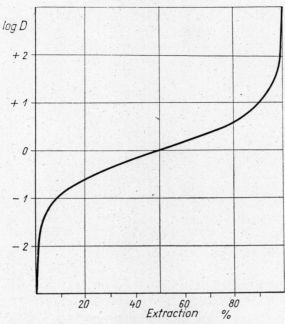

Fig. 3.15. Plot of the logarithm of the distribution coefficient *vs.* the degree of extraction for a 1 : 1 phase ratio

From this the degree of extraction

$$E = \frac{m_{\text{org}}}{m_0} \times 100 = \frac{100\,D}{D + \dfrac{V_{\text{aq}}}{V_{\text{org}}}}\% \qquad (3.150)$$

The relationship between the degree of extraction and logarithm of the distribution coefficient for a 1 : 1 phase ratio is given in Fig. 3.15.

From the analytical point of view the extraction can be considered as quantitative if 99·9% of the metal ion is transferred into the organic phase. If $V_{\text{aq}} = V_{\text{org}}$ and only one extraction is performed then for the extraction to be quantitative it is necessary that $D \geq 10^3$. For no extraction to occur $D \leq 10^{-2}$.

To separate two metal ions quantitatively it is necessary that the ratio of their distribution coefficients is at least 10^5.

266   Analytical applications   [Ch. 3

In estimating the error, the fraction remaining in the aqueous phase should be given. This can be deduced from Eq. (3.149); for one extraction operation the error is given by

$$\Delta_1 = \frac{m_0 - m_{\text{org}}}{m_0} \times 100 = \frac{100}{1 + \frac{V_{\text{org}}}{V_{\text{aq}}} D} \% \qquad (3.151)$$

For n successive extractions, if the phase ratio is always the same, the fraction remaining in the aqueous phase is

$$\Delta_n = 100 \left[\frac{1}{1 + \frac{V_{\text{org}}}{V_{\text{aq}}} D} \right]^n \% \qquad (3.152)$$

Hence it is more efficient to use a given volume of solvent in several small portions instead of all at once.

3.8.4. Worked examples

69. Calculate the thiocyanate concentration necessary to obtain the best extraction of iron(III) into ether from a solution at pH 1. Presumably iron(III) is present in ether as the complex acid $HFe(SCN)_4$. The overall stability constants of the thiocyanato iron(III) complexes are $\log \beta_1 = 2.3$, $\log \beta_2 = 4.2$, $\log \beta_3 = 5.6$, $\log \beta_4 = 6.4$ and $\log \beta_5 = 6.4$. The protonation constant of both the tetrathiocyanato iron(III) complex ion and thiocyanate is $\log K < 0$.

The distribution coefficient of iron(III), from Eq. (3.135), is

$$D_{\text{Fe}} = d\Phi_4 [H^+] K$$

where d is the distribution constant of $HFe(SCN)_4$, K the protonation constant of $Fe(SCN)_4^-$; and Φ_4 the mole fraction of the tetrathiocyanato-iron(III) complex.

As d, K and $[H^+]$ are constants, the optimum extraction of iron(III) can be expected at a thiocyanate concentration where Φ_4 is a maximum.
If $[SCN^-] = 0.1 M$,

$$\Phi_4 = \frac{10^{-4} \times 10^{6.4}}{1 + 10^{-1} \times 10^{2.3} + 10^{-2} \times 10^{4.2} + 10^{-3} \times 10^{5.6} + 10^{-4} \times 10^{6.4} + 10^{-5} \times 10^{6.4}}$$

$$= \frac{10^{2.4}}{1 + 10^{1.3} + 10^{2.2} + 10^{2.6} + 10^{2.4} + 10^{1.4}} = 0.294$$

(Of the terms in the denominator the fourth is the greatest, i.e. the concentration of the species with ligand number 3 is the greatest.)
If $[SCN^-] = 0.5M$,

$$\Phi_4 = \frac{10^{5.2}}{1 + 10^2 + 10^{3.6} + 10^{4.7} + 10^{5.2} + 10^{4.9}} = 0.54$$

As shown by the terms in the denominator, the relative concentration of the complex with ligand number 4 is greatest.
If $[SCN] = 1M$

$$\Phi_4 = \frac{10^{6.4}}{1 + 10^{2.3} + 10^{4.2} + 10^{5.6} + 10^{6.4} + 10^{6.4}} = 0.46$$

(The species with ligand numbers 4 and 5 are present in the same amount.)
According to the calculations Φ_4, and consequently the distribution coefficient of iron(III), is greatest at a thiocyanate concentration of about $0.5M$. The calculated concentration agrees well with the experimentally found optimum concentration [158].

70. Determine the optimum pH in the extraction of a small amount of aluminium with an 8-hydroxyquinoline (oxine) solution in chloroform. The ratio of the volume of the organic phase to that of the aqueous phase is 1 : 5, the concentration of oxine is $0.05M$, and its distribution constant is $\log d_{OxH} = 2.6$. The composition of aluminium oxinate is $Al(Ox)_3$, and its extraction constant $\log K_e = -5.22$. The protonation constants of oxinate ion are $\log K_1 = 9.9$ and $\log K_2 = 5.0$.

To determine the optimum pH, the error for a single extraction step is calculated for pH 4, 5 and 10. The calculation is done by using Eqs. (3.142), (3.146), (3.148) and (3.151).

At pH 4

$$\alpha_{OxH(aq)} = 1 + [10^{-4} \times 10^5 + 10^4 \times 10^{-9.9}]10^{-2.6} \times 5 = 1.126$$

$$\log \alpha_{OxH(aq)} = 0.05$$

$$\log \alpha_{Al(OH)} = 0.1 \text{ (from Fig. 1.10)}$$

$$\log D'_{Al} = \log K_e - \log \alpha_{Al} + 3 \log (OxH) + 3 \text{ pH}$$

$$\log D'_{Al} = -5.22 - 0.1 + 3[-1.3 - 0.05] + 3 \times 4 = 2.63$$

The distribution coefficient $D'_{Al} = 427$. The error is

$$\Delta = \frac{100}{1 + \dfrac{1}{5} \times 427} = 1\cdot 16\%$$

At pH 5

$$\log \alpha_{OxH(aq)} = 0\cdot 00$$

$$\log \alpha_{Al(OH)} = 0\cdot 4$$

$$\log D'_{Al} = -5\cdot 22 - 0\cdot 4 + 3(-1\cdot 3) + 3 \times 5 = 5\cdot 48$$

$$D'_{Al} = 10^{5\cdot 48}$$

$$\Delta = \frac{100}{1 + \dfrac{1}{5} \times 10^{5\cdot 48}} = 0\cdot 0016\%$$

At pH 10

$$\alpha_{OxH(aq)} = 1 + [10^{-10} \times 10^5 + 10^{-9\cdot 9} \times 10^{10}] \times 10^{-2\cdot 6} \times 5 = 1\cdot 016$$

$$\log \alpha_{OxH(aq)} \simeq 0\cdot 0$$

$$\log \alpha_{Al(OH)} = 17\cdot 7$$

$$\log D'_{Al} = -5\cdot 22 - 17\cdot 7 + 3 \times (-1\cdot 3) + 3 \times 10 = 3\cdot 18$$

$$D'_{Al} = 10^{3\cdot 18}$$

$$\Delta = \frac{100}{1 + \dfrac{1}{5} \times 10^{3\cdot 18}} = 0\cdot 33\%$$

The calculations agree well with experimental results [159]: the error is lowest when the extraction is carried out at pH 5.

71. Is it possible to separate a small amount of copper(II) from mercury(II) by extraction with dithizone if bromide is used for masking mercury(II)? The extraction is performed with $2\cdot 5 \times 10^{-5} M$ dithizone in carbon tetrachloride. The composition of the complex extracted is $M(HD)_2$. The extraction constant of copper(II) is $\log K_{eCu} = 10\cdot 5$, that of mercury(II) is $\log K_{eHg} = 26\cdot 8$. The protonation constant of the ligand HD^- is $\log K = 4\cdot 5$. The distribution constant of dithizone is

$\log d_{H_2D} = 4$. The overall stability constants of the bromocopper(II) complexes are $\log \beta_1 = -0.55$ and $\log \beta_2 = -1.84$; those of the bromomercury(II) complexes are $\log \beta_1 = 9.05$, $\log \beta_2 = 17.3$, $\log \beta_3 = 19.7$ and $\log \beta_4 = 21.0$.

The calculation is done by using Eqs. (3.147) and (3.148). For the separation of the two metal ions to be quantitative it is required that the difference $\log D'_{Cu} - \log D'_{Hg}$ is about 5.

From Eq. (3.148), and considering that n is 2 for both metal ions,

$$\log D'_{Cu} - \log D'_{Hg} = \log K'_{eCu} - \log K'_{eHg}$$

If the pH of the solution does not exceed 2, then the formation of hydroxo-complexes can be neglected. If $[Br^-] = 1M$, the α-functions are

$$\alpha_{Cu(Br)} = 1 + 1 \times 10^{-0.55} + 1 \times 10^{-1.85} = 10^{0.1}$$

$$\alpha_{Hg(Br)} = 1 + 1 \times 10^{9.05} + 1 \times 10^{17.3} + 1 \times 10^{19.7} + 1 \times 10^{21} = 10^{21}$$

$$\log K'_{eCu} - \log K'_{eHg} = \log K_{eCu} - \log \alpha_{Cu} - \log K_{eHg} + \log \alpha_{Hg}$$
$$= 10.5 - 0.1 - 26.8 + 21 = 4.6$$

Although the difference between the logarithms of the conditional extraction constants or of the distribution coefficients is somewhat smaller than 5, the separation can be considered as practically quantitative, since trace amounts of substances are involved.

For mercury(II) to remain in the aqueous phase after extraction it is necessary that $D_{Hg} \leq 10^{-2}$. According to Eq. (3.148)

$$\log D'_{Hg} = \log K'_e + 2[\log C'_{H_2D} - \log \alpha_{H_2D(aq)}] + 2\,\text{pH} - 2$$

$$= 5.8 - 2 \times 4.6 + 2\,\text{pH}$$

$$\text{pH} = 0.7$$

The distribution coefficient of copper(II) at pH 0.7 is

$$\log D'_{Cu} = 10.4 - 2 \times 4.6 + 2 \times 0.7 = 2.6$$

i.e.

$$D'_{Cu} = 10^{2.6}$$

If the pH of the solution is 0·7 and the volume ratio of the organic to aqueous phases is 1 : 5, a single extraction is done, and the total concentration of trace metals does not exceed $10^{-6}M$, at equilibrium the degrees of extractions [Eq. (3.150)] are

$$E_{Cu} = \frac{100 \times 10^{2 \cdot 6}}{10^{2 \cdot 6} + 5} \simeq 99\%$$

$$E_{Hg} = \frac{100 \times 10^{-2}}{10^{-2} + 5} = 0 \cdot 2\%$$

3.9. Ion-exchange Separations

The efficiency of ion-exchange separations can be greatly enhanced by the use of complexing agents.

For details of ion-exchange materials (ion-exchange resins, liquid ion exchangers, etc.) and the chemistry and analytical applications of ion exchangers see [53] and [160].

The equilibrium calculations which will be given here refer mainly to ion-exchange processes taking place on the most widely used ion-exchange resins, which can be considered as strong electrolytes, namely sulphonic acid-type, strongly acidic resins (e.g. Dowex 50) and strongly basic ones containing quaternary amino groups (e.g. Dowex 1). Separate sections will be devoted to separation with liquid ion exchangers and to ion-exchange paper chromatography.

3.9.1. Cation-exchange equilibria

If a cation-exchange resin containing ion A^+ is in contact with a solution containing metal ion M^{z+} the following ion-exchange equilibrium is set up between the two phases:

$$M^{z+} + zR-A \rightleftharpoons R_z-M + zA^+ \qquad (3.153)$$

where R_z represents the equivalent amount of the cation-exchange resin. At equilibrium the law of mass-action states, in terms of concentrations, that

$$K_{xMA} = \frac{(M)[A]^z}{[M](A)^z} \qquad (3.154)$$

where round brackets indicate concentrations in the ion exchanger, square brackets those in the solution (expressed in mole/litre) and K_x is the volume ion-exchange constant.

If the metal ions in the solution are not involved in a side reaction (i.e. there is no complex formation), the distribution coefficient of the metal ion M can be obtained from Eq. (3.154) as

$$D_M = \frac{(M)}{[M]} = K_x \left[\frac{(A)}{[A]}\right]^z \tag{3.155}$$

In chromatographic separations the amount of the eluted metal ion is small compared with that of the eluting ion, that is $[M^{z+}] \ll [A^+]$ and $(M^{z+}) < (A^+)$. In these cases, to a good approximation K_x can be considered to be a constant, and (A^+), which is practically equal to the volume capacity of the resin, Q, is also a constant. The following expression is suitable for calculating the approximate value of the distribution coefficient from the equilibrium constant, volume capacity of the resin and concentration of the eluting ion:

$$D_M \simeq K_x Q^z [A^+]^{-z}$$

$$\log D_M \simeq \log K_x + z \log Q - z \log [A^+] \tag{3.156}$$

It is seen that the distribution coefficient of the metal ion decreases with increasing concentration of the eluting ion. Plotting $\log D_M$ vs. $\log [A^+]$ gives a straight line with slope z (strictly, the ratio of the charge of M to that of the eluting ion A). From the intercept, K_x can be calculated, provided that Q is known.

In Fig. 3.16 the logarithms of the distribution coefficients found experimentally for some metal ions are plotted vs. the logarithm of the concentration of the eluent [161]. The slopes of the lines are -1, -2 and -3, respectively, depending on the ratio of the charges.

The value of the distribution coefficient is of great practical importance, since in chromatographic separations it determines the feasibility of the elution and the rate of movement of an ion on the ion-exchange column (see later). The ion-exchange equilibrium constants given in the literature are usually referred to the same ion, namely hydrogen ion, and the concentration in the ion-exchanger phase is given in units of mmole/g of dry resin. To get 'volume distribution coefficients' which are best for calculations concerning chromatographic separations, 'volume equilibrium constants' should be used in which the concentration is given

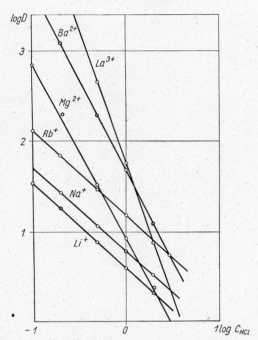

Fig. 3.16. Logarithm of the distribution coefficients of some metal ions on a cation-exchange resin as a function of the logarithm of the eluent concentration. By permission from Strelow [161]

in the resin phase in units of mole/l of resin column. The general form of the relationship between the weight constant K_w and the volume ion-exchange constant K_x is

$$K_{xMA} = \frac{K_{wMA}}{\sigma^{(z-p)}} \qquad (3.157)$$

where σ is the dry weight (g) of resin in 1 ml column volume, and z and p are the charge numbers of the ions M and A, respectively (σ is usually 0·3–0·4 g/ml). Care should be taken in using Eq. (3.157) as the value of σ depends on the nature of the ions and also on the concentration of the solution.

Ion-exchange equilibrium constants are usually dependent on the ratio of the amounts of the ions taking part in the ion-exchange equilibrium.

Sec. 3.9]   Ion-exchange separations

The change in the value of K_x is especially great in the case of ions having different charges. On the basis of thermodynamic considerations [162, 163] and chromatographic practice it is expedient to consider K_x as the value which is valid if a trace of the ion in question is in equilibrium with a large amount of another ion (H^+, Na^+ or Cl^-) which is taken as the reference.

A collection of ion-exchange equilibrium constants taken from the literature, which can be used in exploratory calculations, is given at the end of this book (Section 4.5).

The α-functions have been used by Ringbom [15] for describing ion-exchange equilibria. In the presence of a complexing ion L^{a-}, if the ion M^{z+} forms only uncharged or negatively charged complexes which are not bound by a cation-exchange column, Eq. (3.156) is modified to

$$D_M = \frac{(M)}{[M']} \simeq \frac{K_x}{\alpha_{M(L)}} Q^z [A]^{-z} \quad (3.158)$$

$$\log D_M \simeq \log K_x - \log \alpha_{M(L)} + z \log Q - z \log [A]$$

$$= \log K'_x + z \log Q - z \log [A]$$

The distribution coefficient is smaller than in the absence of the complexing agent, as $[M'] > [M]$ and K'_x, the conditional ion-exchange equilibrium constant, is less than K_x. If the metal ion M forms only one complex of composition ML_n, and $[L]^n \beta_n > 1$, then Eq. (3.158) can be written as

$$\log D_M \simeq \log K_x - \log \beta_n - n \log [L] + z \log Q - z \log [A] \quad (3.159)$$

3.9.1.1. *Chromatographic separation of metal ions on cation-exchange columns*

A small amount of the metal ions to be separated is bound at the top of the column and then eluted with a solution of an electrolyte of suitable composition. During elution the ions travel through the column at different rates depending on their distribution coefficients, and appear successively in the eluate.

The rate of migration of the eluted ion in the column can be described by the basic chromatographic equation

$$\left(\frac{dx}{dv}\right)_M = \frac{[M]}{(M)} = \frac{1}{D_M} \quad (3.160)$$

The rate of migration of an ion M is inversely proportional to its distribution coefficient, where dx is the infinitesimal volume of the column through which the ion M moves on addition of dv ml of eluent. Separating the variables, integrating between suitable limits, and solving the equation for v give the volume of eluent necessary to elute the ion in question from a column of given size, or, conversely, solving for x gives the path length of the ion in the column on passage of a given volume of the eluent.

$$\int_0^v \frac{1}{D}\,dv = \int_0^x dx \qquad (3.161)$$

If the upper limit on the right-hand side is equal to the column volume, $x = X$, then the net eluent volume will be

$$v = XD \qquad (3.162)$$

On addition of dead volume of the column (v_0) to the net eluent volume, the total eluent volume (retention volume) used in practice, v_{max}, is obtained.

$$v_{max} = v + v_0 = XD + Xa = X(D + a) \qquad (3.163)$$

where X is the column volume in ml and a is the void fraction (~ 0.4). Then v_{max} ml of the eluent are necessary for the ion in question to appear in maximum concentration in the eluate (see Fig. 3.17).

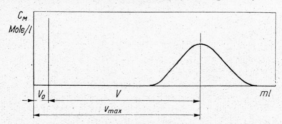

Fig. 3.17. Chromatographic elution curve. v_0 is the dead volume, v is the net eluent volume and v_{max} is the peak eluent volume

The efficiency of the separation of two metal ions is dependent first on the position of the chemical equilibria involved (i.e. on the distribution coefficients), and secondly on the rate of attainment of the equilibrium, that is, on kinetic parameters. Looking at the efficiency of the

Sec. 3.9]   Ion-exchange separations

separation from the viewpoint of chemical equilibria, the selectivity of separation of two similarly charged ions depends on the ratio of the distribution coefficients of the two ions. For ions M and N

$$K_d = \frac{D_N}{D_M} \qquad (3.164)$$

where K_d is the separation factor.

If $K_d > 100$, the separation can be done on a simple ion-exchange column, by selective sorption or selective elution. If $K_d < 10$, then strictly prescribed experimental conditions have to be maintained for the chromatographic separation to give adequate purity of the products. In such cases the resolution of the column (number of theoretical plates) and kinetic factors become important.

In practice not only the ratio but the absolute values of the distribution coefficients are important in determining the efficiency of the separation. It is important that the distribution coefficient of the ion eluted first should be small (<1). At the same time the distribution coefficient of the ion eluted next should be greater than 1 [otherwise, according to Eq. (3.163) this could also pass through the column without being bound]. It is better to use the ratio

$$\theta = \frac{v_{\max, N}}{v_{\max, M}} = \frac{D_N + a}{D_M + a} \qquad (3.165)$$

for estimating the efficiency of the separation.

In separation of metal ions with the same charge [cf. 164, 165] the ratio can be calculated from the equation:

$$\frac{D_N + a}{D_M + a} = \frac{K'_{xNA} Q^z [A]^{-z} + a}{K'_{xMA} Q^z [A]^{-z} + a} \qquad (3.166)$$

[see Eq. (3.158)].

As shown by Eq. (3.166), the efficiency of the separation depends on the K'_x, values, and can be only slightly influenced by the concentration of the eluting ion.

For ions with the same charge the K_x values do not differ significantly. However, in the presence of complexing agents the conditional constants should be used instead of the K_x values, and the selectivity of the separation can be remarkably increased. If the first terms in both the numerator and denominator are much greater than 0·4, then

$$\theta \simeq K_d \simeq \frac{K'_{xNA}}{K'_{xMA}} = \frac{K_{xNA}\alpha_{M(L)}}{K_{xMA}\alpha_{N(L)}} \qquad (3.167)$$

If a suitable complexing agent is used, the value of the quotient defined by Eq. (3.165) increases with the ligand concentration, reaches a maximum and then gradually decreases to approximately 1 [164].

The free ligand concentration can be increased by increasing the total concentration of the complexing agent, or, if it is a weak acid, by changing the pH of the solution. When the total concentration of the complexing agent rises, the concentration of the counter-ion of the ligand (sodium, hydrogen ion, etc.) also increases, resulting in an enhanced displacement effect. Since for metal ions of the same charge this displacement effect changes similarly with concentration [see Eq. (3.156)], beyond a certain limit the increase in the total concentration of the complexing agent leads to a decrease in the selectivity.

However, the free ligand concentration can generally be changed advantageously and sensitively by changing the pH of the eluting solution. As the concentration of the metal ion eluted is usually much lower than the total concentration of the complexing agent, the free ligand concentration can be calculated simply by using Eq. (1.77) if the protonation constants of the ligand are known. That is

$$[\alpha_{M(L)}]^r = 1 + \left[\frac{C_L}{\alpha_{L(H)}}\right]\beta_1 + \left[\frac{C_L}{\alpha_{L(H)}}\right]^2 \beta_2 + \ldots \qquad (3.168)$$

If the distribution coefficient is to be calculated for various pH values at the same total concentration of the complexing agent it should be noted that as the pH rises, the nature of the counter-ion A also changes. Instead of hydrogen ions alkali metal (or possibly ammonium) ions will be present in amounts depending on the degree of neutralization of the weak acid.

The efficiency of the separation of calcium and magnesium has been calculated for elution with $10^{-2}M$ EDTA solutions at different pH values by using Eqs. (3.158), (3.166) and (3.168) and equilibrium constant data. The results of the calculation are shown in logarithmic form in Fig. 3.18. At low pH values both $\alpha_{Ca(EDTA)}$ and $\alpha_{Mg(EDTA)}$ are approximately 1, and the efficiency of the separation is determined by the K_x values. As the pH rises, the degree of complex formation increases and accordingly the ratio of the α-values becomes the predominant factor.

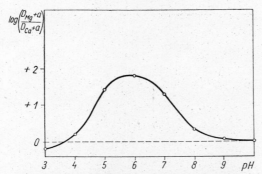

Fig. 3.18. The efficiency of the separation in the elution of calcium and magnesium ions with $0.01M$ EDTA as a function of pH. $K_{xCaNa} = 2.54$; $K_{xMgNa} = 1.65$.

At still higher pH values the distribution coefficients decrease until their values become comparable with the value of the void fraction. The efficiency of separation rapidly decreases and the value of the term calculated by using Eq. (3.165) approximates to 1. Optimum separation is achieved at pH 6. The calculated result agrees well with experimental data [166].

In the separation of metal ions of different oxidation states the separation factor is dependent on the concentration of the eluting ion. For example in the case of the separation of the trivalent ion N^{3+} and the divalent ion M^{2+}

$$K_{dNM} = \frac{D_N}{D_M} = \frac{K_{xNA}\alpha_{M(L)}}{K_{xMA}\alpha_{N(L)}} \times Q[A]^{-1} \qquad (3.169)$$

It has already been mentioned that equilibrium calculations provide only approximate values of distribution coefficients, and an estimate of the position of chromatographic bands on a column of given size, but do not give information about the shape and tailing of the bands which depend on kinetic and hydrodynamic parameters. The curves describing the concentration distribution within the bands can be obtained from the plate theory of chromatography.

The size of the ion-exchange column necessary for the quantitative separation of two components of similar properties can be calculated from the following considerations.

If the distribution coefficients of the two ions to be separated are D_I and D_{II}, $(D_{II} > D_I)$, the minimum number of theoretical plates necessary for separation [cf. 167] is:

$$N \geq 2\pi \left[\frac{\dfrac{D_{II} + a}{D_I + a} + 1}{\dfrac{D_{II} + a}{D_I + a} - 1} \right]^2 \qquad (3.170)$$

The height equivalent to one theoretical plate, h, can be calculated to a good approximation for simple cases by using the equation deduced by Glueckauf [168]. However, care should be taken in the calculations, since in the presence of large complexant molecules the average diffusion constants for simple inorganic ions cannot be used in the calculations. When a chelating agent is used as an eluent, the apparent diffusion coefficients may differ by 1–2 orders of magnitude from those of simple ions.

The value of h can be determined by an experimental method [168] from the data obtained from an elution curve

$$h = \frac{Lb^2}{8v_{max}^2} \qquad (3.171)$$

where L is the length of the ion-exchange column (cm), v_{max} is the eluent volume at the peak (ml) and b is the peak width (ml) at a height of $0.368\ C_{max}$ (see Fig. 3.17).

It should be borne in mind that h is dependent on the nature of the eluting ion even if the experiments are made under identical conditions.

If h is known, the column length necessary for a separation is obtained by multiplying N calculated from (3.170) by h.

3.9.1.2. *Chromatographic separation of organic bases on cation-exchange columns*

Uncharged organic bases, amines, take up protons in aqueous solution to a degree depending on the pH of the solution

$$B + H^+ \rightleftharpoons BH^+ \qquad (3.172)$$

The binding of an organic base on a cation-exchange resin depends on the degree of protonation, since in principle electrostatic bonding may occur only between the protonated base and the cation exchanger. The distribution coefficient can be described by the equation

$$D_B = \frac{(BH^+)}{C_B} = \frac{K_{xBHA}}{\alpha_{BH(H)}} Q[A]^{-1} \qquad (3.173)$$

where

$$\alpha_{BH(H)} = \frac{C_B}{[BH^+]} = 1 + \frac{1}{K[H^+]} \qquad (3.174)$$

K is the protonation constant of the base and K_{xBHA} is the ion-exchange equilibrium constant of the ions BH^+ and A^+. If the constants are known, D can be calculated for different pH values by using these equations, and the optimum eluent concentration for the separation of amines can be found in the same way as for the separation of metal ions [169, 170]. If pH $> \log K + 2$, practically speaking the amine is not bound on the exchanger.

It must be noted in the calculations that if the ion A^+ taking part in the ion exchange is sodium, then in solutions of low pH the concentration of hydrogen ions becomes commensurable with that of sodium ions, and the constant K_{xBHNa} may not be used. In Fig. 3.19 the distribution coefficients found experimentally for some organic bases are plotted against pH. The concentration of sodium was constant, $2 \times 10^{-2} M$ in each case. The shapes of the experimental curves correspond to Eq. (3.173), but they are usually shifted somewhat at low pH values compared with the calculated curves. The reason for this is that the proton activity in the resin is not exactly equal to that in the solution. In spite of this, Eq. (3.173) can generally be used in planning chromatographic separations (see Example 79).

The behaviour of amino acids in acidic solutions is similar to that of amines. In acidic solution they are present as cations, the proportion depending on the protonation constant (with neutral or basic amino acids on the second protonation constant), and can be bound on a cation-exchange resin, and thus chromatographically separated with an eluent of appropriate pH. This is the basic principle of the usual chromatographic separation of amino acids.

The selectivity of the separation of amines and amino acids can be increased by previously loading the cation-exchange column with complex-forming metal ions, e.g. zinc(II) or nickel(II) [172].

Whereas in the case dealt with previously the binding of an amine on the ion-exchange resin is enhanced by protonation, in the presence of a complex-forming metal ion the degree of binding depends on the

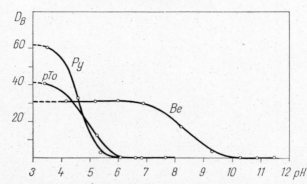

Fig. 3.19. Distribution coefficients of pyridine (Py), p-tolidine (pTo) and benzylamine (Be) on Lewatit SP-100 Na-form resin, in a 50% ethanolic solution of sodium chloride, as a function of pH. $[Na^+] = 2 \times 10^{-2} M$. The logarithms of the protonation constants of the bases in 50% ethanolic solution are 5, 5·1 and 9, respectively [171]

complex formation reaction of the non-protonated amine and the metal ion.

$$RM + B \rightleftharpoons RMB \qquad (3.175)$$

Accordingly, the distribution coefficient of the amine is that of the base, and can be given as

$$D_B = \frac{(MB) + (B)}{C_B} = \frac{(M)(B)\overline{\beta}_1 + (B)}{[M][B]\beta_1 + [B]} \qquad (3.176)$$

Since the resin is saturated with metal ion M, as a first approximation the capacity Q can be written in place of (M). If the base is bound as ligand, then $Q\overline{\beta}_1 \gg 1$ and Eq. (3.176) can be simplified [171]:

$$D_B \simeq Qd_B \frac{\overline{\beta}_1}{\alpha_{B(M)}} \qquad (pH > 1 + \log K) \qquad (3.177)$$

where d_B is the distribution constant of the non-protonated, neutral base molecule, given by $d_B = (B)/[B]$, $\overline{\beta}_1$ is the complex-formation equilibrium constant in the resin phase (to a first approximation $\overline{\beta}_1 \simeq \beta_1$) and $\alpha_{B(M)}$ is the side-reaction function of the base in the aqueous phase and is given by

$$\alpha_{B(M)} = 1 + [M]\beta_1 \qquad (3.178)$$

Usually $d_B \leq 1$.

From Eqs. (3.177) and (3.178) it can be seen that the distribution coefficient can be altered by changing $\alpha_{B(M)}$, i.e. the concentration of the ion M. If, however, $[M]\beta_1 \ll 1$, that is $[M] \ll 1/\beta_1$, then $\alpha_{B(M)} \simeq 1$ and the distribution coefficient is independent of the concentration of the metal ion. It has been assumed above that only the 1 : 1 complex species is formed, since $C_M \gg C_B$.

Equations (3.177) and (3.178) are suitable for the calculation, to a fairly good approximation, of the distribution coefficient of an organic base on a cation-exchange resin containing a complex-forming ion M, if $pH > 1 + \log K$ (see Example 80). On acidification of the solution the complex decomposes, the amine is protonated, and the distribution is determined by the ion-exchange equilibrium of the protonated amine (cation) and the metal ion M. A rise in the pH favours the formation of the complex. However, care should be taken, since at higher pH values the hydroxo-complex formation by the metal ion may be enhanced.

3.9.2. ANION-EXCHANGE EQUILIBRIA

On a strongly basic anion-exchange resin, if an anion L^{z-} and an anion B^- change places, the equilibrium can be given as

$$zR-B + L^{z-} \rightleftharpoons R_z-L + zB^- \qquad (3.179)$$

The ion-exchange equilibrium constant given in terms of concentrations is

$$K_x = \frac{(L)[B]^z}{[L](B)^z} \qquad (3.180)$$

The charges are omitted for simplicity.

When a chromatographic separation is made, then $(L) \ll (B)$; $[L] \ll [B]$ and $(B) \simeq Q$. The equation for the calculation of the approximate value of the distribution coefficient is similar to Eq. (3.156)

$$D_L = \frac{(L)}{[L]} \simeq K_x Q^z [B]^{-z} \qquad (3.181)$$

If the ion L^{z-} is also involved in a side-reaction (protonation or complex formation with a metal ion) and no new negatively charged ion is formed in the side-reaction, the equation is modified to

$$D_L = \frac{(L)}{[L']} \simeq \frac{K_x}{\alpha_L} Q^z [B]^{-z}$$

$$\log D_L = \log K_x - \log \alpha_L + z \log Q - \log [B]$$
$$= \log K'_x + z \log Q - z \log [B] \quad (3.182)$$

where $\alpha_L = \alpha_{L(H)} + \alpha_{L(M)} - 1$.

In the case of the protonation of multivalent anions other ionic species may be formed. For example in a solution containing the anion L^{2-}, HL^- may also be present simultaneously (the proportion depending on the pH and the protonation constant) and both are bound to the resin with different strengths. The distribution coefficient in this case is

$$D_L = \frac{(L^{2-}) + (HL^-)}{[L']} \quad (3.183)$$

which depends on two constants as well as the protonation constants [173], i.e.

$$D_L = \frac{d_L + d_{HL}[H^+]K_1}{1 + [H^+]K_1 + [H^+]^2 K_1 K_2} \quad (3.184)$$

The partition coefficients d_L and d_{LH} can be calculated as follows:

$$d_L = \frac{(L)}{[L]} \simeq K_{xLB} Q^2 [B]^{-2} \quad (3.185)$$

$$d_{HL} = \frac{(HL)}{[HL]} \simeq K_{xHLB} Q [B]^{-1} \quad (3.186)$$

provided that K_{xLB} and K_{xHLB} are known.

In exploratory calculations the following simplifying assumptions can often be made.

(i) It can be predicted from the protonation constants which of the two anions will predominate in a given pH interval, and the other can possibly be left out of consideration.

(ii) In a pH range where the two anions are simultaneously present it is usual that $K_{xHL} > K_{xL}$.

The values of d_L and d_{HL} can be determined experimentally by using Eq. (3.184). If, after determination of the distribution coefficient D_L for solutions of different pH values, and calculation of $\alpha_{L(H)}$, the products $D_L \alpha_{L(H)}$ are plotted against the hydrogen ion concentration, a straight line is obtained the intercept of which is d_L, and the slope $d_{HL}K_1$, K_{xLB} and K_{xHLB} can be calculated if d_L and d_{HL} are known, by using Eqs. (3.185) and (3.186).

3.9.2.1. Separation of anions and organic acids on anion-exchange resin columns

The optimum eluent concentration and pH to be used in the separation of two monobasic weak acids can be calculated from the equation

$$\theta = \frac{D_{II} + a}{D_I + a} = \frac{K'_{xL_{II}B}Q[B]^{-1} + a}{K'_{xLIB}Q[B]^{-1} + a} \quad (3.187)$$

The values of θ calculated for the separation of acetic acid and trichloroacetic acid by using Eqs. (3.182) and (3.187) are plotted on a logarithmic scale in Fig. 3.20. According to the calculations, the best separation can be expected with a $0.1M$ sodium chloride solution at pH 2.4 as eluent.

The optimum conditions for the separation of multivalent anions or organic acids can be predicted by similar calculations. The value of θ is found from the calculated distribution coefficients, by using Eq. (3.182) or (3.184).

The efficiency of the separation of simple inorganic anions (e.g. thiocyanate, sulphate etc.) and of polybasic organic acids can be increased significantly by using, in addition to eluents at various pH values, appropriate complex-forming metal ions in the eluent (see Example 78) [174].

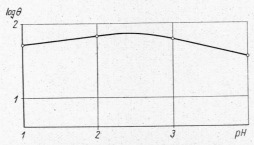

Fig. 3.20. Efficiency of separation of acetic acid and trichloroacetic acid with $0.1M$ sodium chloride solutions of different pH as eluents. (See Example 77)

3.9.2.2. *Chromatographic separation of metal ions on anion-exchange resin columns*

Metal ions can be bound by an anion-exchange resin in the presence of a complexant. The conditions for bonding are the formation of negatively charged complexes and that the size of the complex does not hinder the absorption.

Most ion-exchange methods developed for the separation of metal ions are based on the different stabilities of the complexes of metal ions with halides. The advantages of the use of halide ions are as follows.

(i) Metal halide-complex ions are absorbed by the resin with great selectivity compared with the corresponding halide ion (i.e. K_x is large), and thus a high ligand concentration is allowable.

(ii) Owing to the relatively small sizes of halide ions and of the halide-complexes the rate of the attainment of equilibrium is favourable.

When M^{z+} forms mainly the species $ML_n^{(n-z)-}$ with the halide L^-, the distribution coefficient of the metal ion can be expressed as

$$D_M = \frac{(ML_n)}{[M']} = K_x \Phi_n \left[\frac{(L)}{[L]} \right]^{n-z} \tag{3.188}$$

where K_x is the equilibrium constant of the ion exchange involving $ML_n^{(n-z)-}$ and L^-:

$$K_x = \frac{(ML_n)[L]^{n-z}}{[ML_n](L)^{n-z}} \tag{3.189}$$

and Φ_n is the mole fraction of the complex $ML_n^{(n-z)}$ in the solution, which can be calculated from the halide concentration and stability constants by using Eq. (1.20).

If $C_M \ll C_L$, then

$$D_M \simeq K_x \Phi_n Q^{n-z}[L]^{-(n-z)} \tag{3.190}$$

If the absorption of more than one negatively charged species is to be considered, then the gross distribution coefficient will be the sum of the distribution coefficients of the single complexes, thus

$$D_M = \frac{(ML_n)}{[M']} + \frac{(ML_{n-1})}{[M']} + \ldots \tag{3.191}$$

However, usually one term is predominant, and the others can be neglected in the calculations.

In cation-exchange chromatographic separations an increase in the concentration of the complexing ligand as well as in the concentration of the counter-ion leads to a reduction in the distribution coefficient. However, in anion-exchange chromatography, two opposing effects occur: the amount of anionic metal complexes which are absorbed by the resin increases with an increase in the concentration of the ligand, but the displacing effect of the ligand as a counter-ion also becomes greater. Therefore, the variation of distribution coefficient curve with ligand concentration usually exhibits a maximum (see Fig. 2.29). The maximum distribution coefficient occurs at a ligand concentration where the charge on the metal ion is just equal to the average ligand number, that is $z = \bar{n}$ and the neutral complex is predominant.

The optimum conditions for the separation of metal ions by anion exchange can be predicted by using Eqs. (3.165) and (3.190) in the same way as for cation-exchange separations. However, it should be noted that the uncertainty of the calculations is greater in the case of anion-exchange separations and the results are to be considered as exploratory only.

Most commercially available strongly acidic cross-linked (8% DVB) ion-exchange resins behave similarly irrespective of the manufacturer, but this is not the case with anion exchangers and two types of the most widely used strongly basic anion exchangers are available. One contains quaternary trimethylammonium groups (Dowex 1, Amberlite 400), the other quaternary ethoxydimethylammonium groups (Dowex 2, Amberlite 410). The selectivities of the two types, and accordingly, the ion-exchange equilibrium constants of ions are greatly different (see Section 4.5). When solutions with a concentration exceeding $1M$ are used, the effect of the electrolyte penetration is pronounced and ion-association reactions may occur, which falsify the results of the approximating calculations described.

3.9.3. SEPARATION OF MORE THAN TWO SIMILARLY CHARGED IONS

During the chromatographic separation of mixtures containing more than two components an increase in the efficiency of the separation may be obtained by changing the concentration of the eluent in a stepwise manner or continuously. Changing the composition of the eluent is justified also in cases where only two ions are to be separated and the distribution coefficients differ by some orders of magnitude.

3.9.3.1. *Separation by a stepwise change in eluent*

In using stepwise elution, the composition of the eluent is changed so as to reduce the distribution coefficient of the ion to be eluted to below 2 while that of the ion next to be eluted still exceeds 5. The eluent composition used in the successive steps can be calculated as given earlier [174, 175] (see p. 273).

Since not only the ion eluted but also the other ions remaining on the column migrate during elution, the migration due to previous eluents should also be taken into account when calculating the eluent volume (v_{max}) for elution of the second, third ions, etc. This can be calculated by

$$x = \frac{v}{D + a} \quad (3.192)$$

The ion travels through x ml of column volume during passage of v ml of eluent. This column volume should be subtracted from the total column volume in subsequent calculations (see Example 78).

3.9.3.2. *Separation by gradient elution*

From theoretical considerations and practical experience the separation of components of similar behaviour in multicomponent systems can be carried out most effectively by the gradient elution technique.

The principle of gradient elution is that the composition of the eluent is changed continuously during the separation. As the composition of the eluent changes, the distribution coefficients of the ions to be separated also change continuously during chromatography [175a].

Calculations concerning only the simplest cases will be described here, where similarly charged metal ions are separated on a cation-exchange column, and the concentration or pH of the eluent is changed linearly (i.e. the gradient is constant during elution).

When divalent metal ions are separated on a cation-exchange column, and the elution is done by continuously increasing the concentration of the monovalent eluting ions the eluent volumes of the ions to be separated may be calculated as follows.

The concentration of the eluting ion A rises linearly from zero with the eluent volume:

$$[A] = vb \quad (3.193)$$

where v is the volume of the eluent in ml and b is the constant concentration gradient. Therefore

$$\frac{\Delta [A]}{\Delta v} = b \quad \text{(concentration change/ml)}$$

From Eqs. (3.193) and (3.156),

$$\frac{1}{D_M} = \frac{b^2 v^2}{K_{xMA} Q^2} \quad (3.194)$$

To calculate the eluent volume, Eq. (3.161) is integrated between the limits $v = 0$ and $v = V$, and $x = 0$ and $x = X$, taking into account that D_M is a function of the eluent volume as given by Eq. (3.194).

$$\frac{b^2}{K_{xMA} Q^2} \int_0^V v^2 \, dv = \int_0^X dx$$

$$\frac{b^2}{K_{xMA} Q^2} \cdot \frac{V^3}{3} = X \quad (3.195)$$

Adding the dead volume to V from Eq. (3.195) gives the total eluent volume

$$v_{max} = \left(\frac{3 X K_{xMA} Q^2}{b^2} \right)^{1/3} + Xa \quad (3.196)$$

The greater the concentration gradient, the smaller the eluent volume required to elute the ion. If K_x for the ion in question and the data of the column are known, v_{max} can be calculated from Eq. (3.196).

In a similar way, equations can be deduced for the calculation of the eluent volume of monovalent and trivalent ions, etc.

However, ions of the same charge can only rarely be separated by changing the concentration of the eluting ion without using complexing agents. The K_x values of the ions to be separated rarely differ significantly. An increase or decrease in the concentration gradient b has hardly any influence on the selectivity of separation.

Consider now the separation of divalent metal ions on a cation-exchange column, with an eluent containing a complexing agent, and where the pH is charged during elution.

If the complexing ligand is a weak base, the relationship between the distribution coefficient and the pH is somewhat complicated [see Eqs.

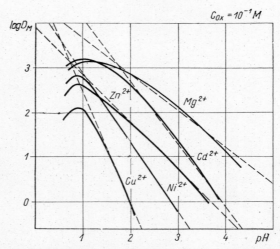

Fig. 3.21. Logarithms of the distribution coefficients of some divalent metal ions calculated by Eq. (3.158) as a function of pH, for $0 \cdot 1 M$ oxalic acid–sodium oxalate as eluent [176]

(3.158) and (3.168)] even if the total concentration of the ligand is constant.

If, however, the distribution coefficients of the ions to be separated are calculated for constant total ligand concentration but different pH values by using these equations, and then plotted against pH, all the curves have a portion which, to a first approximation, is a straight line.

The distribution coefficients of some divalent metal ions as a function of pH, calculated by using Eqs. (3.158) and (3.168) for oxalic acid–sodium oxalate solution as eluent, are shown in Fig. 3.21 [175, 176]. It is easily seen from the figure that the parts of the curves that are most valuable from the point of view of separation can be approximated by straight lines.

The equations of the lines can be given from the intercepts (B) and slopes (A) of the straight lines in the figure, i.e.

$$\log D_M = B - A \, \text{pH} \tag{3.197}$$

If the separation is to be made with an eluent of linearly increasing pH, and the initial pH is 2, then

$$\text{pH} = 2 + bv \tag{3.198}$$

Sec. 3.9] Ion-exchange separations

where v is the volume (ml) of eluent added to the column and b is the pH gradient (ΔpH/ml), the latter being constant.

Equations (3.197) and (3.198) can be combined to give

$$\frac{1}{D_M} = 10^{bAv} \times 10^{(2A-B)} \tag{3.199}$$

Substitution in Eq. (3.161) and integrating between appropriate limits gives

$$10^{(2A-B)} \int_0^V 10^{bAv} \, dv = \int_0^X dx \tag{3.200}$$

Solving for V, then adding $V_0 = Xa$ gives

$$v_{\max,M} = \frac{\log(2\cdot 3 \times 10^B \times 10^{-2A} bAX + 1)}{bA} + Xa \tag{3.201}$$

Equation (3.201) can be used to calculate the eluent volumes for different pH gradients before any experimental work need be done. To calculate the optimum pH gradient, it is necessary that the condition

$$N \geq 2\pi \left[\frac{\dfrac{v_{\max,II}}{v_{\max,I}} + 1}{\dfrac{v_{\max,II}}{v_{\max,I}} - 1} \right]^2 \tag{3.202}$$

where N is the number of theoretical plates of the column, is fulfilled for the eluent volumes of the two components which follow each other most closely [167].

It should be noted that ion-exchange processes are sometimes accompanied in practice by adsorption phenomena, which may be particularly important in the ion exchange of solutions containing organic ions or molecules. The calculations above are restricted to ion-exchange equilibria only, and accompanying adsorption phenomena may falsify the results.

3.9.4. Extraction with liquid ion-exchangers

Liquid ion exchangers are solutions of high molecular weight organic acids (cation exchangers) or bases (anion exchangers) which are sparingly soluble in water but readily soluble in organic solvents and by which ions can be extracted from aqueous solution. (For more details see [53].)

A significant difference between liquid ion exchangers and ion-exchange resins, in addition to the difference in their physical states, is that liquid ion exchangers usually have weakly acidic or weakly basic character (i.e. their capacity depends on the pH), and also ion-pair formation, and ionic and molecular association occur to a degree depending on the nature and dielectric constant of the organic solvent.

3.9.4.1. *Cation-exchange equilibria*

One of the most widely used liquid cation exchangers is di(2-ethylhexyl) phosphoric acid (HDEHP) solution in toluene. HDEHP is present in dimeric form in the solvent. The equilibrium constant for the ion exchange according to the equation

$$2(HL)_{2(org)} + M^{2+}_{(aq)} \rightleftharpoons M(HL_2)_{2(org)} + 2H^+_{(aq)} \qquad (3.203)$$

may be written in terms of concentrations as

$$K_x = \frac{(M(HL_2)_2)[H^+]^2}{[M^{2+}](H_2L_2)^2} \qquad (3.204)$$

If the amount of the metal ion M is small compared with that of the ion exchanger, then half of the total concentration of the ion exchanger $(Q/2)$ can be written in place of (H_2L_2) and D can be expressed as

$$D_M = \frac{(M(HL_2)_2)}{[M^{2+}]} \simeq K_x \left(\frac{Q}{2}\right)^2 [H^+]^{-2} \qquad (3.205)$$

$$\log D_M \simeq \log K_x + 2 \log (Q/2) + 2 \, pH$$

[see Eq. (3.156)].

From Eq. (3.205)

$$K_x = \frac{\bar{K}_M d_M}{\bar{K}_H^2 d_H^2} \qquad (3.206)$$

where $\bar{K}_M$ and $\bar{K}_H$ are the stability constants of the ion-association complexes formed in the organic phase and d_M and d_H are the distribution constants of the metal and hydrogen ions, respectively. The value of the ion-exchange equilibrium constant is effectively determined by the complex stability constants of the ions involved in the ion exchange, which in turn depend on the solvent.

If a complexing agent X which forms a non-extractable complex with M is added to the aqueous phase, Eq. (3.205) giving the distribution coefficient will be modified to

$$\log D_M \simeq \log K_x - \log \alpha_{M(X)} + 2\log(Q/2) + 2\,\mathrm{pH}$$

$$\log D_M \simeq \log K'_x + 2\log(Q/2) + 2\,\mathrm{pH} \qquad (3.207)$$

The optimum conditions for the separation of metal ions can be calculated by using Eqs. (3.205) and (3.207) in the same way as for extraction separations from K_x values and α-functions.

3.9.4.2. *Anion-exchange equilibria*

Between an anion exchanger, such as an organic secondary amine (e.g. Amberlite LA-2) and a strong acid HL dissolved in water the following equilibrium occurs

$$\mathrm{RNH_{(org)} + H^+_{(aq)} + L^-_{(aq)} \rightleftharpoons RNH_2L_{(org)}} \qquad (3.208)$$

The equilibrium constant given in terms of concentrations is

$$K_L = \frac{(\mathrm{HL})}{(\mathrm{R})[\mathrm{H^+}][\mathrm{L^-}]} = \bar{K}_{HL} d_H d_L \qquad (3.209)$$

where $\bar{K}_{HL}$ is the stability constant of the ion-association complex in the organic phase.

From Eq. (3.209) the distribution coefficient is given by

$$D_L = \frac{(\mathrm{HL})}{[\mathrm{L^-}]} = K_L(\mathrm{R})[\mathrm{H^+}] \qquad (3.210)$$

where (R) is the concentration of the free amine. Also,

$$Q = (\mathrm{R}) + (\mathrm{HL}) \qquad (3.211)$$

It should be noted that Eq. (3.210) is only valid if $(\mathrm{HL}) < Q$; thus if $(\mathrm{HL}) = Q$, then

$$D_L = \frac{Q}{[\mathrm{L^-}]} \qquad (3.212)$$

Equation (3.210) is suitable for the calculation of K_L values characteristic of the bond strengths of protonated amine–anion complexes from experimentally determined distribution coefficients. It is reasonable to do the calculation by using the $pH_{1/2}$ values for $D = 1$.

If two strong monobasic acids HB and HL are simultaneously present in the aqueous solution, the distribution equilibrium can be given as

$$K_{xLB} = \frac{D_L}{D_B} = \frac{(HL)[B]}{[HB][L]} = \frac{K_L}{K_B} \qquad (3.213)$$

The ion-exchange equilibrium constant is the ratio of the equilibrium constants defined by Eq. (3.209).

If the anions are involved in a side-reaction in which a non-extractable product is formed in the aqueous phase, the equation will be modified to

$$K'_{xLB} = \frac{D'_L}{D'_B} = \frac{K_L \alpha_B}{K_B \alpha_L} \qquad (3.214)$$

It should be noted that sometimes, mainly at higher electrolyte concentrations ($>0.1\ M$), ionic or molecular aggregates are formed in the organic phase. In such cases the equation and equilibrium constant describing the process are as follows

$$RNH_{(org)} + nH^+_{(aq)} + nB^-_{(aq)} \rightleftarrows RNH(HB)_{n(org)} \qquad (3.215)$$

$$K_{B(n)} = \frac{(HB)^n}{(R)[H^+]^n[B^-]^n} \qquad (3.216)$$

From the analytical point of view, in the selective separation of metal ions the important case is that where the complexing anion and the metal ion capable of forming extractable negatively charged complexes capable of proton uptake are simultaneously present in the aqueous phase. Usually in this case $C_M \ll C_L$.

If a negatively charged ligand L (e.g. chloride) and a divalent metal ion M are present in the solution, and a complex ion of the composition ML_4^{2-} is mainly formed, the equation for the equilibrium constant of the anion exchange will be similar to Eq. (3.213):

$$K_x = \frac{K_{MLH}}{K_L^2} = \frac{(H_2ML_4)[L^-]^2}{[ML_4^{2-}](HL)^2} \qquad (3.217)$$

where K_{ML_4} is the equilibrium constant of the reaction

$$2RNH_{(org)}^+ + 2H_{(aq)}^+ + ML_{4(aq)}^{2-}$$
$$\rightleftharpoons (RNH_2)_2ML_{4(org)}$$

[see Eqs. (3.208), (3.209)]. Usually, $K_{ML_4} > K_L$.

Multiplying the numerator and denominator on the right-hand side of Eq. (3.217) by C_M gives

$$K_x = \frac{C_M}{[ML_4]} \cdot \frac{(H_2ML_4)}{C_M} \cdot \frac{[L]^2}{(HL)^2} \quad (3.218)$$

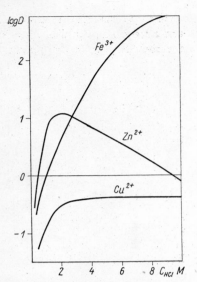

Fig. 3.22. Extraction of metal ions with a liquid anion-exchanger (20% solution of Amberlite LA-2 in xylene) from hydrochloric acid of various concentrations [177]

The first factor on the right-hand side of Eq. (3.218) is the reciprocal value of the mole-fraction of the complex species with ligand number four, the second is the distribution coefficient of the metal ion, and the third is the square of the reciprocal value of the distribution coefficient of the ligand [see Eq. (3.210)].

From the equation

$$D_M = K_x \Phi_4 D_L^2 \quad (3.219)$$

if $C_L \gg C_M$ and $(HL) \simeq Q$ (that is, the ion exchanger is saturated), then considering Eq. (3.212) the distribution coefficient of the metal ion M can be approximated as

$$D_M \simeq K_x \Phi_4 Q^2 [L]^{-2} \quad (3.220)$$

Equation (3.220) is similar in form to Eq. (3.190) deduced for anion-exchange resins. The variation of distribution coefficient with C_{HL} often exhibits a maximum (see Fig. 3.22).

The selective separation of trace amounts of a metal ion can also be carried out with a chelating agent, by means of a liquid anion exchanger, if the metal ion forms a negatively charged chelate with the ligand, and this is strongly bound in the organic phase, e.g. the metal complexes of pyrocatechol disulphonic acid (tiron).

Suppose that the solution contains the chelate-forming ion H_2L^{2-} and a trace amount of the bivalent metal ion M^{2+}. The latter reacts with the chelating ion in the reaction

$$M^{2+} + H_2L^{2-} \rightleftharpoons ML^{2-} + 2H^+ \qquad (3.221)$$

The equilibrium constant of the reaction is the ratio of the stability constant of the complex formed to the product of the protonation constants. In logarithmic form:

$$\log K = \log \beta_1 - \log K_1 - \log K_2$$

Both H_2L^{2-} and ML^{2-} can be bound in the organic ion-exchanger phase in a process accompanied by proton uptake. The ion-exchange equilibrium constant is given by the ratio of the equilibrium constants of the distribution processes, similar to that given by Eq. (3.208) [see Eqs. (3.213) and (3.217)].

$$K_{xMLH_2L} = \frac{K_{ML}}{K_{H_2L}} = \frac{(H_2ML)[H_2L^{2-}]}{(H_4L)[ML^{2-}]} \qquad (3.222)$$

$$K_{ML} > K_{H_2L}, \quad \text{hence} \quad K_x > 1$$

The distribution coefficient of the metal ion can be calculated from an equation similar to Eq. (3.219).

$$D_M = K_x \Phi_{ML} D_{H_2L} \qquad (3.223)$$

or in logarithmic form:

$$\log D_M = \log K_x + \log \Phi_{ML} + \log D_{H_2L}$$

When a chelating agent is applied it may happen that $(H_2ML) + (H_4L) < Q$, that is, the capacity of the ion exchanger is only partially utilized. In such cases the distribution coefficient D_{H_2L} is given by

$$D_{H_2L} = K_{H_2L}(R)^2[H^+]^2 \qquad (3.224)$$

instead of Eq. (3.212) [see Eq. (3.210)].

It should be noted in connection with the application of Eq. (3.223) that the distribution of H_2L^{2-} should also be taken into account when calculating the free ligand concentration necessary to obtain Φ_{ML}.

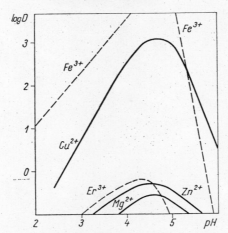

Fig. 3.23. Extraction of metal ions with a 0·2M solution of Amberlite LA II anion-exchanger in xylene in the presence of tiron (T). $C_M = 10^{-3}\ M$; $C_T = 10^{-2}\ M$ and $C_{Cl} = 10^{-1}M$ [178].

That is

$$[L^{2-}] = \frac{C_{H_2L}}{\alpha_{L(H)}} \cdot \frac{1}{(1 + D_{H_2L})} \tag{3.225}$$

If the total amount of the chelating agent is kept constant and the pH of the solution is varied, the distribution coefficient of the metal ion exhibits a maximum, since an increase in pH favours the formation of the complex ML, but simultaneously reduces D_{H_2L} in view of Eq. (3.224) [see Eq. (3.223)].

For the metal ion to be extractable it is necessary that in the pH range where the distribution of the ligand and the complex ion takes place, the complex formation also proceeds to a considerable extent according to Eq. (3.221).

It should be noted that the relationships given can be used in exploratory calculations even if another ion (e.g. chloride) which is absorbed less strongly than the chelating agent, is also present (though in not too high a concentration). Thus, if $K_{Cl} \ll K_{H_2L} < K_{ML}$, chloride has but little influence on the position of the equilibrium.

In Fig. 3.23 the logarithms of the experimentally determined distribution coefficients of some metal ions are plotted *vs.* pH. The total concentrations of tiron and chloride were constant [178].

The selectivity of the extraction of the chosen metal ion can be increased by using a masking agent which forms non-extractable complexes with the interfering metal ions. For example, by means of tiron and a liquid anion exchanger copper(II) and molybdenum(VI) are both extracted from a solution of pH 5; however, if EDTA is added, copper(II) remains in the aqueous phase.

For the interfering metal ion not to be extracted it is necessary that its distribution coefficient $D \leq 10^{-2}$. The concentration of the complexing agent required for masking can be calculated by using Eq. (3.223) if the conditional ion-exchange constant is inserted in place of K_x:

$$K'_x = \frac{K_x}{\alpha_{M(Y)}} \qquad (3.226)$$

where Y represents the masking ligand (see Example 81).

When liquid anion exchangers containing quaternary ammonium groups (e.g. Aliquat 336) are used, the distribution equilibria are similar to those dealt with in the case of strongly basic ion-exchange resins (see p. 281). The capacity of the ion exchanger is practically constant in the pH range 1–14. However, it must be stressed that molecular association in the organic phase appears as a frequent side-reaction, as well as the partition of some uncharged species.

3.9.5. Ion-exchange paper chromatography

Ion-exchange chromatography can be performed on papers impregnated with ion-exchange resin or on papers made of cellulose containing ion-exchange groups. For more details see [53].

According to Martin and Synge [179], in separations by paper chromatography the relationship

$$\frac{1}{R_f} = \frac{m_s}{m_0} D + 1 \qquad (3.227)$$

holds between the retention factor and distribution coefficient of an ion, where m_s/m_0 is the ratio of the mass of the solid phase to that of the mobile solution and R_f is the retention factor (its value can be obtained from the chromatogram, by dividing the distance between the centre of the spot and the starting point by that of the solvent front from the starting point).

Sec. 3.9]   Ion-exchange separations   297

In principle, Eq. (3.227) can be used for calculating R_f values expected in separations by ion-exchange paper chromatography, from distribution coefficients. The distribution coefficients, which depend on the composition of the eluent can be calculated from Eqs. (3.156) and (3.158) for cation-exchange papers, and by Eqs. (3.181), (3.182) and (3.190) for anion-exchange papers. (See Example 82.)

In practice the results of calculations are exploratory only, as m_s/m_0 can only be determined with some uncertainty, and the adsorption phenomena, which often accompany ion-exchange processes, have not been taken into account in the derivation of the equations used.

3.9.6. Worked examples

72. Calculate the approximate value of the volume equilibrium constant $K_{x\,CaNa}$, given that the weight ion-exchange equilibrium constants are $K_{w\,CaH} = 1\cdot75$ and $K_{w\,NaH} = 1\cdot40$. The density of the ion-exchange resin column is $\sigma = 0\cdot35$ g/ml.

Equation (3.157) is used:

$$K_{w\,CaNa} = \frac{K_{w\,CaH}}{K_{w\,NaH}^2} = \frac{1\cdot75}{(1\cdot4)^2} = 0\cdot89$$

$$K_{x\,CaNa} = \frac{K_{x\,CaNa}}{\sigma} = \frac{0\cdot89}{0\cdot35} = 2\cdot54$$

73. Determine whether iron(III) can be separated from manganese(II) by selective sorption on a sodium-form cation-exchange column by addition of sodium pyrophosphate to the solution if the concentration of both metal ions is $10^{-2}M$. Calculate the necessary concentration of pyrophosphate and the optimum pH. The overall stability constant of $[Fe(HP_2O_7)_2]^{3-}$ is $\log \beta_2 = 22\cdot2$; the protonation constants of pyrophosphate are $\log K_1 = 8\cdot5$, $\log K_2 = 6\cdot1$, $\log K_3 = 2\cdot5$ and $\log K_4 = 1$. The ion-exchange equilibrium constants are $K_{x\,FeNa} = 3\cdot65$; $K_{x\,MnNa} = 1\cdot94$.

The complexing ligand is $HP_2O_7^{3-}$ containing one proton, and so the Φ_1 function can be used instead of the α-function for calculating the free ligand concentration and the conditional overall stability constant.

$$\frac{[HP_2O_7^-]}{[HP_2O_7^{3-}]} = \frac{1}{\Phi_1} = \alpha_{LH(H)}$$

On the basis of the protonation constants of $P_2O_7^{4-}$, the mole-fraction of $HP_2O_7^{3-}$ has its maximum value at a pH of about

$$\frac{\log K_1 + \log K_2}{2} = \frac{8\cdot 5 + 6\cdot 1}{2} = 7\cdot 3$$

In a solution of pH 7·3, however, the formation of iron(III) hydroxo-complexes is considerable. Therefore the conditional overall stability constants at pH 7 and 3 are calculated.

At pH 7

$$\Phi_1 = \frac{10^{-7} \times 10^{8\cdot 5}}{1 + 10^{-7} \times 10^{8\cdot 5} + 10^{-14} \times 10^{14\cdot 6} + 10^{-21} \times 10^{17\cdot 1} + 10^{-28} \times 10^{18\cdot 1}} = 0\cdot 86$$

$$\log \alpha_{Fe(OH)} \simeq 8 \quad \text{(see Fig. 1.10)}$$

$$\log \beta_2' = \log \beta_2 - \log \alpha_{Fe(OH)} - 2\log \frac{1}{\Phi_1}$$

$$= 22\cdot 2 - 8 - 2 \times 0\cdot 065 = 14\cdot 07 .$$

At pH 3

$$\Phi_1 = \frac{10^{-3} \times 10^{8\cdot 5}}{1 + 10^{-3} \times 10^{8\cdot 5} + 10^{-6} \times 10^{14\cdot 6} + 10^{-9} \times 10^{11\cdot 7} + 10^{-12} \times 10^{18\cdot 1}} = 10^{-3\cdot 2}$$

$$\log \alpha_{Fe(OH)} = 0\cdot 4 \quad \text{(see Fig. 1.10)}$$

$$\log \beta_2' = 22\cdot 2 - 0\cdot 4 - 2 \times 3\cdot 2 = 15\cdot 4$$

Thus, it is better to work at pH 3. The distribution coefficient of iron(III) is calculated by using Eq. (3.159). If the concentration of sodium pyrophosphate is $0\cdot 04M$, $[Na^+] \simeq 0\cdot 16M$. The concentration of pyrophosphate in excess is ca. $0\cdot 02M$. The volumetric capacity of the cation exchanger is $Q \simeq 2$ meq/ml.

$$\log D_{Fe} = \log K_x + 3\log Q - \log \beta_2' - 2\log [HP_2O_7'] - 3\log [Na^+]$$

$$= 0\cdot 56 + 3 \times 0\cdot 3 - 15\cdot 4 + 2 \times 1\cdot 7 + 3 \times 0\cdot 8 = -8\cdot 14$$

$$D_{Fe} = 7\cdot 2 \times 10^{-9}$$

Manganese(II) is not complexed appreciably by pyrophosphate under these conditions ($\alpha_{Mn(HP_2O_7)} \simeq 1$). The distribution coefficient can be

Sec. 3.9] Ion-exchange separations 299

calculated by using Eq. (3.156). (If pH = 3, then $\alpha_{Mn(OH)} = 1$.)

$$\log D_{Mn} = \log K_{x\,MnNa} + 2 \log Q - 2 \log [Na^+]$$
$$= 0.29 + 2 \times 0.3 + 2 \times 0.8 = 2.49$$
$$D_{Mn} = 309$$

Hence the separation can be carried out. Iron(III) passes through the column practically without being bound, whereas manganese(II) is adsorbed. A solution of pH 3 and of similar concentration to the value calculated has been suggested by Ryabtshikov and Osipova [180] for the separation of iron(III) and manganese(II).

74. Determine whether small amounts of copper(II) and nickel(II) can be separated on a cation-exchange column, with $0.5M$ sulphosalicylic acid as eluent, and if so, what pH should be used?

$$K_{x\,CuNa} = 2.04; \quad K_{x\,NiNa} = 2.0$$

The overall stability constants of the sulphosalicylate–copper(II) complexes are $\log \beta_1 = 9.5$ and $\log \beta_2 = 16.5$. Those of the nickel(II) complexes are $\log \beta_1 = 6.4$ and $\log \beta_2 = 10.2$. The protonation constants of sulphosalicylate are $\log K_1 = 11.6$, $\log K_2 = 2.6$ and $\log K_3 < 1$.

The distribution coefficients are calculated for pH 3, 5 and 7 by using Eqs. (3.158) and (3.168) and then the ratio $(D_{Ni} + a)/(D_{Cu} + a)$. It can be established on the basis of the protonation constants that in the pH range 4–10 $[Na^+] \simeq 1M$ and the capacity of the cation-exchange resin is $Q = 2$ meq/ml.

At pH 3,

$$\log \alpha_{SS(H)} = 8.75 \quad \text{(see Fig. 1.13)}$$

$$\alpha_{Cu(SS)} = 1 + \left[\frac{0.5}{10^{8.75}}\right] 10^{9.5} + \left[\frac{0.5}{10^{8.75}}\right]^2 10^{16.5} = 3.8$$

$$D_{Cu} = \frac{2.04}{3.8} \left(\frac{2}{0.8}\right)^2 = 3.36$$

$$\alpha_{Ni(SS)} = 1 + \left[\frac{0.5}{10^{8.75}}\right] 10^{6.4} + \left[\frac{0.5}{10^{8.75}}\right]^2 10^{10.2} \simeq 1$$

$$D_{Ni} = 2 \times 2^2 (0.8)^{-2} = 12.5$$

$$\frac{D_{Ni} + a}{D_{Cu} + a} = \frac{12.5 + 0.4}{3.36 + 0.4} = 3.43$$

At pH 5,
$$\log \alpha_{SS(H)} = 6.6$$

$$\alpha_{Cu(SS)} = 1 + \left[\frac{0.5}{10^{6.6}}\right] 10^{9.5} + \left[\frac{0.5}{10^{6.6}}\right]^2 10^{16.5} = 9 \times 10^2$$

$$D_{Cu} = \frac{2 \times 04}{9 \times 10^2} \left(\frac{2}{1}\right)^2 = 9.1 \times 10^{-3}$$

$$\alpha_{Ni(SS)} = 1 + \left[\frac{0.5}{10^{6.6}}\right] 10^{6.4} + \left[\frac{0.5}{10^{6.6}}\right]^2 10^{10.2} = 1.3$$

$$D_{Ni} = \frac{2}{1.3} \left(\frac{2}{1}\right)^2 = 6.16$$

$$\frac{D_{Ni} + a}{D_{Cu} + a} = \frac{6.56}{0.41} = 16.0.$$

At pH 7,
$$\log \alpha_{SS(H)} = 4.6$$

$$\alpha_{Cu(SS)} = 1 + 10^{-4.9} \times 10^{9.5} + 10^{-9.8} \times 10^{16.5} = 5 \times 10^6$$

$$D_{Cu} = \frac{2.04}{5 \times 10^6} \left(\frac{2}{1}\right)^2 = 1.6 \times 10^{-6}$$

$$\alpha_{Ni(SS)} = 1 + 10^{-4.9} \times 10^{6.4} + 10^{-9.8} \times 10^{10.2} = 35.1$$

$$D_{Ni} = \frac{2}{35.1} \left(\frac{2}{1}\right)^2 = 0.228$$

$$\frac{D_{Ni} + a}{D_{Cu} + a} = \frac{0.628}{0.40} = 1.57$$

Best separation can thus be achieved with a solution at pH 5. If the length of the ion-exchange column is 10 cm, its diameter 6 mm ($X = 2.83$ ml), then copper(II) will appear in maximum concentration in the effluent when

$$v_{max} = X(D + a) = 2.83(0.01 + 0.4) \simeq 1.2 \text{ ml}$$

of eluent has passed through the column and nickel(II) at

$$v_{max} = 2.83(6.16 + 0.4) \simeq 18.6 \text{ ml}$$

Experiments made by the author have yielded results similar to those calculated above [181].

75. Consider the possibility of separation of 4-aminoantipyrine (A) and 4-dimethylaminoantipyrine (amidazophen, DA) by ion-exchange chromatography from the following data. The protonation constant of A is $\log K_1 = 3·9$. The ion-exchange equilibrium constant of the protonated base is $\log K_{xAHNa} = 0·32$. The corresponding data for DA are $\log K_1 = 4·9$, $K_{xDAHNa} = 0·25$. In view of the large size of the molecules of the substances to be separated, a resin with large pore size is used (e.g. Lewatit SP-100) with $Q = 1·9$ meq/ml; grain-size 0·2 mm, at 60°C. In order to increase the solubility, the eluent contains 20% alcohol. The inner diffusion coefficient of the base cations at 60 °C is $d_r = 3 \times 10^{-7}$ cm²/sec.

It can be established on the basis of the protonation constants that the optimum separation can be expected at pH 4–6. To ensure appropriate distribution coefficients, the concentration of sodium in the eluent is adjusted to $10^{-1} M$. The distribution coefficients of the bases for pH 4·5, 5 and 5·5 are calculated by using Eqs. (3.173) and (3.174), as well as the value of θ characteristic of the efficiency of the separation [see Eq. (3.165)].

At pH 4·5,

$$\alpha_{AH(H)} = 1 + \frac{1}{10^{3·9} \times 10^{-4·5}} = 1 + 10^{0·6} = 5·0$$

$$D_A = \frac{0·32}{5} \times \frac{1·9}{0·1} = 1·22$$

$$\alpha_{DAH(H)} = 1 + \frac{1}{10^{4·9} 10^{-4·5}} = 1 + 10^{-0·4} = 1·4$$

$$D_{DA} = \frac{0·25}{1·4} \times \frac{1·9}{0·1} = 3·39$$

$$\theta_{pH 4·5} = \frac{3·39 + 0·4}{1·22 + 0·4} = 2·34$$

At pH 5,

$$\alpha_{AH(H)} = 1 + \frac{1}{10^{3·9} 10^{-5}} = 1 + 10^{1·1} = 13·6$$

$$D_A = \frac{0·32}{13·6} \times \frac{1·9}{0·1} = 0·45$$

$$\alpha_{DAH(H)} = 1 + \frac{1}{10^{4 \cdot 9} \times 10^{-5}} = 1 + 10^{0 \cdot 1} = 2 \cdot 26$$

$$D_{DA} = \frac{0 \cdot 25}{2 \cdot 26} \times \frac{1 \cdot 9}{0 \cdot 1} = 2 \cdot 10$$

$$\theta_{pH\,5} = \frac{2 \cdot 10 + 0 \cdot 4}{0 \cdot 45 + 0 \cdot 4} = 2 \cdot 94$$

At pH 5·5,

$$\alpha_{AH(H)} = 1 + \frac{1}{10^{3 \cdot 9} \times 10^{-5 \cdot 5}} = 1 + 10^{1 \cdot 6} = 40 \cdot 8$$

$$D_A = \frac{0 \cdot 32}{40 \cdot 8} \times \frac{1 \cdot 9}{0 \cdot 1} = 0 \cdot 15$$

$$\alpha_{DAH(H)} = 1 + \frac{1}{10^{4 \cdot 9} \times 10^{-5 \cdot 5}} = 1 + 10^{0 \cdot 6} = 5 \cdot 0$$

$$D_{DA} = \frac{0 \cdot 25}{5 \cdot 0} \times \frac{1 \cdot 9}{0 \cdot 1} = 0 \cdot 95$$

$$\theta_{pH\,5 \cdot 5} = \frac{0 \cdot 95 + 0 \cdot 4}{0 \cdot 15 + 0 \cdot 4} = 2 \cdot 45.$$

The best separation can thus be obtained with a solution of pH 5. A sodium acetate–acetic acid buffer of pH 5 is prepared in which the concentration of sodium ions is $10^{-1}M$. The number of theoretical plates necessary for quantitative separation is calculated from Eq. (3.202):

$$N \geq 6 \cdot 28 \left[\frac{2 \cdot 94 + 1}{2 \cdot 94 - 1} \right]^2 = 25 \cdot 9$$

The height equivalent to one theoretical plate is calculated by the equation proposed by Glueckauf [168], taking into consideration that the inner diffusion is the rate-controlling process, owing to the large size of the ions:

$$h = 1 \cdot 64\,r + \frac{D}{(D + a)^2} \times \frac{0 \cdot 142\,r^2 F}{d_r}$$

where r is the radius of a resin particle (cm), F is the linear flow rate of the eluent (cm/sec) and d_r is the inner diffusion coefficient (cm²/sec).

Let F be 100 cm/h (approx. 0·028 cm/sec), and using the mean value of the distribution coefficients of the two bases: then

$$h = 1·64 \times 0·01 + \frac{1·27}{(1·27 + 0·4)^2} \times \frac{0·142(0·01)^2 \times 0·028}{3·10^{-7}} = 0·62 \text{ cm}$$

The necessary length L of the column is therefore

$$L = Nh = 25·9 \times 0·62 = 16·1 \text{ cm}$$

To be on the safe side a column 18 cm long is chosen. If the inner diameter of the column is 6·5 mm, then the volume of the resin bed is

$$X = 18 \left(\frac{0·65}{2}\right)^2 \pi = 5·97 \text{ ml}$$

4-Aminoantipyrine appears in the effluent in maximum concentration after

$$v_{max, A} = 5·97(0·45 + 0·4) = 5·1 \text{ ml}$$

of eluent, and dimethylaminoantipyrine after

$$v_{max, DA} = 5·97(2·10 + 0·4) = 14·9 \text{ ml}$$

of eluent have been passed through the column [see Eq. (3.163)].

To complete the elution

$$v_{end} = 14·9 \sqrt{\frac{14·9}{5·1}} \simeq 25·5 \text{ ml}$$

of eluent are required. The time of separation is

$$t = \frac{v_{end}}{AF} = \frac{25·5}{0·332 \times 1·67} = 46 \text{ min}$$

(A is the cross-sectional area of the column). Experimental results are in agreement with these calculated values [170].

76. The protonation constants of pyridine and *p*-tolidine are very close to each other (log K_{Py} = 5·0 and log K_{To} = 5·1 in a 1 : 1 water–alcohol mixture). Their chromatographic separation is thus difficult to

achieve by simply adjusting the pH. Consider whether the separation is improved if a nickel-form column is employed and a $10^{-2}M$ nickel(II) chloride solution is used as eluent. The stability constant of the pyridine–nickel(II) complex is $\beta_1 = 10^{1.78}$; p-tolidine does not form a complex with nickel(II). The capacity of the ion-exchange resin is $Q = 1.9$ meq/ml and the distribution constant of both free bases is $d_B \simeq 0.44$ in 50% aqueous alcohol.

For both bases to be present in the non-protonated form it is necessary that the pH of the solution is at least $5.1 + 1.5 = 6.6$. Above pH 7, however, hydroxo-complexes of nickel(II) are formed. The distribution coefficient is calculated to a first approximation by using Eq. (3.177), the calculation being done with β_1 instead of $\bar{\beta}_1$.

$$\alpha_{Py(Ni)} = 1 + 10^{-2} \times 10^{1.78} = 1.6$$

$$\alpha_{To(Ni)} \simeq 1$$

$$D_{Py} \simeq 1.9 \times 0.44 \frac{10^{1.78}}{1.6} = 31.4$$

$$D_{To} \simeq d_{To} = 0.44$$

If the separation is carried out on a column 7.6 cm long and 6.5 mm in diameter ($X = 2.5$ ml), p-tolidine appears in maximum concentration in the eluate after $v_{max, To} = 2.5(0.44 + 0.4) \sim 2$ ml of eluent have been added and pyridine after $v_{max, Py} = 2.5(31.4 + 0.4) \sim 80$ ml of the eluent have been added to the column.

According to experiments, good separation can be achieved even on a small column (see Fig. 3.24).

77. Calculate the pH of a $0.1M$ sodium chloride (or hydrochloric acid) solution to be used as eluent in the separation of acetic acid (Ac) and trichloroacetic acid (Tac). A chloride-form Varion AD ion-exchange column is used for the separation. The capacity of the ion-exchange resin is $Q = 1.2$ meq/ml; $K_{x\,AcCl} = 0.17$ and $K_{x\,TacCl} = 2.3$. The protonation constant of acetate is $\log K = 4.65$, that of trichloroacetate is $\log K = 0.5$ ($I = 0.1$).

To calculate the optimum pH the distribution coefficients are calculated at pH 1, 2, 3 and 4.

At pH 1,

$$\alpha_{Ac(H)} = 1 + 10^{-1} \times 10^{4.65} \simeq 4.5 \times 10^3$$

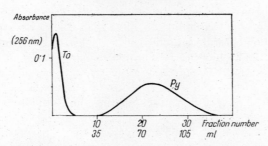

Fig. 3.24. Separation of 5 μmole of *p*-tolidine and 1 μmole of pyridine on an Ni-form cation-exchange column, with $10^{-2}M$ nickel(II) chloride (pH 6·6) in 50% ethanol as eluent. Ion-exchange resin: Lewatit SP-100 (0·2–0·4 mm). Column: 6·5×76 mm; flow-rate 0·3 ml/min; fraction volume 3·5 ml [171]

$$D_{Ac} = \frac{0·17}{4·5 \times 10^3} \times \frac{1·2}{0·1} \simeq 4·5 \times 10^{-4}$$

$$\alpha_{Tac(H)} = 1 + 10^{-1} \times 10^{0·5} = 1·3$$

$$D_{Tac} = \frac{2·3}{1·3} \times \frac{1·2}{0·1} = 21·3.$$

The distribution coefficients obtained by similar calculations for the other three pH values are:

at pH 2, $D_{Ac} = 4·5 \times 10^{-3}$; $D_{Tac} = 26·8$

at pH 3, $D_{Ac} = 4·5 \times 10^{-2}$; $D_{Tac} = 27·6$

at pH 4, $D_{Ac} = 0·37$; $D_{Tac} = 27·6$

From these values and the void fraction of the ion-exchange column ($a = 0·4$), the logarithms of the ratios $(D_{Tac} + a)/(D_{Ac} + a)$ are found and plotted against the pH (see Fig. 3.20). The pH at the maximum of the curve (about 2·4) will provide the best separation. At pH 2·4 $D_{Ac} = 1·1 \times 10^{-2}$ and $D_{Tac} = 27·6$. It follows from the calculated values of the distribution coefficients that it the pH of the solution containing acetic and trichloroacetic acids is adjusted to 2·4, and sodium chloride is added to give a chloride concentration of $0·1M$ the separation can be carried out by selective sorption. Acetic acid passes through the column without being sorbed, whereas trichloroacetic acid is bound and can be eluted subsequently with a more concentrated sodium chloride solution.

78. Phthalic acid and dinitrophthalic acid are to be separated by an anion-exchange resin column. Calculate the necessary volume of eluent if $0.25M$ lithium chloride is used to elute the less strongly bound phthalic acid and $0.125M$ nickel(II) chloride solution containing complex-forming ions is used to elute the more strongly bound dinitrophthalic acid. Both solutions are in 50% aqueous alcohol and at pH 5·5. The ion-exchange column is 180 mm in length and 5 mm in diameter ($X = 3.52$ ml). The capacity of the ion-exchange resin is $Q = 0.91$ meq/ml and the ion-exchange equilibrium constants are $K_{x\,PCl} = 0.48$; $K_{x\,DNPCl} = 2.9$. The stability constant of the dinitrophthalato–nickel(II) complex is $\log K = 1.78$.

The protonation constants of the conjugated bases are in both cases smaller than 10^4, hence at pH 5·5 the acids are completely dissociated giving the doubly charged anions.

The distribution coefficient of the phthalic acid eluted first with $0.25M$ lithium chloride is calculated by using Eq. (3.181), and the eluent volume by using Eq. (3.163):

$$D_P = 0.48(0.91)^2(0.25)^{-2} = 6.35$$

$$v_{\max,P} = 3.52(6.35 + 0.4) \simeq 23.8 \text{ ml}$$

To elute the phthalic acid completely, the elution is made with 32 ml of eluent. The migration of dinitrophthalic acid caused by 32 ml of eluent is calculated by taking the distribution coefficient into consideration [see Eq. (3.192)].

$$D_{DNP}^{(1)} = 2.9(0.91)^2(0.25)^{-2} = 38.4$$

$$x = \frac{32}{38.4 + 0.4} \simeq 0.83 \text{ ml}$$

The distribution coefficient of dinitrophthalic acid when the nickel(II) chloride solution is used as eluent is [see Eq. (3.182)]:

$$\alpha_{DNP(Ni)} = 1 + [Ni^{2+}]K = 1 + 0.125 \times 10^{1.78} = 8.5$$

$$D_{DNP}^{(2)} = \frac{K_x}{\alpha_{DNP(Ni)}} Q^2 [Cl^-]^{-2} = \frac{2.9}{8.5}(0.91)^2(0.25)^{-2} = 4.5$$

From the distribution coefficient and the remaining column volume $(X - x)$ the eluent volume necessary to obtain the maximum concentra-

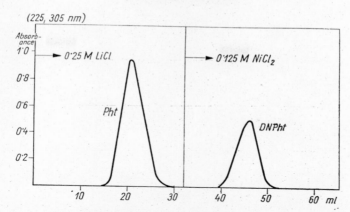

Fig. 3.25. Ion-exchange chromatographic separation of phthalic acid and dinitrophthalic acid with stepwise elution. Ion-exchange column: Lewatit MP 500 (150–300 mesh); diameter 5 mm; length 100 mm; flow-rate 6 ml/hr [174]

tion of dinitrophthalic acid in the effluent is calculated to be

$$v_{\max, \text{DNP}} = 32 + (3 \cdot 52 - 0 \cdot 83)(4 \cdot 5 + 0 \cdot 4) \simeq 45 \cdot 2 \text{ ml}$$

Thus phthalic acid is in maximum concentration in the effluent when the eluate volume is 23·8 ml and is obtained quantitatively when the eluate volume is 32 ml. If the eluent composition is changed at this point, dinitrophthalic acid appears in maximum concentration in the effluent when the eluate volume is 45·2 ml.

The experimental results are in good agreement with the calculated values [174] (see Fig. 3.25).

79. Calculate the distribution coefficient of mercury(II) in $5M$ hydrochloric acid on Dowex 1×10 ion-exchange resin. The ion-exchange equilibrium constant of the tetrachloro-mercury(II) complex ion, referred to chloride, is $K_x = 3 \times 10^4$. The overall stability constants of the chlorocomplexes of mercury(II) are $\log \beta_1 = 6 \cdot 7$, $\log \beta_2 = 13 \cdot 2$, $\log \beta_3 = 14 \cdot 1$ and $\log \beta_4 = 15 \cdot 1$. The capacity of the resin is $Q = 1 \cdot 4$ meq/ml.

The calculation is done by using Eqs. (1.20) and (3.190).

$$\Phi_4 = \frac{10^{2 \cdot 8} \times 10^{15 \cdot 1}}{1 + 10^{0 \cdot 7} \times 10^{6 \cdot 7} + 10^{1 \cdot 4} \times 10^{13 \cdot 2} + 10^{2 \cdot 1} \times 10^{14 \cdot 1} + 10^{2 \cdot 8} \times 10^{15 \cdot 1}} = 0 \cdot 98$$

$$D_{\text{Hg}} = K_x \Phi_4 Q^2 [\text{Cl}^-]^{-2} = 3 \times 10^4 \times 0 \cdot 98 (1 \cdot 4)^2 (5)^{-2} \sim 2 \cdot 3 \times 10^4$$

$$\log D_{\text{Hg}} = 3 \cdot 36.$$

This value is in good agreement with that in ref. [878].

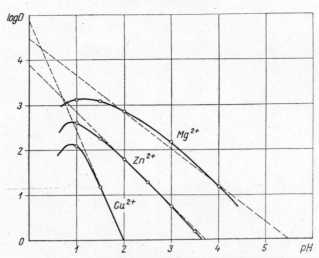

Fig. 3.26. Logarithms of the distribution coefficients of copper(II), zinc(II) and magnesium(II) calculated from Eq. (3.158) as a function of pH, in the presence of 0·1M oxalic acid–sodium oxalate. (See Example 80 [175])

80. In the separation of copper(II), zinc(II) and magnesium(II) on a cation-exchange column calculate the eluent volumes at which each ion is at maximum concentration in the effluent if 0·1M oxalic acid is used with the pH changing linearly from 2 upwards. The pH-gradient is $b = 0·012$ pH/ml. The ion-exchange column is 80 mm in height, and 7 mm in diameter ($X = 3·08$ ml). The capacity of the ion-exchange resin is $Q = 2·0$ meq/ml and $K_{x\,CuH} = 4$, $K_{x\,CuNa} = 2·04$, $K_{x\,ZnH} = 3·6$, $K_{x\,ZnNa} = 1·84$, $K_{x\,MgH} = 3·3$ and $K_{x\,MgNa} = 1·66$.

From the stability constants of the oxalato-complexes and the protonation constants of oxalate the $\alpha_{M(Ox)}$ functions are calculated for the metal ions in 0·1M oxalic acid–sodium oxalate solutions of different pH values and then the distribution coefficients are calculated by using Eq. (3.158). Plotting the logarithms of the calculated distribution coefficients *vs.* pH gives the curves shown in Fig. 3.26.

For example, the distribution coefficient of magnesium at pH 3 is

$$\log D_{Mg} = 0·22 - 0·66 + 2 \times 0·3 + 2 \times 1 = 2·16$$

since

$$\log \alpha_{Ox(H)} = 0·85 \quad \text{(see Fig. 1.13)}$$

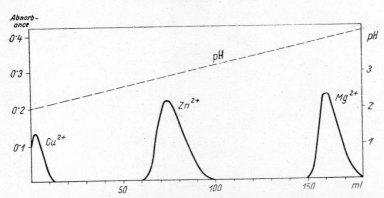

Fig. 3.27. Ion-exchange separation of copper(II), zinc(II) and magnesium(II) by gradient elution. Ion-exchange column: Amberlite CG 120 (150–300 mesh); diameter 7 mm, length 80 mm; eluent: 0·1M oxalic acid–sodium oxalate of continuously increasing pH [175]

and

$$\log \alpha_{\text{Mg(Ox)}} = 1 + 10^{-1\cdot85} \times 10^{2\cdot4} = 10^{0\cdot66}$$

The equations of the straight lines (drawn in dotted lines in the figure) are:

$$\log D_{\text{Cu}} = 4\cdot9 - 2\cdot5 \ \text{pH}$$
$$\log D_{\text{Zn}} = 3\cdot9 - 1\cdot04 \ \text{pH}$$
$$\log D_{\text{Mg}} = 4\cdot5 - 0\cdot82 \ \text{pH}$$

From the equations for the distribution coefficients, the eluent volumes are calculated by using Eq. (3.201).

$$v_{\text{max, Cu}} = \frac{\log(2\cdot3 \times 10^{4\cdot9} \times 10^{-5} \times 12 \times 10^{-3} \times 2\cdot5 \times 3\cdot08 + 1)}{12 \times 10^{-3} \times 2\cdot5}$$

$$+ 3\cdot08 \times 0\cdot4 = \frac{\log(0\cdot17 + 1)}{0\cdot03} + 1\cdot23 \simeq 3\cdot5 \ \text{ml}$$

$$v_{\text{max, Zn}} = \frac{\log(2\cdot3 \times 10^{3\cdot9} \times 10^{-2\cdot08} \times 12 \times 10^{-3} \times 1\cdot04 \times 3\cdot08 + 1)}{1\cdot04 \times 12 \times 10^{-3}} + 1\cdot23$$

$$= \frac{\log(5\cdot84 + 1)}{0\cdot0125} + 1\cdot23 \simeq 68 \ \text{ml}$$

$$v_{max,Mg} = \frac{\log(2 \cdot 3 \times 10^{4 \cdot 5} \times 10^{-1 \cdot 64} \times 12 \times 10^{-3} \times 0 \cdot 82 \times 3 \cdot 08 + 1)}{0 \cdot 82 \times 12 \times 10^{-3}}$$

$$+ 1 \cdot 23 = \frac{\log(50 \cdot 5 + 1)}{0 \cdot 00985} + 1 \cdot 23 \simeq 175 \text{ ml}$$

The author's experiments gave results similar to the values calculated above (see Fig. 3.27) [175, 176].

81. In the extraction of copper(II) with a liquid ion exchanger ($0 \cdot 2M$ Amberlite LA-2 in xylene), with tiron as a complexing agent, calculate the distribution coefficient of copper, if the total concentration of tiron is $10^{-2}M$, that of copper (II) $10^{-3}M$ in the original solution, and the pH of the solution after extraction is 5. The volume ratio of the aqueous and organic phase is 1 : 1. The ion-exchange equilibrium constant is $\log K_{xCuT-H_2T} = 1 \cdot 6$ and the extraction equilibrium constant of tiron is $\log K_{H_2T} = 13 \cdot 2$. The stability constant of the tiron–copper(II) complex is $\log \beta_1 = 14 \cdot 5$, the protonation constants of tiron are $\log K_1 = 12 \cdot 7$, $\log K_2 = 7 \cdot 7$, $\log K_3 < 0$ and $\log K_4 < 0$.

To calculate the distribution coefficient of copper(II) by using Eq. (3.223) D_{H_2T} and Φ_{CuT} must first be found. Taking into account that $Q = 0 \cdot 20$ and $C_{H_2T} = 0 \cdot 01M$, the concentration of the free amine, (R) should be between $0 \cdot 2$ and $0 \cdot 18$. If (R) $\simeq 0 \cdot 19$, then according to Eq. (3.224)

$$D_{H_2T} = 10^{13 \cdot 2}(0 \cdot 19)^2(10^{-5})^2 = 10^{1 \cdot 76}$$

The free ligand concentration is calculated by using Eq. (3.225):

$$\alpha_{T(H)} = 1 + 10^{-5} \times 10^{12 \cdot 7} + 10^{-10} \times 10^{20 \cdot 4} \simeq 10^{10 \cdot 4}$$

$$[T^{4-}] = \frac{10^{-2}}{10^{10 \cdot 4}} \times \frac{1}{1 + 10^{1 \cdot 76}} \simeq 10^{-14 \cdot 2}$$

Equation (1.20) gives

$$\Phi_{CuT} = \frac{[T]\beta_1}{1 + [T]\beta_1} = \frac{10^{-14 \cdot 2} \times 10^{14 \cdot 5}}{1 + 10^{-14 \cdot 2} \times 10^{14 \cdot 5}} = 10^{-0 \cdot 18}$$

The distribution coefficient of copper(II) is

$$\log D_{Cu} = 1 \cdot 6 - 0 \cdot 18 + 1 \cdot 76 = 3 \cdot 18$$

$$D_{Cu} = 1 \cdot 5 \times 10^3$$

Now the concentration of EDTA required to prevent the extraction of $10^{-3}M$ copper(II) is calculated. For the distribution coefficient of copper(II) to be reduced to 10^{-2} it is necessary that

$$\alpha_{Cu(EDTA)} = 1 + \frac{[EDTA']}{\alpha_{EDTA(H)}} K_{Cu\,EDTA} > 10^{5 \cdot 18}$$

Considering that at pH 5, $\log \alpha_{EDTA(H)} = 6 \cdot 6$ (see Fig. 1.14) and $\log K_{Cu\,EDTA} = 18 \cdot 8$, then

$$1 + \frac{[EDTA']}{10^{6 \cdot 6}} \times 10^{18 \cdot 8} > 10^{5 \cdot 18}$$

From this

$$[EDTA'] > 10^{-7} M$$

Thus it is sufficient to add enough EDTA to give a concentration $10^{-7}M$ in excess of the amount stoichiometric with copper(II).

82. Copper(II) and cadmium(II) are absorbed to a similar degree on a cation-exchange resin. The weight ion-exchange equilibrium constants, referred to hydrogen ion, are $K_{w\,CuH} = 1 \cdot 60$ and $K_{w\,CdH} = 1 \cdot 64$. It is expected, however, that the two ions can be separated on cation-exchange paper with dilute hydrochloric acid as eluent, as their tendencies to form chloro-complexes are different. The overall stability constants of the cadmium chloro-complexes are $\log \beta_1 = 1 \cdot 42$; $\log \beta_2 = 1 \cdot 92$; $\log \beta_3 = 1 \cdot 76$ and $\log \beta_4 = 1 \cdot 06$ while those of the copper(II) complexes are $\log \beta_1 = 0 \cdot 98$, $\log \beta_2 = 0 \cdot 69$, $\log \beta_3 = 0 \cdot 55$ and $\log \beta_4 = 0$. The ion-exchange paper used is Amberlite SA-1, which contains Amberlite IR-120 cation-exchange resin ($Q_w = 4 \cdot 2$ meq/g). The ratio of the weight of the solvent to that of the ion-exchange resin impregnated with dilute hydrochloric acid is $3 \cdot 5 : 1$.

If the retention factor R_f of cadmium(II), which moves faster on the cation-exchange paper since it forms the more stable chloride complexes, is chosen to be $0 \cdot 60$ then the corresponding distribution coefficient may be calculated from Eq. (3.227)

$$D_{w\,Cd} = \left(\frac{1}{R_f} - 1\right) \frac{m_0}{m_s} = \left(\frac{1}{0 \cdot 6} - 1\right) 3 \cdot 5 = 2 \cdot 34$$

where $D_{w\,Cd}$ is the weight distribution coefficient expressed with concentration referred to dry weight:

$$D_{Cd} = D_{w\,Cd}\, \sigma$$

To obtain this distribution coefficient, in view of Eq. (3.158),

$$2\cdot 34 = \frac{1\cdot 64}{\alpha_{Cd(Cl)}} \times \frac{(4\cdot 2)^2}{[H^+]^2}$$

it is necessary that the value of the product $\alpha_{Cd(Cl)}[H^+]^2$ should be approximately $12\cdot 4$. If the concentration of hydrochloric acid is $0\cdot 1 M$, then

$$\alpha_{Cd(Cl)} = 1 + 10^{-1} \times 10^{1\cdot 42} + 10^{-2} \times 10^{1\cdot 92} + 10^{-3} \times 10^{1\cdot 76}$$
$$+ 10^{-4} \times 10^{1\cdot 06} \simeq 4\cdot 5$$

and

$$\alpha_{Cd(Cl)}[H^+]^2 = 4\cdot 5 \times 10^{-2}$$

which is not large enough. If the hydrochloric acid concentration is $0\cdot 5M$ then

$$\alpha_{Cd(Cl)} = 1 + 10^{-0\cdot 3} \times 10^{1\cdot 42} + 10^{-0\cdot 6} \times 10^{1\cdot 92}$$
$$+ 10^{-0\cdot 9} \times 10^{1\cdot 76} + 10^{-1\cdot 2} \times 10^{1\cdot 06} \simeq 43$$

The product $\alpha_{Cd(Cl)}[H^+]^2$ is then $43(0\cdot 5)^2 \simeq 10\cdot 8$ which is satisfactory. If $0\cdot 5M$ hydrochloric acid is used, the distribution coefficient and R_f value of cadmium(II) will be

$$D_{wCd} = \frac{1\cdot 64}{43}\left(\frac{4\cdot 2}{0\cdot 5}\right)^2 = 2\cdot 68$$

$$2\cdot 68 = \left(\frac{1}{R_f} - 1\right) \times 3\cdot 5$$

$$R_{f(Cd)} = 0\cdot 57$$

The corresponding values for copper(II) are

$$\alpha_{Cu(Cl)} = 1 + 10^{-0\cdot 3} \times 10^{0\cdot 98} + 10^{-0\cdot 6} \times 10^{0\cdot 69}$$
$$+ 10^{-0\cdot 9} \times 10^{0\cdot 55} + 10^{-1\cdot 2} \times 10^{0\cdot 00} = 7\cdot 5$$

$$D_{wCu} = \frac{1\cdot 60}{7\cdot 5}\left(\frac{4\cdot 2}{0\cdot 5}\right)^2 \simeq 15$$

$$15 = \left(\frac{1}{R_f} - 1\right)3\cdot 5; \quad R_{f(Cu)} \simeq 0\cdot 19$$

Grimaldi and Liberti [182] using $0\cdot 5M$ hydrochloric acid have found experimentally R_f values similar to those calculated above.

3.10. Electrophoresis

The basic principle of electrophoretic methods is that in an electric field the ions in a solution move in a direction determined by their sign of charge and at a velocity determined by the magnitude of their partial charge and mobility and, when in contact with a solid phase, by their distribution. If the velocities of two components are different, they can be separated. For more details of this technique see [183].

Paper electrophoresis is the most widely used technique. The sample solution is spotted onto the centre of a paper strip impregnated with a buffer solution and a voltage of 500–2000 V is applied to the ends of the wet paper strip. It is necessary to cool the apparatus as significant heating occurs. The velocity of migration (cm/min) due to the potential difference applied is composed of electrophoretic and osmotic terms:

$$v = v_e + v_0 \qquad (3.228)$$

Uncharged particles also move but at lower velocity, because of the electro-osmosis. For charged particles, however, v_e is predominant. The velocity of the electrophoretic migration is given by the equation

$$v_e = \mu \frac{V}{L} f \frac{1}{D+1} \qquad (3.229)$$

where μ is the mobility of the ion, V is the potential difference between the ends of the paper (in volts), L is the length of the paper strip between the two electrodes (cm), f is a tortuosity factor allowing for the random stationary-phase path length available to the ions, and D is the distribution coefficient of the ion between the two phases. The mobility of the ion (cm^2 volt^{-1} min^{-1}) gives the path length (cm) travelled by the ion in 1 min when 1 V/cm potential drop is applied, in the absence of the porous solid carrier (i.e. $f = 1$ and $D = 0$). The quantity μ is thus a constant, characteristic of an ion for a given solvent, temperature and ionic strength and in the absence of side-reactions. In many cases there is no appreciable interaction between the solute and stationary phase and $D = 0$ can be assumed.

The direction of migration of an ion depends on the sign of its charge. In paper electrophoresis with aqueous solutions the mobility of cations is increased by electro-osmosis, whereas that of anions is reduced.

If a divalent metal ion M^{2+} forms an uncharged complex ML with a doubly charged ligand L^{2-}, and $D_M = 0$, then Eq. (3.229) is modified

$$v_{e,M} = \mu_M \frac{V}{L} f \Phi_0 \qquad (3.230)$$

to where Φ_0 is the mole-fraction of the free metal ion:

$$\Phi_0 = \frac{1}{1 + [L]\beta_1}$$

According to Eq. (3.230), the velocity of migration of the metal ion decreases with increasing ligand concentration. At high ligand concentrations where Φ_0 tends towards zero, the velocity is given by the velocity due to electro-osmosis alone [see Eq. (3.228)]. If a negatively charged complex ML_2^{2-} can also be formed by successive ligand uptake, the velocity of electrophoretic migration will be

$$v_e = \frac{V}{L} f \left(\mu_M \Phi_0 - \mu_{ML_2} \Phi_2 \right) \qquad (3.231)$$

where Φ_0 and Φ_2 are the respective mole-fractions. From Eq. (3.231) it can be seen that the velocity reaches a minimum as the ligand concentration increases, then increases in the opposite direction.

If the relative mobilities of the free metal ion M^{2+} and of the complex ion ML_2 are known for the given experimental conditions, as well as β_1 and β_2, then the direction and velocity of migration can be calculated for any ligand concentration, from Eq. (3.231).

When metal ions of similar behaviour are to be separated, the concentration of complexant and pH necessary to obtain the best separation can be calculated.

On the other hand from runs made under similar experimental conditions but at different ligand concentrations, related values of [L] and Φ can be obtained from which β_1 and β_2 can be calculated (see p. 125).

If an anion A^- gives an uncharged molecule HA by proton uptake and a positively charged species H_2A^+ on taking up a second proton, then the equation of the velocity of electrophoretic migration will be given by

$$v_{e,A} = \frac{V}{L} f \left[\mu_{H_2A} \Phi_{H_2A} - \mu_A \Phi_A \right] \qquad (3.232)$$

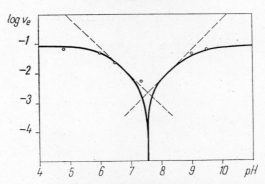

Fig. 3.28. Logarithmic diagram of the electrophoretic migration velocity of histidine. The points marked are the values measured by McDonald [184]

In alkaline solution, if only A^- is present, the migration will be towards the anode. As the pH decreases, Φ_A decreases and Φ_{H_2A} increases. At a certain pH the terms on the right-hand side of Eq. (3.232) become equal and thus v_e will be zero. As the pH decreases further the direction of migration changes, and the cation H_2A^+ will migrate towards the cathode.

In the case of amino acids, peptides and proteins the mobilities and distribution coefficients of positively and negatively charged particles are nearly identical (that is, $\mu_{H_2A} \simeq \mu_A$ and $D_{H_2A} \simeq D_A$), and Eq. (3.232) can be simplified to

$$v_e = B[\Phi_{H_2A} - \Phi_A] \quad (3.233)$$

where

$$B = \mu \frac{V}{L} f \frac{1}{D+1} \quad (3.234)$$

In such cases the pH at which $v_e = 0$ coincides with the isoelectric point.

The logarithm of the velocity of migration of histidine [calculated from Eq. (3.233)], as a function of the pH is shown in Fig. 3.28. The values determined by McDonald [184] are also shown in the figure. It can be seen from the figure that the calculated curve is in good agreement with the experimental points (the large deviation close to the isoelectric point is due to the electro-osmotic effect).

The equations of the dotted lines in the figure are

$$\log v'_e = \log B + \log K_2 - \text{pH} \qquad (3.235)$$

$$\log v''_e = \log B - \log K_1 + \text{pH} \qquad (3.236)$$

The pH at the point of intersection of the two lines (if $v_0 \simeq 0$) gives the pH of the isoelectric point. For histidine, $\log K_1 = 9\cdot 2$ and $\log K_2 = 6\cdot 1$. Under the experimental conditions applied $\log B = -1\cdot 2$ and

$$\text{pH}_{iso} = \frac{\log K_1 + \log K_2}{2} = 7\cdot 6 \qquad (3.237)$$

If similar diagrams are constructed from calculated data the pH and ligand concentration etc. required to obtain the best separation can be found.

It should be noted that for two similarly charged ions (I and II), a quantitative separation by one-dimensional electrophoresis is only obtained when

$$\frac{v_{II}}{v_I} > 1\cdot 05 \qquad (3.238)$$

Chapter 4

TABLES OF EQUILIBRIUM CONSTANTS

4.1. Protonation and Complex Formation Equilibrium Constants

The tables contain the logarithms of the successive protonation and complex formation constants, for room temperature, defined by Eqs. (1.40) and (1.11), respectively. For protonated and mixed-ligand complexes with composition different from that of the normal complexes, the logarithm of the overall stability constants and formulae giving the composition are presented. That is

$$\beta_{M_mH_eL_nY_p} = \frac{[M_mH_eL_nY_p]}{[M]^m[H]^e[L]^n[Y]^p}$$

From the overall stability constants the successive complex formation (protonation) constants can be calculated by subtracting the logarithms of the constants [see Eqs. (1.85) and (1.86)].

For inorganic ligands the formula of the ligand is given and for organic ligands the name of the complex-forming compound and the formula indicating the composition of the molecule (ligand and proton) are given.

For ease of reference inorganic ligands are arranged in alphabetical order of symbols of formulae and organic ligands in alphabetical order of names. The metal ions are arranged in alphabetical order of symbols, after values for protonation constants.

I = ionic strength; v = variable. The abbreviations of the methods used for the determinations are: pot = potentiometry; pol = polarography; sp = spectrophotometry; ex = extraction; i = ion exchange; cond = conductometry; k = kinetic; sol = solubility; oth = other methods.

318 Tables of equilibrium constants [Ch. 4

Inorganic ligands

Ligand	Metal ion	Method	I	$\log \beta$	Ref.
$As(OH)_4^-$	H^+	pot	0.1	HL 9.38	[185]
AsO_4^{3-}	H^+	pot	0.1	HL 11.2; H_2L 17.9; H_3L 20	[186]
$B(OH)_4^-$	H^+	pot	0.1	HL 9.1	[187]
Br^-	Ag^+		0.1	AgL 4.15; AgL_2 7.1; AgL_3 7.95; AgL_4 8.9; Ag_2L 9.7	[188] [62]
	Bi^{3+}	pot	2	BiL 2.3; BiL_2 4.45; BiL_3 6.3; BiL_4 7.7; BiL_5 9.3; BiL_6 9.4	[189]
	Cd^{2+}	pol	0.75	CdL 1.56; CdL_2 2.1; CdL_3 2.16; CdL_4 2.53 $CdBrI$ 3.32; $CdBrI_2$ 4.51; $CdBrI_3$ 5.83; $CdBrI$ 3.75; $CdBr_2I_2$ 5.33; $CdBr_3I$ 4.18	[190] [191]
	Cu^{2+}	sp	2	CuL -0.55; CuL_2 -1.84	[192]
	Fe^{3+}	sp	1	FeL -0.21; FeL_2 -0.7	[193]
	Hg^{2+}	pot	0.5	HgL 9.05; HgL_2 17.3; HgL_3 19.7; HgL_4 21; HgBrCN 26.97	[194] [195]
	In^{3+}	i	1	InL 1.2; InL_2 1.8; InL_3 2.5	[19]
	Pb^{2+}	pol	1	PbL 1.56; PbL_2 2.1; PbL_3 2.16; PbL_4 2.53	[190]
	Sn^{2+}	pot	3	SnL 0.73; SnL_2 1.14; SnL_3 1.34	[196]
	Tl^+	sol	v	TlL 0.92; TlL_2 0.92; TlL_3 0.40	[197]
	Tl^{3+}	pot	0.4	TlL 8.3; TlL_2 14.6; TlL_3 19.2; TlL_4 22.3; TlL_5 24.8; TlL_6 26.5	[198]
	Zn^{2+}	pot	4.5	ZnL -0.6; ZnL_2 -0.97; ZnL_3 -1.70; ZnL_4 -2.14	[199]
Cl^-	Ag^+	sol	v	AgL 3.4; AgL_2 5.3; AgL_3 5.48; AgL_3 5.4; Ag_2L 6.7	[200] [201]
	Au^{3+}	pot	v	AuL_4 26	[202]
	Bi^{3+}	pot	2	BiL 2.4; BiL_2 3.5; BiL_3 5.4; BiL_4 6.1; BiL_5 6.7; BiL_6 6.6	[189]
	Cd^{2+}	i	0	CdL 1.42; CdL_2 1.92; CdL_3 1.76; CdL_4 1.06	[203]
	Cu^{2+}	i	0.69	CuL 0.98; CuL_2 0.69; CuL_3 0.55; CuL_4 0.0	[204]
	Fe^{2+}	sp	2	FeL 0.36; FeL_2 0.4	[205]
	Fe^{3+}	sp	2	FeL 0.76; FeL_2 1.06; FeL_3 1.0	[205]

Sec. 4.1] Protonation and complex formation equilibrium constants

Inorganic ligands (continued)

Ligand	Metal ion	Method	I	$\log \beta$	Ref.
Cl^-	Hg^{2+}	pot	0.5	HgL 6.74; HgL_2 13.22; HgL_3 14.07; HgL_4 15.07	[194]
				$HgClCN$ 28.2	[195]
	In^{3+}	i	1	InL 1.42; InL_2 2.23; InL_3 3.23	[19]
	Mn^{2+}	i	0.69	MnL 0.59; MnL_2 0.26; MnL_3 -0.36	[206]
	Pb^{2+}	i	0	PbL 1.6; PbL_2 1.78; PbL_3 1.68; PbL_4 1.38	[207]
	Pd^{2+}	sp	0	PdL 3.88; PdL_2 6.94; PdL_3 9.08 PdL_4 10.42	[208]
	Sn^{2+}	pot	3	SnL 1.15; SnL_2 1.7; SnL_3 1.68	[209]
	Th^{4+}	ex	4	ThL 0.23; ThL_2 -0.85; ThL_3 -1.0; ThL_4 -1.74	[210]
	Tl^+	pol	0	TlL 0.46	[211]
	Tl^{3+}	pot	0	TlL 6.25; TlL_2 11.4; TlL_3 14.5; TlL_4 17; TlL_5 19.15	[212]
	U^{4+}	ex	2	UL 0.52	[213]
	UO_2^{2+}	sp	1.2	UO_2L 1.6	[214]
	Zn^{2+}	ex	v	ZnL -0.72; ZnL_2 -0.85; ZnL_3 -1.50; ZnL_4 -1.75	[215]
ClO^-	H^+	pot	0.1	HL 7.4	[216]
CN^-	H^+	pot	0.1	HL 9.2	[217]
	Ag^+		0.2	AgL_2 21.1; AgL_3 21.9; AgL_4 20.7	[120]
	Au^+		0	AuL_2 38.3	[11]
	Au^{3+}			AuL_4 56	[218]
	Cd^{2+}	pot	3	CdL 5.5; CdL_2 10.6; CdL_3 15.3; CdL_4 18.9	[5]
	Cu^+		0	CuL_2 24; CuL_3 28.6; CuL_4 30.3	[62]
	Fe^{2+}		0	FeL_6 35.4	[219]
	Fe^{3+}		0	FeL_6 43.6	[219]
	Hg^{2+}		0.1	HgL 18.0; HgL_2 34.7; HgL_3 38.5; HgL_4 41.5	[220]
	Ni^{2+}		0.1	NiL_4 31.3	[221]
	Pb^{2+}	pol	1	PbL_4 10.3	[222]
	Pd^{2+}	pot	0	PdL_4 42.4; PdL_5 45.3	[223]
	Tl^{3+}		v	TlL_4 35	[224]
	Zn^{2+}		0	ZnL_4 16.72	[225]
CNO^-	H^+	pot	0.1	HL 3.6	[226]
	Ag^+	cond	0	AgL_2 5.0	[227]

Tables of equilibrium constants [Ch. 4

Inorganic ligands (continued)

Ligand	Metal ion	Method	I	$\log \beta$	Ref.
CNO^-	Cu^{2+}	sp	v	CuL 2·7; CuL_2 4·7; CuL_3 6·1; CuL_4 7·4	[228]
	Ni^{2+}	sp	v	NiL 1·97; NiL_2 3·53; NiL_3 4·90; NiL_4 6·2	[228]
CO_3^{2-}	H^+	pot	0·1	HL 10·1; H_2L 16·4	
	UO_2^{2+}	sol	0·2	UO_2L 15·57; UO_2L_2 20·70	[229]
CrO_4^{2-}	H^+	pot	0·1	HL 6·2; H_2L 6·9; H_2L_2 12·4	[230]
F^-	H^+	pot	0·1	HL 3·15	[321]
	Al^{3+}		0·53	AlL 6·16; AlL_2 11·2; AlL_3 15·1; AlL_4 17·8; AlL_5 19·2; AlL_6 19·24	[232]
	Be^{2+}		0·5	BeL 5·1; BeL_2 8·8; BeL_3 11·8	[233]
	Cr^{3+}		0·5	CrL 4·4; CrL_2 7·7; CrL_3 10·2	[234]
	Cu^{2+}		0·5	CuL 0·95	[235]
	Fe^{2+}	oth	v	$FeL < 1·5$	[236]
	Fe^{3+}		0·5	FeL 5·21; FeL_2 9·16; FeL_3 11·86	[236]
	Ga^{3+}	sp	0·5	GaL 5·1	[234]
	Hg^{2+}		0·5	HgL 1·03	[237]
	In^{3+}		1	InL 3·7; InL_2 6·3; InL_3 8·6; InL_4 9·7	[238]
	La^{3+}	pot	0·5	LaL 2·7	[239]
	Mg^{2+}	pot	0·5	MgL 1·3	[240]
	Ni^{2+}	pot	1	NiL 0·7	[235]
	Pb^{2+}	pot	0·5	$PbL < 0·3$	[237]
	SbO^+	pot	0·1	$SbOL$ 5·5	[241]
	Sc^{3+}	pot	0·5	ScL 6·2; ScL_2 11·5; ScL_3 15·5	[239]
	Sn^{4+}	pol	v	SnL_6 25	[242]
	Th^{4+}		0·5	ThL 7·7; ThL_2 13·5; ThL_3 18·0	[236]
	TiO^{2+}	pot	3	$TiOL$ 5·4; $TiOL_2$ 9·8; $TiOL_3$ 13·7; $TiOL_4$ 17·4	[243]
	UO_2^{2+}	pot	1	UO_2L 4·5; UO_2L_2 7·9; UO_2L_3 10·5; UO_2L_4 11·8	[244]
	Zn^{2+}	pot	0·5	ZnL 0·73	[237]
	Zr^{4+}		2	ZrL 8·8; ZrL_2 16·1; ZrL_3 21·9	[245]
$Fe(CN)_6^{4-}$	H^+	pot	0	HL 4·28; H_2L 6·58; $H_3L \simeq 6·58$	[246]
	K^+	cond	0	KL 2·3	[247]
	Mg^{2+}	sp	0	MgL 3·81	[248]
	La^{3+}	sp	0	LaL 5·06	[249]

Inorganic ligands (continued)

Ligand	Metal ion	Method	I	$\log \beta$	Ref.
$Fe(CN)_6^{3-}$	H^+	pot		$HL < 1$	[250]
	K^+	cond		KL 1·4	[251]
	Mg^{2+}	cond		MgL 2·79	[252]
	La^{3+}	cond		LaL 3·74	[253]
I^-	Ag^+	pot	4	AgL_3 13·85; AgL_4 14·28;	[62]
				Ag_3L 14·15	[254]
	Bi^{3+}	sol	2	BiL_4 15·0; BiL_5 16·8; BiL_6 18·8	[189]
	Cd^{2+}	pot	v	CdL 2·4; CdL_2 3·4; CdL_3 5·0;	
				CdL_4 6·15	[255]
	Hg^{2+}		0·5	HgL 12·87; HgL_2 23·8; HgL_3	
				27·6; HgL_4 29·8	[256]
				$HgICN$ 29·3	[195]
	I_2	ex	v	I_2L 2·9	[257]
	In^{3+}	i	0·69	InL 1·64; InL_2 2·56; InL_3 2·48	[258]
	Pb^{2+}	pol.	1	PbL 1·3; PbL_2 2·8; PbL_3 3·4;	
				PbL_4 3·9	[259]
IO_3^-	H^+	sp	0	HL 0·78	[260]
	Th^{4+}	ex	0·5	ThL 2·9; ThL_2 4·8; ThL_3 7·15	[261]
MoO_4^{2-}	H^+	pot	3	HL 3·9; HL_2 7·50; H_8L_7 57·7;	
				H_9L_7 62·14; $H_{10}L_7$ 65·7;	
				$H_{11}L_7$ 68·2	[262]
NH_3	H^+	pot	0·1	HL 9·35	[26]
	Ag^+	pot	0·1	AgL 3·4; AgL_2 7·40	[16]
	Au^+	pot	v	AuL_2 27	[224]
	Au^{3+}	pot		AuL_4 30	[12]
	Ca^{2+}	pot	2	CaL $-0·2$; CaL_2 $-0·8$;	
				CaL_3 $-1·6$; CaL_4 $-2·7$	[16]
	Cd^{2+}	pot	0·1	CdL 2·6; CdL_2 4·65; CdL_3	
				6·04; CdL_4 6·92; CdL_5 6·6;	
				CdL_6 4·9	[16]
	Co^{2+}	pot	0·1	CoL 2·05; CoL_2 3·62; CoL_3	
				4·61; CoL_4 5·31; CoL_5 5·43;	
				CoL_6 4·75	[16]
	Co^{3+}	pot	2	CoL 7·3; CoL_2 14·0; CoL_3 20·1;	
				CoL_4 25·7; CoL_5 30·8; CoL_6	
				35·2	[16]
	Cu^+	pot	2	CuL 5·90; CuL_2 10·80	[16]
	Cu^{2+}	pot	0·1	CuL 4·13; CuL_2 7·61; CuL_3	
				10·48; CuL_4 12·59	[16]

Inorganic ligands (continued)

Ligand	Metal ion	Method	I	$\log \beta$	Ref.
NH_3	Fe^{2+}		0	FeL 1·4; FeL_2 2·2; FeL_4 3·7	[263]
	Hg^{2+}	pot	2	HgL 8·80; HgL_2 17·50; HgL_3 18·5; HgL_4 19·4	[16]
	Mg^{2+}	pot	2	MgL 0·23; MgL_2 0·08; MgL_3 $-0·36$; MgL_4 $-1·1$	[16]
	Mn^{2+}	pot	v	MnL 0·8; MnL_2 1·3	[62]
	Ni^{2+}	pot	0·1	NiL 2·75; NiL_2 4·95; NiL_3 6·64; NiL_4 7·79; NiL_5 8·50; NiL_6 8·49	[16]
	Tl^+	pot	v	TlL $-0·9$	[16]
	Tl^{3+}	pot	v	TlL_4 17	[224]
	Zn^{2+}	pot	0·1	ZnL 2·27; ZnL_2 4·61; ZnL_3 7·01; ZnL_4 9·06	[16]
NH_2OH	H^+	pot	0·1	HL 6·2	[264]
	Ag	pot	0·5	AgL 1·9	[265]
	Co^{2+}	pot	0·5	CoL 0·9	[265]
	Cu^{2+}	pot	0·5	CuL 2·4; CuL_2 4·1	[265]
	Zn^{2+}	pol	1	ZnL 0·40; ZnL_2 1·0	[266]
N_2H_4	H^+	pot	0·1	HL 8·1	[267]
	Cd^{2+}	pot	0·5	CdL 2·25; CdL_2 2·4; CdL_3 2·78; CdL_4 3·89	[268]
	Co^{2+}	pot	1	CoL 1·78; CoL_2 3·34	[269]
	Cu^{2+}	pot	1	CuL 6·67	[269]
	Mn^{2+}	pot	1	MnL 4·76	[269]
	Ni^{2+}	pot	1	NiL 3·18	[269]
	Zn^{2+}	pot	1	ZnL 3·69; ZnL_2 6·69	[269]
NO_2^-	H^+	cond	0·1	HL 3·2	[270]
	Cu^{2+}	sp	1	CuL 1·2; CuL_2 1·42; CuL_3 0·64	[271]
	Hg^{2+}	pot	v	HgL_4 13·54	[272]
	Pb^{2+}	pol	1	PbL 1·93; PbL_2 2·36; PbL_3 2·13	[273]
NO_3^-	Ba^{2+}	pot	0	BaL 0·94	[274]
	Bi^{3+}	i	1	BiL 0·96; BiL_2 0·62; BiL_3 0·35; BiL_4 0·07	[275]
	Ca^{2+}	cond	0	CaL 0·31	[274]
	Ce^{3+}	ex	1	CeL 0·21	[276]
	Ce^{4+}	sp	3·5	CeL 0·33	[277]
	Eu^{3+}	i	1	EuL 0·15; EuL_2 $-0·4$	[278]
	Pb^{2+}	pol	2	PbL 0·3; PnL_2 0·4	[279]
	Sc^{3+}	i	0·5	ScL 0·55; ScL_2 0·08	[280]

Sec. 4.1] Protonation and complex formation equilibrium constants 323

Inorganic ligands (continued)

Ligand	Metal ion	Method	I	$\log \beta$	Ref.
NO_3^-	Sr^{2+}	cond	0	SrL 0·54	[274]
	Th^{4+}	i	2	ThL 1·22; ThL$_2$ 1·53; ThL$_3$ 1·1	[281]
	Tl^{3+}	pot	3	TlL 0·9; TlL$_2$ 0·12; TlL$_3$ 1·1	[282]
OH^-	H^+	pot	0	HL 14·0	[39]
	Ag^+	sol	0	AgL 2·3; AgL$_2$ 3·6; AgL$_3$ 4·8	[283]
	Al^{3+}	pot	2	AlL$_4$ 33·3; Al$_6$L$_{15}$ 163	[284]
	Ba^{2+}	pot	0	BaL 0·7	[285]
	Be^{2+}	pot	3	BeL$_2$ 3·1; Be$_2$L 10·8; Be$_3$L$_3$ 33·3	[286]
	Bi^{3+}		3	BiL$_3$ 12·4; Bi$_6$L$_{12}$ 168·3; Bi$_9$L$_{20}$ 277	[287]
	Ca^{2+}	sol	0	CaL 1·3	[288]
	Cd^{2+}	ex	3	CdL 4·3; CdL$_2$ 7·7; CdL$_3$ 10·3; CdL$_4$ 12·0	[29]
	Ce^{3+}	pot		CeL 5	[289]
	Ce^{4+}	pot	v	CeL 13·3; Ce$_2$L$_3$ 40·3; Ce$_2$L$_4$ 53·7	[290]
	Co^{2+}	pot	0·1	CoL 4·1; CoL$_2$ 9·2	[291]
	Co^{3+}	oth	3	CoL 13·3	[292]
	Cr^{3+}	pot	0·1	CrL 10·2; CrL$_2$ 18·3	[293]
	Cu^{2+}	pot	0	CuL 6·0; Cu$_2$L$_2$ 17·1	[294]
	Fe^{2+}	pot	1	FeL 4·5	[295]
	Fe^{3+}	pot	3	FeL 11·0; FeL$_2$ 21·7; Fe$_2$L$_2$ 25·1	[296]
	Ga^{3+}	sp	0·5	GaL 11·1	[22]
	Hg_2^{2+}	pot	0·5	Hg$_2$L 9	[297]
	Hg^{2+}	pot	0·5	HgL 10·3; HgL$_2$ 21·7	[298]
	In^{3+}	pot	3	InL 7·0; In$_2$L$_2$ 17·9	[299]
	La^{3+}	pot	3	LaL 3·9; LaL$_2$ 4·1; La$_5$L$_9$ 54·6	[300]
	Li^+	pot	0	LiL 0·2	[301]
	Mg^{2+}	pot	0	MgL 2·6	[285]
	Mn^{2+}	pot	0·1	MnL 3·4	[302]
	Ni^{2+}	pot	0·1	NiL 4·6	[21]
	Pb^{2+}	pot	0·3	PbL 6·2; PbL$_2$ 10·3; PbL$_3$ 13·3; Pb$_2$L 7·6; Pb$_4$L$_4$ 36·1; Pb$_6$L$_8$ 69·3	[21]
	Sc^{3+}	pot	1	ScL 9·1; ScL$_2$ 18·2; Sc$_2$L$_2$ 21·8	[287]
	Sn^{2+}	pot	3	SnL 10·1; Sn$_2$L$_2$ 23·5	[303]
	Sr^{2+}	pot	0	SrL 0·8	[304]
	Th^{4+}	pot	1	ThL 9·7; Th$_2$L 11·1; Th$_2$L$_2$ 22·9	[285]
	Ti^{3+}	pot	0·5	TiL 11·8	[305]
	TiO^{2+}	i	1	TiOL 13·7	[306]
					[307]

21*

Inorganic ligands (continued)

Ligand	Metal ion	Method	I	$\log \beta$	Ref.
OH^-	Tl^+	k	0	TlL 0·8	[308]
	Tl^{3+}		3	TlL 12·9; TlL$_2$ 25·4	[309]
	U^{4+}	pot	3	UL 12	[310]
	UO_2^{2+}	pot	1	$(UO_2)_2L$ 10·3; $(UO_2)_2L_2$ 22·0	[311]
	VO^{2+}	pot	3	VOL 8·0; $(VO)_2L_2$ 21·1	[312]
	Zn^{2+}	pot	2	ZnL 4·9; ZnL$_4$ 13·3; Zn$_2$L 6·5; Zn$_2$L$_6$ 26·8	[313]
	Zr^{4+}	sol	4	ZrL 13·8; ZrL$_2$ 27·2; ZrL$_3$ 40·2; ZrL$_4$ 53	[314]
OOH^-	H^+	ex	0	HL 11·75	[315]
	Co^{3+}	k	v	CoL 13·9	[316]
	Fe^{3+}	pot	0·1	FeL 9·3	[317]
H_2O_2	TiO^{2+}	sp	v	TiOL 4·0	[318]
	VO_2^+	pot	v	VO$_2$L 4·5;	[319]
HPO_3^{2-}	H^+	pot	v	HL 6·58; H$_2$L 8·58	[320]
PO_4^{3-}	H^+	pot	0·1	HL 11·7; H$_2$L 18·6; H$_3$L 20·6	[321]
	Ca^{2+}	pot	0·2	CaHL 13·4	[322]
	Co^{2+}	pot	0·1	CoHL 13·9	[323]
	Cu^{2+}	pot	0·1	CuHL 14·9	[323]
	Fe^{3+}	sp	0·66	FeHL 21·0	[324]
	Mg^{2+}	pot	0·2	MgHL 13·6	[322]
	Mn^{2+}	pot	0·2	MnHL 14·3	[322]
	Ni^{2+}	pot	0·1	NiHL 13·8	[323]
	Sr^{2+}	i	0·15	SrL 4·2; SrHL 12·9; SrH$_2$L 18·85	[325]
	Zn^{2+}	pot	0·1	ZnHL 14·1	[323]
$P_2O_7^{4-}$	H^+	pot	0·1	HL 8·5; H$_2$L 14·6; H$_3$L 17·1; H$_4$L 18·1	[326]
	Ca^{2+}	pot	1	CaL 5·0; CaHL 10·8	[327]
	Cd^{2+}	pot	0	CdL 8·7; Cd(OH)L 11·8	[328]
	Cu^{2+}	sol	1	CuL 6·7; CuL$_2$ 9·0	[62]
	Fe^{3+}	sol	v	FeH$_2$L$_2$ 39·2	[329]
	Hg_2^{2+}	pot	0·75	Hg$_2$(OH)L 15·6	[330]
	Hg^{2+}	pot	0·75	Hg(OH)L 17·45	[330]
	K^+	pot	0	KL 2·3	[328]
	Li^+	pot	0	LiL 3·1	[328]
	Mg^{2+}	oth	0·02	MgL 5·7	[62]
	Na^+	pot	0	NaL 2·3	[328]

Sec. 4.1]   Protonation and complex formation equilibrium constants   325

Inorganic ligands (continued)

Ligand	Metal ion	Method	I	$\log \beta$	Ref.
$P_2O_7^{4-}$	Ni^{2+}	sol	0.1	NiL 5.8; NiL_2 7.2	[62]
	Pb^{2+}	cond	v	PbL_2 5.32	[331]
	Sr^{2+}	i	0.15	SrL 3.26	[325]
	Tl^+	pol	v	TlL 1.7; TlL_2 1.9	[332]
	Zn^{2+}	pot	0	ZnL 8.7; ZnL_2 11.0; $Zn(OH)L$ 13.1	[328]
$P_3O_{10}^{5-}$	H^+	pot	0.1	HL 8.82; H_2L 14.75; H_3L 16.95	[333]
	Ba^{2+}	pot	0.1	BaL 6.3	[328]
	Ca^{2+}	pot	0.1	CaL 6.31; $CaHL$ 12.82	[333]
	Cd^{2+}	pot	0.1	CdL 8.1; $CdHL$ 13.79	[333]
	Co^{2+}	pot	0.1	CoL 7.95; $CoHL$ 13.75	[333]
	Cu^{2+}	pot	0.1	CuL 9.3; $CuHL$ 14.9	[333]
	Fe^{2+}	pot	1	FeL 2.54; FeH_2L 15.9	[334]
	Fe^{3+}	sp	1	FeH_2L 18.8; FeH_2L_2 34.6	[335]
	Hg_2^{2+}	pot	0.75	Hg_2L_2 11.2; $Hg_2(OH)L$ 15.0	[330]
	K^+	pot	0	KL 2.8	[328]
	La^{3+}			LaL 6.56; $LaHL$ 11.78	[336]
	Li^+		0	LiL 3.9	[328]
	Mg^{2+}	pot	0.1	MgL 7.05; $MgHL$ 13.27	[333]
	Mn^{2+}	pot	0.1	MnL 8.04; $MnHL$ 13.90	[333]
	Ni^{2+}	pot	0.1	NiL 7.8; $NiHL$ 13.7	[333]
	Sr^{2+}	pot	0.1	SrL 5.46; $SrHL$ 12.38	[333]
	Zn^{2+}	pot	0.1	ZnL 8.35; $ZnHL$ 13.9	[333]
$P_4O_1^{6-}$	H^+	pot	1	HL 8.34; H_2L 14.97	[337]
	Ca^{2+}	pot	1	CaL 5.46; $CaHL$ 11.88	[338]
	Cu^{2+}	pot	1	CuL 9.44; Cu_2L 10.6; $Cu(OH)L$ 13.30	[339]
	K^+	pot	1	KHL 9.45	[340]
	La^{3+}	pot	0.1	LaL 6.59; $LaHL$ 12.13	[336]
	Li^+	pot	1	$LiHL$ 9.93	[340]
	Mg^{2+}	pot	1	MgL 6.04; $MgHL$ 12.08	[338]
	Na^+	pot	1	$NaHL$ 9.44	[340]
	Sr^{2+}	pot	1	SrL 4.82; $SrHL$ 11.83; Sr_2L 8.24	[338]
S^{2-}	H^+	pot	0	HL 12.92; H_2L 19.97	[15]
	Ag^+	pot	0.1	AgL 16.8; $AgHL$ 26.2; AgH_2L_2 43.5	[341]
	Hg^{2+}	pot	v	HgL_2 53; HgH_2L_2 66.8	[342]
SCN^-	Ag^+	sol	2.2	AgL_2 8.2; AgL_3 9.5; AgL_4 10.0	[343]
	Au^+		v	AuL_2 25	[202]

Inorganic ligands (continued)

Ligand	Metal ion	Method	I	$\log \beta$	Ref.
SCN$^-$	Au^{3+}		v	AuL$_2$ 42	[202]
	Bi^{3+}	pot	0.4	BiL 0.8; BiL$_2$ 1.9; BiL$_3$ 2.7; BiL$_4$ 3.5; BiL$_5$ 3.25; BiL$_6$ 3.2	[344]
	Cd^{2+}	pol	2	CdL 1.4; CdL$_2$ 1.88; CdL$_3$ 1.93 CdL$_4$ 2.38	[345]
	Co^{2+}	sp	1	CoL 1.01	[346]
	Cr^{3+}		v	CrL 2.52; CrL$_2$ 3.76; CrL$_3$ 4.42; CrL$_5$ 4.62; CrL$_6$ 4.23 (50 °C)	[4]
	Cu$^+$	sol	5	CuL$_2$ 11.0	[347]
	Cu^{2+}	sp	0.5	CuL 1.7; CuL$_2$ 2.5; CuL$_3$ 2.7; CuL$_4$ 3.0	[348]
	Fe^{2+}	sp	v	FeL 1.0	[349]
	Fe^{3+}	sp	v	FeL 2.3; FeL$_2$ 4.2; FeL$_3$ 5.6; FeL$_4$ 6.4; FeL$_5$ 6.4	[350]
	Hg^{2+}	pol	1	HgL$_2$ 16.1; HgL$_3$ 19.0; HgL$_4$ 20.9	[351]
	In^{3+}	pot	2	InL 2.6; InL$_2$ 3.6; InL$_3$ 4.6	[238]
	Mn^{2+}	sp	0	MnL 1.23	[352]
	Ni^{2+}	i	1.5	NiL 1.2; NiL$_2$ 1.6; NiL$_3$ 1.8	[20]
	Pb^{2+}	pol	2	PbL 0.5; PbL$_2$ 1.4; PbL$_3$ 0.4; PbL$_4$ 1.3	[353]
	Tl$^+$	pol	2	TlL 0.4	[353]
	Zn^{2+}	pol	2	ZnL 0.5; ZnL$_2$ 1.32; ZnL$_3$ 1.32; ZnL$_4$ 2.62	[354]
SO$_3^{2-}$	H$^+$	pot	0.1	HL 6.8; H$_2$L 8.6	
SO$_4^{2-}$	H$^+$	pot	0.1	HL 1.8	
	Ca^{2+}	sol	0	CaL 2.3	[355]
	Cd^{2+}	pot	3	CdL 0.85	[356]
	Ce^{3+}	i	1	CeL 1.63; CeL$_2$ 2.34; CeL$_3$ 3.08	[357]
	Ce^{4+}	sp	2	CeL 3.5; CeL$_2$ 8.0; CeL$_3$ 10.4	[358]
	Co^{2+}	cond	0	CoL 2.47	[359]
	Cr^{3+}	pol	0.1	CrL 1.76	[360]
	Cu^{2+}	pot	1	CuL 1.0; CuL$_2$ 1.1; CuL$_3$ 2.3	[361]
	Eu^{3+}	ex	1	EuL 1.54; EuL$_2$ 2.69	[362]
	Fe^{2+}	k	1	FeL 1.0	[334]
	Fe^{3+}	sp	1.2	FeL 2.23; FeL$_2$ 4.23; FeHL 2.6	[363]
	In^{3+}	ex	1	InL 1.85; InL$_2$ 2.6; InL$_3$ 3.0	[364]
	K$^+$	pot	0.1	KL 0.4	[365]
	La^{3+}	ex	1	LaL 1.45; LaL$_2$ 2.46	[366]
	Lu^{3+}	ex	1	LuL 1.29; LuL$_2$ < 2.5; LuL$_3$ 3.36	[367]

Sec. 4.1] Protonation and complex formation equilibrium constants

Inorganic ligands (continued)

Ligand	Metal ion	Method	I	$\log \beta$	Ref.
SO_4^{2-}	Mg^{2+}	pot	0	MgL 2·25	[368]
	Mn^{2+}	cond	0	MnL 2·3	[369]
	Ni^{2+}	cond	0	NiL 2·3	[359]
	Sc^{3+}	i	0·5	ScL 1·66; ScL_2 3·04; ScL_3 4·0	[370]
	U^{4+}	ex	2	UL 3·6; UL_2 6·0	[213]
	UO_2^{2+}	sp	0	UO_2L 2·96; UO_2L_2 4·0	[371]
	Th^{4+}	ex	2	ThL 3·32; ThL_2 5·6	[372]
	Y^{3+}	pot	3	YL 2·0; YL_2 3·4; YL_3 4·36	[373]
	Zn^{2+}	cond	0	ZnL 2·31	[374]
	Zr^{4+}	ex	2	ZrL 3·7; ZrL_2 6·5; ZrL_3 7·6	[245]
$S_2O_3^{2-}$	H^+	pot	0	HL 1·72; H_2L 2·32	[375]
	Ag^+	pot	0	AgL 8·82; AgL_2 13·46; AgL_3 14·15	[376]
	Ba^{2+}	sol	0	BaL 2·33	[377]
	Ca^{2+}	sp	0	CaL 1·91	[378]
	Cd^{2+}	sp	0	CdL 3·94	[378]
	Co^{2+}	sol	0	CoL 2·05	[377]
	Cu^+	pol	2	CuL 10·3; CuL_2 12·2; CuL_3 13·8	[379]
	Fe^{2+}		0·48	FeL 0·92 (6·1 °C)	[380]
	Fe^{3+}	sp	0·47	FeL 2·10	[380]
	Hg^{2+}	pot	0	HgL_2 29·86; HgL_3 32·26; HgL_4 33·61	[381]
	Mg^{2+}	sp	0	MgL 1·79	
	Mn^{2+}	sol	0	MnL 1·95	[378]
	Ni^{2+}	sol	0	NiL 2·06	[377]
	Pb^{2+}	sol	v	PbL 5·1; PbL_2 6·4	[382]
	Sr^{2+}	sol	0	SrL 2·04	[377]
	Tl^+	pol	0	TlL 1·91	[383]
	Zn^{2+}	sp	0	ZnL 2·29	[378]
Se^{2-}	H^+	pot	0	HL 11·0; H_2L 14·89	[384]
SeO_3^{2-}	H^+	pot	0	HL 8·32; H_2L 10·94	[385]
SeO_4^{2-}	H^+	pot	0	HL 1·88	[386]
$SiO_2(OH)_2^{2-}$	H^+	cond	0	HL 11·81; H_2L 21·27	
	Fe^{3+}	sp	0·1	FeHL 21·03	
TeO_4^{2-}	H^+	pot	0	HL 11·04; H_2L 18·74	[389]

Organic complexants

Metal ion	Method	I	$\log \beta$	Ref.
	Acetic acid HL			
H^+	pot	0·1	HL 4·65	
Ag^+	pot	0	AgL_2 0·64; Ag_2L 1·14	[390]
Ba^{2+}	pot	0	BaL 1·15	[391]
Ca^{2+}	pot	0	CaL 1·24	[391]
Cd^{2+}	pot	0·1	CdL 1·61	[392]
Ce^{3+}	pot	0·1	CeL 2·09; CeL_2 3·53	[392]
Co^{2+}	pot	0	CoL 1·5; CoL_2 1·9	[393]
Cu^{2+}	pot	1	CuL 1·67; CuL_2 2·65; CuL_3 3·07; CuL_4 2·88	[361]
Dy^{3+}	pot	0·1	DyL 2·03; DyL_2 3·64	[392]
Er^{3+}	pot	0·1	ErL 2·01; ErL_2 3·60	[392]
Eu^{3+}	pot	0·1	EuL 2·31; EuL_2 3·91	[392]
Fe^{3+}	pot	0·1	FeL 3·2; FeL_2 6·1; FeL_3 8·3	[394]
Gd^{3+}	pot	0·1	GdL 2·16; GdL_2 3·76	[392]
La^{3+}	pot	0·1	LaL 2·02; LaL_2 3·26	[392]
Lu^{3+}	pot	0·1	LuL 2·05; LuL_2 3·69	[392]
Mg^{2+}	pot	0	MgL 1·25	[391]
Mn^{2+}	pot	0	MnL 1·40	[395]
Nd^{3+}	pot	0·1	NdL 2·22; NdL_2 3·76	[392]
Ni^{2+}	pot	0	NiL 1·43	[395]
Pb^{2+}	pot	0·1	PbL 2·20; PbL_2 3·59	[392]
Pr^{3+}	pot	0·1	PrL 2·18; PrL_2 3·63	[392]
Sm^{3+}	pot	0·1	SmL 2·30; SmL_2 3·88	[392]
Sr^{2+}	pot	0	SrL 1·19	[391]
Tb^{3+}	pot	0·1	TbL 2·07; TbL_2 3·66	[392]
Tl^{3+}		0·2	TlL_4 15·4	[396]
Tm^{3+}	pot	0·1	TmL 2·02; TmL_2 3·61	[392]
UO_2^{2+}	ex	0·1	UO_2L 2·61; UO_2L_2 4·9; UO_2L_3 6·3	[397]
Y^{3+}	pot	0·1	YL 1·97; YL_2 3·60	[392]
Yb^{3+}	pot	0·1	YbL 2·03; YbL_3 3·67	[392]
Zn^{2+}	pot	0·1	ZnL 1·28; ZnL_2 2·09	[392]
	Acetylacetone HL			
H^+	pot	0·2	HL 8·9	[398]
Al^{3+}	pot	0	AlL 8·6; AlL_2 16·5; AlL_3 22·3	[17]
Be^{2+}	pot	0	BeL 7·8; BeL_2 14·5	[399]
Cd^{2+}	pot	0	CdL 3·84; CdL_2 6·7	[399]
Ce^{3+}	pot	0	CeL 5·3; CeL_2 9·27; CeL_3 12·65	[399]
Co^{2+}	pot	0	CoL 5·4; CoL_2 9·57	[399]
Cu^{2+}	pot	0	CuL 8·31; CuL_2 15·6	[399]
Dy^{3+}	pot	0·1	DyL 6·03; DyL_2 10·70; DyL_3 14·04	[400]
Er^{3+}	pot	0·1	ErL 5·99; ErL_2 10·67; ErL_3 14·05	[400]
Eu^{3+}	pot	0·1	EuL 5·87; EuL_2 10·35; EuL_3 13·64	[400]

Sec. 4.1] Protonation and complex formation equilibrium constants 329

Organic complexants (continued)

Metal ion	Method	I	$\log \beta$	Ref.
Fe^{2+}	pot	0	FeL 5·07; FeL$_2$ 8·67	[399]
Fe^{3+}	pot	0	FeL 9·8; FeL$_2$ 18·8; FeL$_3$ 26·4	[17]
Ga^{3+}	pot	0	GaL 9·5; GaL$_2$ 17·4; GaL$_3$ 23·1	[17]
Gd^{3+}	pot	0·1	GdL 5·9; GdL$_2$ 10·38; GdL$_3$ 13·79	[400]
Hf^{4+}	pot	0·1	HfL 8·7; HfL$_2$ 15·4; HfL$_3$ 21·8; HfL$_4$ 28·1	[401]
Ho^{3+}	pot	0·1	HoL 6·05; HoL$_2$ 10·73; HoL$_3$ 14·13	[400]
In^{3+}	pot	0	InL 8·0; InL$_2$ 15·1	[17]
La^{3+}	pot	0·1	LaL 4·96; LaL$_2$ 8·41; LaL$_3$ 10·91	[400]
Lu^{3+}	pot	0·1	LuL 6·23; LuL$_2$ 11·0; LuL$_3$ 14·63	[400]
Mg^{2+}	pot	0	MgL 3·67; MgL$_2$ 6·38	[399]
Mn^{2+}	pot	0	MnL 4·24; MnL$_2$ 7·35	[399]
Nd^{3+}	pot	0·1	NdL 5·3; NdL$_2$ 9·4; NdL$_3$ 12·6	[400]
Ni^{2+}	pot	0	NiL 6·06; NiL$_2$ 10·77; NiL$_3$ 13·09	[17]
Pb^{2+}	pot	0·1	PbL 4·2; PbL$_2$ 6·6	[401]
Pd^{2+}	pot	0	PdL 16·7; PdL$_2$ 27·6	[402]
Pr^{3+}	pot	0·1	PrL 5·27; PrL$_2$ 9·17; PrL$_3$ 12·7	[400]
Sc^{3+}	pot	0	ScL 0·0; ScL$_2$ 15·2	[17]
Sm^{3+}	pot	0·1	SmL 5·59; SmL$_2$ 10·05; SmL$_3$ 12·95	[400]
Tb^{3+}	pot	0·1	TbL 6·02; TbL$_2$ 10·63; TbL$_3$ 14·04	[400]
Th^{4+}	pot	0	ThL 8·8; ThL$_2$ 16·2; ThL$_3$ 22·5; ThL$_4$ 26·7	[17]
Tm^{3+}	pot	0·1	TmL 6·09; TmL$_2$ 10·85; TmL$_3$ 14·33	[400]
U^{4+}	ex	0·1	UL 8·6; UL$_2$ 17; UL$_3$ 23·4; UL$_4$ 29·5	[403]
UO_2^{2+}	pot	0	UO$_2$L 7·66; UO$_2$L$_2$ 14·15	[399]
Y^{3+}	pot	0	YL 6·4; YL$_2$ 11·1; YL$_3$ 13·9	[17]
Yb^{3+}	pot	0·1	YbL 6·18; YbL$_2$ 11·04; YbL$_3$ 14·64	[400]
Zn^{2+}	pot	0	ZnL 5·07; ZnL$_2$ 9·02	[399]
Zr^{4+}	pot	0·1	ZrL 8·4; ZrL$_2$ 16·0; ZrL$_3$ 23·2; ZrL$_4$ 30·1	[401]

N-acetyl cysteine H$_2$L

Metal ion	Method	I	$\log \beta$	Ref.
H^+	pot	0·1	HL 9·75; H$_2$L 12·95; H$_3$L 14·65	[404]
Ni^{2+}	pot	0·1	NiL 5·10; NiL$_2$ 9·25	[404]
Zn^{2+}	pot	0·1	ZnL 6·35; ZnL$_2$ 12.11	[404]

Adenosine-5-triphosphate (ATP) H$_4$L

Metal ion	Method	I	$\log \beta$	Ref.
H^+	pot	0·1	HL 6·54; H$_2$L 10·68	[405]
Ba^{2+}	pot	0·1	BaL 3·42; BaHL 8·46	[405]
Ca^{2+}	pot	0·1	CaL 3·99; CaHL 8·75	[405]
Co^{2+}	pot	0·1	CoL 4·69; CoHL 8·93	[405]
Cu^{2+}	pot	0·1	CuL 6·13; CuHL 9·74	[405]
Mg^{2+}	pot	0·1	MgL 4·22; MgHL 8·70	[405]
Mn^{2+}	pot	0·1	MnL 4·78; MnHL 9·02	[405]
Ni^{2+}	pot	0·1	NiL 5·02; NiHL 9·34	[405]
Zn^{2+}	pot	0·1	ZnL 4·88; ZnHL 9·27	[405]

Organic complexants (continued)

Metal ion	Method	I	$\log \beta$	Ref.

α-Alanine HL

Metal ion	Method	I	$\log \beta$	Ref.
H^+	pot	0.1	HL 9.8; H_2L 12.1	[406]
Ag^+	pot	0	AgL 3.64; AgL_2 7.18	[407]
Ba^{2+}	sol	0	BaL 0.8	[407]
Ca^{2+}	pot	0	CaL 1.24	[408]
Cd^{2+}	pol	2	CdL 5.13; CdL_2 7.82; CdL_3 9.16	[409]
Co^{2+}	pot	0	CoL 4.82; CoL_2 8.48	[407]
Cu^{2+}	pot	0	CuL 8.51; CuL_2 15.37	[407]
Fe^{2+}	pot	0.01	FeL 7.3	[410]
Fe^{3+}	pot	1	FeL 10.4	[411]
Mn^{2+}	pot	0.01	MnL 3.24; MnL_2 6.05	[412]
Ni^{2+}	pot	0	NiL 5.96; NiL_2 10.66	[407]
Pb^{2+}	pot	0	PbL 5.0; PbL_2 8.24	[407]
Sr^{2+}	sol	0	SrL 0.73	[413]
Zn^{2+}	pot	0	ZnL 5.21; ZnL_2 9.54	[407]

β-Alanine HL

Metal ion	Method	I	$\log \beta$	Ref.
H^+	pot	0.5	HL 10.21; H_2L 13.83	[414]
Ag^+	pot	0.5	AgL 3.44; AgL_2 7.25	[415]
Cd^{2+}	pol	1	CdL_2 5.70; CdL_3 6.78; $CdL_3(OH)$ 7.20; CdL_2CO_3 6.60; $CdL_2(NH_3)_4$ 7.98	[416]
Co^{2+}	pot	0.2	CoL 3.58	[417]
Cu^{2+}	pot	0.2	CuL 7.10	[417]
Ni^{2+}	pot	0.5	NiL 4.46; NiL_2 7.84; NiL_3 9.55; NiL (pyr) 8.34; NiL_2 (pyr) 11.95; NiL_2 (pyr)$_2$ 15.17; (pyr = pyruvate)	[414]
Pb^{2+}	pol	1	$PbL_2(OH)_2$ 12.11	[416]
Zn^{2+}	pot	0.5	ZnL 3.9; ZnL(pyr) 7.08; ZnL_2(pyr)$_2$ 12.1 (pyr = pyruvate)	[414]

Alizarin Red S H_3L

Metal ion	Method	I	$\log \beta$	Ref.
H^+	pot		HL 11.1; H_2L 17.17	
Be^{2+}	pot	0.1	BeL 10.96	[418]
Zr^{4+}	sp	0.1	$Zr(OH)_2L$ 49.0	[419]

Organic complexants (continued)

Metal ion	Method	I	$\log \beta$	Ref.
			Aniline L	
H^+	pot	1	HL 4·78	[420]
Hg^{2+}	pot		HgL 4·61; HgL_2 9·21	[420]
			Anthranilic acid HL	
H^+	pot	0·1	HL 4·9; H_2L 7·0	[421]
Ag^+	pot	0	AgL 1·86	[421]
Ce^{3+}	pot	0·1	CeL 3·18	[422]
Cd^{2+}	pot	0	CdL 1·83	[421]
Co^{2+}	pot	0	CoL 1·56	[421]
Cu^{2+}	pot	0	CuL 4·25	[421]
La^{3+}	pot	0·1	LaL 3·14	[422]
Nd^{3+}	pot	0·1	NdL 3·23	[422]
Ni^{2+}	pot	0	NiL 2·12; NiL_2 3·59	[421]
Pb^{2+}	pot	0	PbL 2·82	[321]
Pr^{3+}	pot	0·1	PrL 3·22	[422]
Zn^{2+}	pot	0	ZnL 2·57	[421]
			Antipyrine (1-phenyl-2,3-dimethyl-3-pyrazolone) L	
H^+	pot	0·1	HL 1·38	[170]
			Arsenazo III. H_8L	

H^+	sp	0·2	HL 12·33; H_2L 19·81; H_3L 25·16; H_4L 27·57; H_5L 29·98	[423]
Dy^{3+}	sp	0·2	Dy_2L_2 83·0	[423]
Gd^{3+}	sp	0·2	Gd_2L_2 80·5	[423]
La^{3+}	sp	0·2	La_2L_2 81·2; La_2H_4L 42·5; $La_2H_8L_2$ 83·5	[423]
Sm^{3+}	sp	0·2	Sm_2L_2 82·1	[423]
Yb^{3+}	sp	0·2	Yb_2L_2 81·9	[423]

Organic complexants (continued)

Metal ion	Method	I	$\log \beta$	Ref.

Ascorbic acid H_2L

Metal ion	Method	I	$\log \beta$	Ref.
H^+	pot	0	HL 11·56; H_2L 15·73	[424]
TiO^{2+}	sp	0·1	$TiOH_2L$ 18·81; $TiOH_2L_2$ 47·94; $TiOH_4L_2$ 37·67	[425]

Benzoic acid HL

Metal ion	Method	I	$\log \beta$	Ref.
H^+	pot	0·01	HL 4·12	
Ag^+	pot	1	AgL 3·4; AgL_2 4·2	[426]
Cd^{2+}	pol	0·1	CdL 1·08; CdL_2 1·18; CdL_3 1·64; CdL_4 1·87	[427]
Cu^{2+}	pot	0·1	CuL 3·30 (50% dioxane)	[428]
Pb^{2+}	pol	1	PbL_2 3·30	[429]
UO_2^{2+}	pot	0·1	UO_2L 2·59	[430]
Zn^{2+}	pot	0·1	ZnL 2·35	[431]

2,2'-Bipyridyl L

Metal ion	Method	I	$\log \beta$	Ref.
H^+	pot	0·1	HL 4·47	[432]
Ag^+	pot	0·1	AgL 3·03; AgL_2 6·67	[433]
Cd^{2+}	ex	0·1	CdL 4·12; CdL_2 7·62; CdL_3 10·22	[432]
Co^{2+}	pot	0·1	CoL 6·06; CoL_2 11·42; CoL_3 16·02	[434]
Cu^{2+}	pot	0·1	CuL 8·0; CuL_2 13·6; CuL_3 17·08	[434]
Fe^{2+}	pot	0·1	FeL 4·4; FeL_2 8·0; FeL_3 17·6	[434]
Hg^{2+}	pot	0·1	HgL 9·64; HgL_2 16·74; HgL_3 19·54	[434]
Mn^{2+}	pot	0·1	MnL 2·6; MnL_2 4·6; MnL_3 6·3	[434]
Ni^{2+}	pot	0·1	NiL 7·13; NiL_2 14·01; Ni_3L 20·54	[434]
Pb^{2+}	pot	0·1	PbL 2·9	[434]
Zn^{2+}	pot	0·1	ZnL 5·3; ZnL_2 9·83; ZnL_3 13·63	[434]

Calcon (see Eriochrome Black R)

Calmagite H_3L

Metal ion	Method	I	$\log \beta$	Ref.
H^+	pot		HL 12·4; H_2L 20·5	[435]
Ca^{2+}	sp		CaL 6·1	[435]
Mg^{2+}	sp		MgL 8·1	[435]

Sec. 4.1]   **Protonation and complex formation equilibrium constants**   333

Organic complexants (continued)

Metal ion	Method	I	$\log \beta$	Ref.

Chloranilic acid H_2L

H^+	sp	0·1	HL 2·72; H_2L 3·53	[436]
Fe^{3+}	sp	0·1	FeL 5·81; FeL_2 9·84	[436]
Hf^{3+}	sp	3	HfL 6·45; HfL_3 19·79	[437]
Ni^{2+}	sp	0·1	NiL 4·02	[436]

Chrome Azurol S H_4L

H^+	sp	0·1	HL 11·81; H_2L 16·52; H_3L 18·77	[438]
Be^{2+}	sp	0·1	BeHL 16·57; Be_2L_2 26·8	[439]
Cu^{2+}	sp	0·1	CuHL 15·83; Cu_2L 13·7	[438]
Fe^{3+}	sp	0·1	FeL 15·6; Fe_2L 20·2; Fe_2L_2 36·2	[440]

Chromotropic acid H_4L

H^+	pot	0·1	HL 15·6; H_2L 20·96	[441]
Al^{3+}	pot	0·1	AlL 17·4; AlL_2 34·26	[442]
Be^{2+}	pot	0·1	BeL 16·34; BeL_2 28·19; BeHL 18·50	[443]

Organic complexants (continued)

Metal ion	Method	I	$\log \beta$	Ref.
Cu^{2+}	pot	0.1	CuL 13.44	[444]
Th^{4+}	pot	0.1	ThL 16.46; ThL$_2$ 29.14	[445]
TiO^{2+}	sp	0.1	TiOL$_2$ 40.5; TiOL$_3$ 56.4	[441]

Citric acid H$_4$L

Metal ion	Method	I	$\log \beta$	Ref.
H^+	pot	0.1	HL 16; H$_2$L 22.1; H$_3$L 26.5; H$_4$L 29.5	
Al^{3+}		0.5	AlL 20; AlHL 23; Al(OH)L 30.6	[446]
Ba^{2+}	i	0.16	BaHL 18.5	[447]
Be^{2+}	i	0.15	BeHL 20.5; BeH$_2$L 24.3; BeH$_3$L 27.9	[448]
Ca^{2+}	i	0	CaHL 20.68; CaH$_2$L 25.1; CaH$_3$L 27.6	[449]
Cd^{2+}	pot	0.15	CdHL 20; CdH$_2$L 24.4	[450]
Co^{2+}	pol	0.15	CoHL 20.8; CoH$_2$L 25.3	[450]
Cu^{2+}	pot	0.1	CuL 18; CuHL 22.3; CuH$_3$L 28.3	[451]
Fe^{2+}	pot	1	FeL 15.5; FeHL 19.1; FeH$_2$L 24.2	[452]
Fe^{3+}	pot	1	FeL 25.0; FeHL 27.8; FeH$_2$L 28.4	[452]
Mg^{2+}	pol	0.15	MgHL 19.29; MgH$_2$L 23.7	[450]
Mn^{2+}	pot	0.15	MnHL 19.7; MnH$_2$L$_2$ 24.2	[450]
Ni^{2+}	pot	0.15	NiL 14.3; NiHL 21.1; NiH$_2$L 25.3	[450]
Pb^{2+}	pot		PbHL 19; PbH$_2$L 27.8	[453]
Sr^{2+}	i	0.16	SrHL 18.8	[453]
UO_2^{2+}	pot	0.15	UO$_2$HL 24.5	[450]
Zn^{2+}	pot	0.15	ZnL 11.4; ZnHL 20.8; ZnH$_2$L 25.0	[450]

Cupferron HL

Metal ion	Method	I	$\log \beta$	Ref.
H^+	pot	0.1	HL 4.16	[454]
La^{3+}	ex	0.1	LaL$_3$ 12.9	[454]
Th^{4+}	ex	0.1	ThL$_2$ 14.6; ThL$_4$ 27.0	[454]
UO_2^{2}	sp	0	UO$_2$L$_2$ 11.0	[455]

Cysteine (2-amino-3-mercaptopropanoic acid) H$_2$L

Metal ion	Method	I	$\log \beta$	Ref.
H^+	pot	0.1	HL 10.11; H$_2$L 18.24; H$_3$L 20.2	[456]
Co^{2+}	pot	0.01	CoL 9.3; CoL$_2$ 17	[457]
Co^{3+}	pot	0.01	CoL 16.2; CoL$_2$ 32.9	[457]
Cu^{2+}	pot	1	CuL 19.2	[458]
Fe^{2+}	pot	0	FeL 6.2; FeL$_2$ 11.7; Fe(OH)L 12.7	[459]
Hg^{2+}	pot	0.1	HgL 14.21	[456]
Mn^{2+}	pot	0.1	MnL 4.56	[456]

Organic complexants (continued)

Metal ion	Method	I	$\log \beta$	Ref.
Ni^{2+}	pot	0.1	NiL 9.64; NiL_2 19.04	[456]
Pb^{2+}	pot	0.1	PbL 11.39	[456]
Zn^{2+}	pot	0.1	ZnL 9.04; ZnL_2 17.54	[456]

DCTA (1,2-diaminocyclohexanetetra-acetic acid) H_4L

Metal ion	Method	I	$\log \beta$	Ref.
H^+	pot		HL 11.78; H_2L 17.98; H_3L 21.58; H_4L 24.09	[25]
Al^{3+}	pot	0.1	AlL 17.6; AlHL 19.6 Al(OH)L 24	[25]
Ba^{2+}	pot	0.1	BaL 8.0; BaHL 14.7	[25]
Bi^{3+}	pol	0.5	BiL 31.2	[460]
Ca^{2+}	pot	0.1	CaL 12.5	[25]
Cd^{2+}	pot	0.1	CdL 19.2; CdHL 22.2	[25]
Ce^{3+}	pot	0.1	CeL 16.8	[25]
Co^{2+}	pot	0.1	CoL 18.9; CoHL 21.8	[25]
Cu^{2+}	pot	0.1	CuL 21.3; CuHL 24.4	[25]
Dy^{3+}	pot	0.1	DyL 19.7	[25]
Er^{3+}	pot	0.1	ErL 20.7	[25]
Eu^{3+}	pot	0.1	EuL 18.6	[25]
Fe^{2+}	pot	0.1	FeL 18.2	[461]
Fe^{3+}	pot	0.1	FeL 29.3; Fe(OH)L 34.0	[461]
			Fe(OOH)HL 32.2	[462]
Ga^{3+}	pot	0.1	GaL 22.9	[25]
Gd^{3+}	pot	0.1	GdL 18.8	[25]
Hg^{2+}	pot	0.1	HgL 24.3; HgHL 27.4; Hg(OH)L 27.8	[25]
La^{3+}	pot	0.1	LaL 16.3; LaHL 18.9	[25]
Lu^{3+}	pot	0.1	LuL 21.5	[25]
Mg^{2+}	pot	0.1	MgL 10.3	[25]
Mn^{2+}	pot	0.1	MnL 16.8; MnHL 19.6	[25]
Nd^{3+}	pot	0.1	NdL 17.7	[25]
Ni^{2+}	pot	0.1	NiL 19.4	[463]
Pb^{2+}	pot	0.1	PbL 19.7; PbHL 22.5	[25]
Pr^{3+}	pot	0.1	PrL 17.3	[25]
Sm^{3+}	pot	0.1	SmL 18.4	[25]
Sr^{2+}	pot	0.1	SrL 10.0	[463]
Tb^{3+}	pot	0.1	TbL 19.5	[25]
Tm^{3+}	pot	0.1	TmL 21.0	[25]
Th^{4+}	pot	0.1	ThL 23.2; Th(OH)L 29.6	[464]
VO^{2+}	pot	0.1	VOL 19.4	[25]
Y^{3+}	pot	0.1	YL 19.2	[25]
Yb^{3+}	pot	0.1	YbL 21.1	[25]
Zn^{2+}	pot	0.1	ZnL 18.7; ZnHL 21.7	[25]

Organic complexants (continued)

Metal ion	Method	I	$\log \beta$	Ref.

1,2-Diaminopropane L

Metal ion	Method	I	$\log \beta$	Ref.
H^+	pot	0·1	HL 9·8; H_2L 16·43	[465]
Cu^{2+}	pot	0·1	CuL 10·5; CuL 19·6	[466]
Hg^{2+}	pot	0·1	HgL 23·51	[467]
Ni^{2+}	pot	0·1	NiL 7·3; NiL_2 13·4	[468]
Zn^{2+}	pot	0·1	ZnL 5·7; ZnL_2 10·6	[465]

1,3-Diaminopropane L

Metal ion	Method	I	$\log \beta$	Ref.
H^+	pot	0·1	HL 10·72; H_2L 19·68	[469]
Ag^+	pot	0·1	AgL 5·85	[469]
Cu^{2+}		0·1	CuL 9·98; CuL_2 17·17	[470]
Ni^{2+}		0·1	NiL 6·39; NiL_2 10·78; NiL_3 12·01	[470]

Dichloroacetic acid HL

Metal ion	Method	I	$\log \beta$	Ref.
H^+	pot	0·1	HL 1·1	

Diethanolamine L

Metal ion	Method	I	$\log \beta$	Ref.
H^+	pot	0·5	HL 8·95	[471]
Ag^+	pot	0	AgL 3·48; AgL_2 5·60	[472]
Cd^{2+}	pol		CdL_2 4·30; CdL_3 5·08	[473]
Cu^{2+}	pol	0·5	$CuL(OH)_2$ 18·2; $CuL_2(OH)_2$ 19·8	[471]
Pb^{2+}	pol	0	PbL_2 8·70; PbL_3 9·0	[473]
Zn^{2+}	pol	0	ZnL_2 6·60; ZnL_3 8·08; ZnL_4 9·11	[473]

Diethyldithiocarbamic acid HL

Metal ion	Method	I	$\log \beta$	Ref.
H^+			HL $\simeq$ 4	[474]
Hg^{2+}	pot	0·1	HgL 22·3; HgL_2 38·1; HgL_3 39·1	[475]
Tl^+	old	0·1	TlL 4·3; TlL_2 5·3	[476]

Diethylglycollic acid HL

Metal ion	Method	I	$\log \beta$	Ref.
H^+	pot	1	HL 3·62	[477]
Er^{3+}	pot	1	ErL 3·11; ErL_2 5·31; ErL_3 6·69; ErL_4 7·41	[477]
Gd^{3+}	pot	1	GdL 2·71; GdL_2 4·67; GdL_3 5·71; GdL_4 6·56	[477]
Ho^{3+}	pot	1	HoL 3·07; HoL_2 5·16; HoL_3 6·46; HoL_4 7·30	[477]
Nd^{3+}	pot	1	NdL 2·28; NdL_2 3·89; NdL_3 5·10; NdL_4 6·04	[477]
Pr^{3+}	pot	1	PrL 2·31; PrL_2 3·8; PrL_3 4·8; PrL_4 5·39	[477]
Tb^{3+}	pot	1	TbL 3·01; TbL_2 5·08; TbL_3 6·45; TbL_4 6·98	[477]
Yb^{3+}	pot	1	YbL 3·1; YbL_2 5·36; YbL_3 6·67; YbL_4 7·76	[477]

Sec. 4.1] Protonation and complex formation equilibrium constants 337

Organic complexants (continued)

Metal ion	Method	I	$\log \beta$	Ref.
			Diethylmalonic acid H_2L	
H^+	pot	0·1	HL 6·98; H_2L 8·94	[478]
Ce^{3+}			CeL 3·78; CeL_2 6·32	[478]
Gd^{3+}			GdL 4·49; GdL_2 7·05	[478]
La^{3+}			LaL 3·61; LaL_2 5·95	[478]
Lu^{3+}			LuL 4·69; LuL_2 7·40	[478]
			Diethylenetriamine L	
H^+	pot	0·1	HL 9·94; H_2L 19·07; H_3L 23·4	[28]
Ag^+	pot	0·1	AgL 6·1; AgHL 13·2	[28]
Cd^{2+}	pot	0·1	CdL 8·45; CdL_2 13·85	[28]
Co^{2+}	pot	0·1	CoL 8·1; CoL_2 14·1	[28]
Cu^{2+}	pot	0·1	CuL 16·0; CuL_2 21·3	[28]
Fe^{2+}	pot	0·1	FeL 6·23; FeL_2 10·36	[479]
Hg^{2+}	pol	0·1	HgL_2 25·06; HgL_3 24·0	[480]
Mn^{2+}	pot	0·1	MnL 3·99; MnL_2 6·82	[479]
Ni^{2+}	pot	0·1	NiL 10·7; NiL_2 18·9	[28]
Zn^{2+}	pot	0·1	ZnL 8·9; ZnL_2 14·5	[28]
			Diethylenetriaminepenta-acetic acid (DTPA) H_5L	
H^+	pot	0·1	HL 10·56; H_2L 19·25; H_3L 23·62; H_4L 26·49; H_5L 28·43	[481]
Ag^+	pot	0·1	AgL 8·70	[482]
Al^{3+}	pot	0·1	AlL 18·51	[483]
Ba^{2+}	pot	0·1	BaL 8·8; BaHL 14·1	[481]
Bi^{3+}	pot	1	BiL 35·4; BiHL 38·2 Bi(OH)L 38·3	[484]
Ca^{2+}	pot	0·1	CaL 10·6; CaHL 17; Ca_2L 12·6	[481] [485]
Cd^{2+}	pot	0·1	CdL 19·0; CdHL 22·9; Cd_2L 22	[481]
Ce^{3+}	pot	0·1	CeL 20·5	[486]
Co^{2+}	pot	0·1	CoL 19·0; CoHL 23·8; Co_2L 22·5	[485]
Cu^{2+}	pot	0·1	CuL 20·5; CuHL 24·5; Cu_2L 26·0	[481]
Dy^{3+}	pot	0·1	DyL 22·8; DyHL 25·0	[486]
Er^{3+}	pot	0·1	ErL 22·7; ErHL 24·7	[486]
Eu^{3+}	pot	0·1	EuL 22·4; EuHL 24·55	[486]
Fe^{2+}	pot	0·1	FeL 16·0; FeHL 21·4; Fe(OH)L 21·0; Fe_2L 19·0	[485]
Fe^{3+}	pot	0·1	FeL 27·5; FeHL 30·9; Fe(OH)L 31·6	[485]
Ga^{3+}		0·1	GaL 25·54; GaHL 29·89; Ga(OH)L 32·06	[484]
Gd^{3+}	pot	0·1	GdL 22·46; GdHL 24·85	[486]
Hg^{2+}	pot	0·1	HgL 27·0; HgHL 30·6	[481]
Ho^{3+}	pot	0·1	HoL 22·78; HoHL 25·03	[486]

22 Inczédy

Organic complexants (continued)

Metal ion	Method	I	$\log \beta$	Ref.
La^{3+}	pot	0·1	LaL 19·43; LaHL 22·03	[487]
Li^{+}	pot	0·1	LiL 3·1	[481]
Lu^{3+}	pot	0·1	LuL 22·44; LuHL 24·62	[486]
Mg^{2+}	pot	0·1	MgL 9·3; MgHL 16·2	[481]
Mn^{2+}	pot	0·1	MnL 15·5; MnHL 20·0; Mn_2L 17·6	[481]
Nd^{3+}	pot	0·1	NdL 21·6; NdHL 24·0	[486]
Ni^{2+}	pot	0·1	NiL 20·0; NiHL 25·6; Ni_2L 25·4	[481]
Pb^{2+}	pot	0·1	PbL 18·9; PbHL 23·4; Pb_2L 22·3	[481]
Pr^{3+}	pot	0·1	PrL 21·07; PrHL 23·45	[486]
Sm^{3+}	pot	0·1	SmL 22·34; SmHL 24·54	[486]
Sr^{2+}	pot	0·1	SrL 9·7; SrHL 15·1	[488]
Tb^{3+}	pot	0·1	TbL 22·71; TbHL 24·85	[486]
Th^{4+}	pot	0·1	ThL 28·78; ThHL 30·94; Th(OH)L 33·68	[484]
Tl^{3+}	pot	1	TlL 46·0	[489]
Tm^{3+}	pot	0·1	TmL 22·72; TmHL 24·62	[486]
Yb^{3+}	pot	0·1	YbL 22·62; YbHL 24·92	[496]
Zn^{2+}	pot	0·1	ZnL 18·0; ZnHL 23·6; Zn_2L 22·4	[481]
Zr^{4+}	pot	1	ZrL 36·9	[484]

5-[4-Dimethylaminobenzylidene]rhodanine HL

H^{+}	ex	0·1	HL 8·20	[490]
Ag^{+}	ex	0·1	AgL 9·15	[490]
Cu^{3+}	ex	0·1	CuL 6·08	[490]

Dimethylglyoxime HL

H^{+}	sol		HL 10·6	[118]
Co^{2+}	pot	0·1	CoL 8·35; CoL_2 16·98	[491]
Cu^{2+}	pot	0·1	CuL 9·05; CuL_2 18·50	[491]
Ni^{2+}	ex	0·1	NiL_2 17·24	[119] [492]
Pd^{2+}	ex	0·1	PdL_2 34·1	[493]

Dithio-oxamide (Rubeanic acid) L

$$H_2N-C=S$$
$$|$$
$$H_2N-C=S$$

H^{+}	pot	1	HL 10·4	[494]
Ru^{3+}	sp	1	RuL 13·38; RuL_2 38·14	[494]

Sec. 4.1] Protonation and complex formation equilibrium constants

Organic complexants (continued)

Metal ion	Method	I	log β	Ref.
			Dithizone (Diphenylthiocarbazone) H(HL)	
H^+	sp	0·1	H(HL) 4·5	[495]
Hg^{2+}	ex		$Hg(HL)_2$ 40·34	[496]
Zn^{2+}			$Zn(HL)_2$ 10·8	[497]
			DTPA (see Diethylenetriaminepenta-acetic acid)	
			EDTA (Ethylenediaminetetra-acetic acid) H_4L	
H^+	pot	0·1	HL 10·34; H_2L 16·58; H_3L 19·33;	[51]
			H_4L 21·40; H_5L 23·0; H_6L 23·9	[498]
Ag^+	pot	0·1	AgL 7·3; AgHL 13·3	[499]
Al^{3+}	pol	0·1	AlL 16·13; AlHL 18·7; Al(OH)L 24·2	[25]
Ba^{2+}	pot	0·1	BaL 7·76; BaHL 12·4	[51]
Be^{2+}	ex	0·1	BeL 9·27	[500]
Bi^{3+}	pol	0·5	BiL 28·2; BiHL 29·6	[501]
Ca^{2+}	pot	0·1	CaL 10·7; CaHL 13·8	[51]
Cd^{2+}	pol	0·1	CdL 16·46; CdHL 19·4	[25]
Ce^{3+}	pot	0·1	CeL 15·98; CeHL 19·05	[25]
				[502]
Co^{2+}	pot	0·1	CoL 16·31; CoHL 19·5	[25]
Co^{3+}	pot	0·1	CoL 36; CoHL 37·3	[503]
Cr^{3+}	pot	0·1	CrL 23; CrHL 25·3; Cr(OH)L 29·6	[504]
Cu^{2+}	pot	0·1	CuL 18·8; CuHL 21·8; Cu(OH)L 21·2	[25]
				[505]
Dy^{3+}	pot	0·1	DyL 18·30; DyHL 21·1	[25]
Er^{3+}	pol	0·1	ErL 18·98; ErHL 21·7	[25]
Eu^{3+}	pol	0·1	EuL 17·35; EuHL 20·0	[25]
Fe^{2+}	pot	0·1	FeL 14·33; FeHL 17·2	[25]
Fe^{3+}	pot	0·1	FeL 25·1; FeHL 26·0; Fe(OH)L 31·6	[110]
				[498]
Ga^{3+}	pot	0·1	GaL 20·25; GaHL 21·6	[506]
Gd^{3+}	pot	0·1	GdL 17·37; GdHL 20·0	[25]
Hg^{2+}	pot	0·1	HgL 21·8; HgHL 24·94; Hg(OH)L 26·9;	
			$Hg(NH_3)L$ 28·5	[25]
Ho^{3+}	pot	0·1	HoL 18·74; HoHL 21·4	[25]
In^{3+}	pot	0·1	InL 24·95; InHL 25·95; In(OH)L 30	[506]
La^{3+}	pot	0·1	LaL 15·5; LaHL 17·5	[25]

Organic complexants (continued)

Metal ion	Method	I	log β	Ref.
Li^+	pot	0·1	LiL 2·8	[51]
Lu^{3+}	pot	0·1	LuL 19·83; LuHL 22·3	[25]
Mg^{2+}	pot	0·1	MgL 8·6; MgHL 12·6	[51]
Mn^{2+}	pol	0·1	MnL 14·04; MnHL 17·2	[25]
Mo^{5+}	sp		MoL 6·36	[507]
Na^+	pot	0·1	NaL 1·66	[51]
Nd^{3+}	pot	0·1	NdL 16·61; NdHL 21·00	[25]
Ni^{2+}	pot	0·1	NiL 18·6 NiHL 21·8	[25]
Pb^{2+}	pol	0·1	PbL 18·0; PbHL 20·9	[25]
Pr^{3+}	pol	0·1	PrL 16·4	[25]
Ra^{2+}		0·1	RaL 7·4	[508]
Sc^{3+}	pol	0·1	ScL 23·1; ScHL 21·2; Sc(OH)L 26·6	[25]
Sm^{3+}	pot	0·1	SmL 17·14; SmHL 19·74	[25]
Sn^{2+}	pot	0·1	SnL 22·1	[509]
Sr^{2+}	pot	0·1	SrL 8·6; SrHL 12·64	[51]
Tb^{3+}	pot	0·1	TbL 17·9; TbHL 20·5	[25]
Th^{4+}	pot	0·1	ThL 23·2; Th(OH)L 30·2	[510]
Ti^{3+}	pot	0·1	TiL 21·3	[511]
TiO^{2+}	pot	0·1	TiOL 17·3	[511]
Tl^{3+}	pot	0·1	TlL 22·5; TlHL 24·8	[512]
Tm^{3+}	pot	0·1	TmL 19·32; TmHL 21·9	[25]
UO_2^{2+}	ex	0·1	UO_2HL 17·66	[513]
V^{2+}	pot	0·1	VL 12·7 V(OH)L 30·4	[504]
V^{3+}	pot	0·1	VL 25·9	[504]
VO^{2+}	pol	0·1	VOL 18·77	[25]
VO_2^+	pot	0·1	VO_2L 18·1; VO_2HL 21·7	[514]
Y^{3+}	pot	0·1	YL 18·1	[25]
Yb^{3+}	pol	0·1	YbL 19·54; YbHL 22·2	[25]
Zn^{2+}	pol	0·1	ZnL 16·5; ZnHL 20·9; Zn(OH)L 19·5	[25]
Zr^{4+}	pot	0·1	ZrL 29·9; Zr(OH)L 37·7	[12]

EGTA [(Ethylenedioxy)diethylenedinitrilo]tetra-acetic acid H_4L

Metal ion	Method	I	log β	Ref.
H^+	pot	0·1	HL 9·46; H_2L 18·31; H_3L 20·96; H_4L 22·96	[515]
Ag^+	pot	0·1	AgL 7·06	[516]
Al^{3+}	pot	0·2	AlL 13·90; AlHL 17·87; Al(OH)L 22·70; $Al(OH)_2L$ 28·28	[517]
Bi^{3+}	sp	0·5	BiL 23·8; BiHL 25·46	[518]
Ca^{2+}	pot	0·1	CaL 10·97; CaHL 14·76	[515]
Cd^{2+}	pot	0·1	CdL 16·1; CdHL 19·60	[515]
Co^{2+}	pot	0·1	CoL 12·28; CoHL 17·44; Co_2L 15·58	[515]
Cu^{2+}	pot	0·1	CuL 17·71; CuHL 22·07	[515]
Fe^{2+}	pot	0·1	FeL 11·81; FeHL 15·86	[515]
Fe^{3+}	sp	0·1	FeL 20·5	[519]
Hg^{2+}	pot	0·1	HgL 23·2; HgHL 26·26	[515]

Sec. 4.1] Protonation and complex formation equilibrium constants

Organic complexants (continued)

Metal ion	Method	I	log β	Ref.
La³⁺	pot	0·1	LaL 15·79	[515]
Mg²⁺	pot	0·1	MgL 5·2; MgHL 12·86	[515]
Mn²⁺	pot	0·1	MnL 12·28; MnHL 16·48	[515]
Na⁺	pot	0·1	NaL 1·38	[520]
Ni²⁺	pot	0·1	NiL 11·82; NiHL 17·76; Ni₂L 16·72	[515]
Pb²⁺	pot	0·1	PbL 11·8; PbHL 16·96; Pb₂L 16·40	[515]
Zn²⁺	pot	0·1	ZnL 12·91; ZnHL 17·88; Zn₂L 16·21	[515]

Eriochrome Black A H_3L

H⁺	sp		HL 13·0; H₂L 19·2	[521]
Ca²⁺	sp		CaL 5·3	[521]
Mg²⁺	sp		MgL 7·2	[521]

Eriochrome Black B H_3L

H⁺	sp	0·1	HL 12·5; H₂L 18·7	[521]
Ca²⁺	sp		CaL 5·7	[521]
Mg²⁺	sp		MgL 7·4	[521]
Zn²⁺	sp		ZnL 12·5; Zn(NH)₃L 16·4	[522] [523]

Eriochrome Black R (Calcon) H_3L

H⁺	sp	0·1	HL 13·5; H₂L 20·5	[521]
Ca²⁺	sp	0·1	CaL 5·3	[521]

Organic complexants (continued)

Metal ion	Method	I	$\log \beta$	Ref.
Mg^{2+}	sp	0·1	MgL 7·6	[521]
Zn^{2+}	sp		ZnL 12·5; $Zn(NH_3)L$ 16·4	[523]

Eriochrome Black T H_3L

H^+	sp	0·1	HL 11·55; H_2L 17·8	[521]
Ba^{2+}	sp		BaL 3·0	[15]
Ca^{2+}	sp		CaL 5·4	[521]
Cd^{2+}	sp	0·1	CdL 12·74	[524]
Co^{2+}	sp	0·1	CoL 20·0	[525]
Cu^{2+}	sp	0·1	CuL 21·38	[526]
Mg^{2+}	sp	0·1	MgL 7·0	[521]
Mn^{2+}	sp	0·1	MnL 9·6; MnL_2 17·6	[15]
Pb^{2+}	sp	0·1	PbL 13·19	[524]
Zn^{2+}	sp	0·1	ZnL 12·9; ZnL_2 20·0	[521]

Eriochrome Cyanine R H_4L

H^+	sp	0·1	HL 11·85; H_2L 17·32; H_3L 19·62	[527]
Be^{2+}	sp	0·1	BeHL 17·34; Be_2L_2 28·3	[527]
Fe^{3+}	sp	0·1	FeL 17·9; Fe_2L 22·5; Fe_2L_2 37·9	[528]

Sec. 4.1] Protonation and complex formation equilibrium constants

Organic complexants (continued)

Metal ion	Method	I	$\log \beta$	Ref.

Ethylenediamine (En) L

Metal ion	Method	I	$\log \beta$	Ref.
H^+	pot	0·1	HL 9·94; H_2L 17·08	
Ag^+	pot	0·1	AgL 4·7; AgL_2 7·7; AgHL 12·3; Ag_2L 6·5; Ag_2L_2 13·23	[469]
Cd^{2+}	pot	0·5	CdL 5·47; CdL_2 10·0; CdL_3 12·1	[529]
Co^{2+}	pot	1	CoL 5·89; CoL_2 10·72; CoL_3 13·82	[16]
Co^{3+}	pot	1	CoL 48·69	[16]
Cr^{3+}	sp	0·1	CrL 16·5; $CrL_2 \sim 26$	[530]
Cu^+	pot		CuL_2 10·8	[531]
Cu^{2+}	pot	0·5	CuL 10·55; CuL_2 19·60	[529]
Fe^{2+}	pot	0·1	FeL 4·28; FeL_2 7·53; FeL_3 9·62	[16]
Hg^{2+}	pol	0·1	HgL 14·3; HgL_2 23·3; Hg(OH)L 23·8; $HgHL_2$ 28·5	[532]
Mg^{2+}	pot	0·1	MgL 0·37	[16]
Mn^{2+}	pot	1	MnL 2·73; MnL_2 4·79; MnL_3 5·67	[16]
Ni^{2+}	pot	1	NiL 7·66; NiL_2 14·06; NiL_3 18·61	[16]
Zn^{2+}	pot	1	ZnL 5·71; ZnL_2 10·37; ZnL_3 12·09	[529]

Ferron H_2L

[Structure: 8-hydroxyquinoline with I at position 7 and SO_3H at position 5]

Metal ion	Method	I	$\log \beta$	Ref.
H^+	pot	0·1	HL 7·11; H_2L 9·61	[533]
Al^{3+}	pot	0·1	AlL 7·6; AlL_2 14·70; AlL_3 20·30; $Al(OH)L_2$ 23·70	[533]
Co^{2+}	pot	0·1	CoL 6·70; CoL_2 10·87	[534]
Cu^{2+}	pot	0·1	CuL 8·33; CuL_2 16·58	[534]
Fe^{3+}	pot	0·1	FeL 8·9; FeL_2 17·30; FeL_3 25·20	[534]
Mn^{2+}	pot	0·1	MnL 4·95; MnL_2 8·10	[534]
Ni^{2+}	pot	0·1	NiL 7·70; NiL_2 13·96	[534]
Zn^{2+}	pot	0·1	ZnL 7·25; ZnL_2 13·40	[534]

Formic acid HL

Metal ion	Method	I	$\log \beta$	Ref.
H^+	pot	0	HL 3·77	
Al^{3+}	i	1	AlL 1·78	[535]
Ba^{2+}	pot	0	BaL 1·38	[536]
Ca^{2+}	pot	0	CaL 1·43	[536]

Organic complexants (continued)

Metal ion	Method	I	$\log \beta$	Ref.
Cd^{2+}	pot	0	CdL 0.65; CdL$_2$ 0.40; CdL$_3$ 1.32	[537]
Cu^{2+}	pol	2	CuL 1.57; CuL$_2$ 2.22; CuL$_3$ 2.05; CuL$_4$ 2.45	[538]
Fe^{3+}	i	1	FeL 1.85; FeL$_2$ 3.61; FeL$_3$ 3.95; FeL$_4$ 5.4	[535]
Mg^{2+}	pot	0	MgL 1.43	[536]
Mn^{2+}	i	1	MnL 0.80	[535]
Pb^{2+}	pol	2	PbL 0.78; PbL$_2$ 1.2; PbL$_3$ 1.43; PbL$_4$ 1.18	[538]
Zn^{2+}	pol	2	ZnL 0.6; ZnL$_2$ 1.55; ZnL$_3$ 2.03; ZnL$_4$ 2.77	[538]

Fumaric acid H_2L

H^+	cond.	0	HL 4.39; H$_2$L 7.41	[539]
Ba^{2+}	cond.	0	BaL 1.59	[539]
Ca^{2+}	cond.	0	CaL 2.0	[539]
Cu^{2+}	pot	0	CuL 2.51	[540]
La^{3+}	pot	0	LaL 3.04	[540]
Sr^{2+}	i	0.16	SrL 0.54	[541]

α-Furildioxime HL

H^+	pot	0.1	HL 11.1; H$_2$L 22.5 (75% dioxan)	[542]
Co^{2+}	pot	0.1	CoL 8.2; CoL$_2$ 15.4 (85% dioxan)	[542]
Cu^{2+}	pot	0.1	CuL$_2$ 18.6 (75% dioxane)	[542]
Ni^{2+}	pot	0.1	NiL 6.9; NiL$_2$ 14.1 (75% dioxan)	[542]

Glutamic acid H_2L

H^+	pot	0.1	HL 9.67; H$_2$L 13.95; H$_3$L 16.25	[543]
Ba^{2+}	pot	0.1	BaL 1.28	[543]
Ca^{2+}	pot	0.1	CaL 1.43	[543]
Cd^{2+}	pot	0.1	CdL 3.9	[543]
Co^{2+}	pot	0.02	CoL 5.06; CoL$_2$ 8.46	[544]
Cu^{2+}	pot	0.02	CuL 7.85; CuL$_2$ 14.40	[544]
Mg^{2+}	pot	0.1	MgL 1.9	[543]
Ni^{2+}	pot	0.02	NiL 5.90; NiL$_2$ 10.34	[544]
Pb^{2+}	pol	1	PbL 4.60; PbL$_2$ 6.22	[545]
Sr^{2+}	pot	0.1	SrL 1.37	[543]
UO_2^{2+}	pot	0.2	UO$_2$HL 12.3	[546]
Zn^{2+}	pot	0.02	ZnL 5.45; ZnL$_2$ 9.46	[544]

Glycine HL

H^+	pot	0.1	HL 9.84; H$_2$L 12.36	[547]
Ag^+	oth	0.1	AgL 3.3; AgL$_2$ 6.8	[407]
Ba^{2+}	oth	0	BaL 0.77	[407]
Ca^{2+}	oth	0	CaL 1.43	[407]

Sec. 4.1] Protonation and complex formation equilibrium constants

Organic complexants (continued)

Metal ion	Method	I	$\log \beta$	Ref.
Cd^{2+}	pot	0.1	CdL 4.14; CdL$_2$ 7.46	[548]
Co^{2+}	pot	0.1	CoL 4.7; CoL$_2$ 8.5; CoL$_3$ 11.0	[549]
Cu^{2+}	pot	0.1	CuL 8.1; CuL$_2$ 15.09	[548]
Fe^{2+}	pot	0.01	FeL 4.3; FeL$_2$ 7.8	[550]
Hg^{2+}	pot	0.5	HgL 10.3; HgL$_2$ 19.2	[551]
Mg^{2+}	pot	0	MgL 3.44	[407]
Mn^{2+}	pot	0.01	MnL 3.2; MnL$_2$ 5.5	[550]
Ni^{2+}	pot	0.1	NiL 5.80; NiL$_2$ 10.70	[547]
Pb^{2+}	pot	0	PbL 5.47; PbL$_2$ 8.9	[407]
Sr^{2+}	pot	0	SrL 0.9	[552]
Zn^{2+}	pot	0	ZnL 5.52; ZnL$_2$ 9.96	[407]

Glycolic acid HL

H^+	pot	0	HL 3.8	[553]
Ba^{2+}	pot	0	BaL 1.0	[554]
Ca^{2+}	pot	0	CaL 1.59	[554]
Cu^{2+}	pot	1	CuL 2.34; CuL$_2$ 3.7; CuL$_3$ 3.99; CuL$_4$ 3.77	[361]
Eu^{3+}	i	0.75	EuL 3.03; EuL$_2$ 5.7; EuL$_3$ 7.79	[555]
In^{3+}	i	0.5	InL 2.93; InL$_2$ 5.4	[556]

HEDTA (Hydroxyethyl-ethylenediaminetriacetic acid) H_3L

H^+	pot	0.1	HL 10.0; H$_2$L 15.4; H$_3$L 17.8	[557]
Ag^+	pot	0.1	AgL 6.71	[558]
Al^{3+}	pot	0.1	AlL 14.4; AlHL 16.8; Al(OH)L 23.7	[559]
Ba^{2+}	pot	0.1	BaL 6.2	[487]
Ca^{2+}	pot	0.1	CaL 8.5	[557]
Cd^{2+}	pot	0.1	CdL 13.0	[560]
Ce^{3+}	pot	0.1	CeL 14.2	[557]
Co^{2+}	pot	0.1	CoL 14.4	[560]
Cu^{2+}	pot	0.1	CuL 17.4	[557]
Dy^{3+}	pot	0.1	DyL 15.34; Dy(OH)L 20.1	[557] [561]
Er^{3+}	pot	0.1	ErL 15.4; Er(OH)L 20.5	[557] [561]
Eu^{3+}	pot	0.1	EuL 15.4; Eu(OH)L 19.4	[557] [561]
Fe^{2+}	pot	0.1	FeL 12.2 Fe(OH)L 17.2	[562]
Fe^{3+}	pot	0.1	FeL 19.8; Fe(OH)L 29.9	[562]
Ga^{3+}	pot	0.1	GaL 16.9; GaHL 21.07	[563]
Gd^{3+}	pot	0.1	GdL 15.3; Gd(OH)L 19.4	[537] [561]

Organic complexants (continued)

Metal ion	Method	I	$\log \beta$	Ref.
Hg^{2+}	pot	0·1	HgL 20·1; Hg(OH)L 25·7; Hg(NH$_3$)L 26·2	[115] [487]
Ho^{3+}	pot	0·1	HoL 15·3; Ho(OH)L 20·4	[557] [561]
La^{3+}	pot	0·1	LaL 13·5; La(OH)L 16·95	[557] [561]
Lu^{3+}	pot	0·1	LuL 15·9; Lu(OH)L 20·0; LuHL 19·1	[557] [561]
Mg^{2+}	pot	0·1	MgL 7·0	[487]
Mn^{2+}	pot	0·1	MnL 10·7	[560]
Nd^{3+}	pot	0·1	NdL 14·9; Nd(OH)L 18·6	[557] [561]
Ni^{2+}	pot	0·1	NiL 17·0	[560]
Pb^{2+}	pot	0·1	PbL 15·5	[487]
Pr^{3+}	pot	0·1	PrL 14·7; Pr(OH)L 18·4	[557] [561]
Sm^{3+}	pot	0·1	SmL 15·4; Sm(OH)L 19·1	[557] [561]
Sr^{2+}	pot	0·1	SrL 6·8; SrHL 11·4	[487]
Tb^{3+}	pot	0·1	Tb 15·4; Tb(OH)L 19·9	[557] [561]
Tm^{3+}	pot	0·1	TmL 15·6; Tm(OH)L 20·7	[557] [561]
Y^{3+}	pot	0·1	YL 14·65	[557]
Yb^{3+}	pot	0·1	YbL 15·9; Yb(OH)L 21·1	[557] [561]
Zn^{2+}	pot	0·1	ZnL 14·5	[487]

Hexamethylenetetramine L

H^+	pot	0·01	HL 5·13	[564]
Ag^+	pot	0·01	AgL$_2$ 3·58	[564]

Histidine HL

H^+	pot	0·1	HL 9·2; H$_2$L 15·3	[565]
Cd^{2+}	pot	0·1	CdL$_2$ 10·2	[565]
Co^{2+}	pot	0·25	CoL 6·77; CoL$_2$ 11·9	[566]
Cu^{2+}	pot	0·15	CuL 9·79; CuL$_2$ 17·41	[567]
Hg^{2+}	pol	0·6	HgL$_2$ 20·62	[568]
Mn^{2+}	pot	0·15	MnL 3·24; MnL$_2$ 6·16	[567]
Ni^{2+}	pot	0·25	NiL 8·50; NiL$_2$ 15·19	[566]
Pb^{2+}	pot	0·15	PbL 5·96; PbL$_2$ 8·96	[567]
Zn^{2+}	pot	0·25	ZnL 6·40; ZnL$_2$ 11·95	[566]

Protonation and complex formation equilibrium constants

Organic complexants (continued)

Metal ion	Method	I	$\log \beta$	Ref.

α-Hydroxyisobutyric acid (2-Hydroxy-2-methylpropanoic acid) HL

Metal ion	Method	I	$\log \beta$	Ref.
H^+	pot	0.2	HL 3.8	[569]
Ba^{2+}	pot	1	BaL 0.36; BaL$_2$ 0.51	[569]
Ca^{2+}	pot	1	CaL 0.92; CaL$_2$ 1.42	[569]
Cd^{2+}	pot	1	CdL 1.24; CdL$_2$ 2.16; CdL$_3$ 2.19	[570]
Co^{2+}	pot	1	CoL 1.46; CoL$_2$ 2.53	[570]
Cu^{2+}	pot	1	CuL 2.74; CuL$_2$ 4.34; CuL$_3$ 4.38	[571]
Dy^{3+}	pot	0.2	DyL 2.94; DyL$_2$ 5.45; DyL$_3$ 7.29; DyL$_4$ 8.5	[27]
Er^{3+}	pot	0.2	ErL 3.0; ErL$_2$ 5.7; ErL$_3$ 7.57; ErL$_4$ 9.0	[27]
Eu^{3+}	pot	0.1	EuL 2.70; EuL$_2$ 4.97; EuL$_3$ 6.5; EuL$_4$ 7.6	[571]
Gd^{3+}	pot	0.2	GdL 2.79; GdL$_2$ 4.98; GdL$_3$ 6.5; GdL$_4$ 7.65	[27]
Ho^{3+}	pot	0.2	HoL 2.98; HoL$_2$ 5.54; HoL$_3$ 7.44; HoL$_4$ 8.74	[27]
Lu^{3+}	pot	0.2	LuL 3.18; LuL$_2$ 6.04; LuL$_3$ 8.07; LuL$_4$ 10.0	[27]
Mg^{2+}	pot	1	MgL 0.81; MgL$_2$ 1.47	[569]
Mn^{2+}	pot	1	MnL 0.96; MnL$_2$ 1.54; MnL$_3$ 1.56	[570]
Ni^{2+}	pot	1	NiL 1.67; NiL$_2$ 2.8; NiL$_3$ 2.84	[570]
Pb^{2+}	pot	1	PbL 2.03; PbL$_2$ 3.2; PbL$_3$ 3.22	[570]
Sm^{3+}	pot	0.2	SmL 2.75; SmL$_2$ 4.77; SmL$_3$ 6.17; SmL$_4$ 7.38	[27]
Sr^{2+}	pot	1	SrL 0.55; SrL$_2$ 0.73	[569]
Tb^{3+}	pot	0.2	TbL 2.92; TbL$_2$ 5.24; TbL$_3$ 6.86; TbL$_4$ 8.09	[27]
Tm^{3+}	pot	0.2	TmL 3.1; TmL$_2$ 5.79; TmL$_3$ 7.71; TmL$_4$ 9.33	[27]
UO_2^{2+}	pot	1	UO$_2$L 3.01; UO$_2$L$_2$ 4.83; UO$_2$L$_3$ 6.38	[570]
Yb^{3+}	pot	0.2	YbL 3.13; YbL$_2$ 5.87; YbL$_3$ 7.94; YbL$_4$ 9.72	[27]
Zn^{2+}	pot	1	ZnL 1.71; ZnL$_2$ 3.01	[570]

8-Hydroxyquinoline (Oxine) HL

Metal ion	Method	I	$\log \beta$	Ref.
H^+	pot	0	HL 9.9; H$_2$L 14.9	[8]
Ba^{2+}	pot	0	BaL 2.07	[8]
Ca^{2+}	pot	0	CaL 3.27	[8]
Cd^{2+}	pot	0.01	CdL 7.2; CdL$_2$ 13.4	[572]
Co^{2+}	pot	0.01	CoL 9.1; CoL$_2$ 17.2	[572]
Cu^{2+}	pot	0.01	CuL 12.2; CuL$_2$ 23.4	[572]
Fe^{2+}	pot	0.01	FeL 8.0; FeL$_2$ 15.0	[572]
Fe^{3+}	pot	0.01	FeL 12.3; FeL$_2$ 23.6; FeL$_3$ 33.9	[573]
La^{3+}	ex	0.1	LaL 5.85; LaL$_3$ 16.95	[454]
Mg^{2+}	pot	0.01	MgL 4.5	[572]
Mn^{2+}	pot	0.01	MnL 6.8; MnL$_2$ 12.6	[572]
Ni^{2+}	pot	0.01	NiL 9.9; NiL$_2$ 18.7	[572]
Pb^{2+}	pot	0	PbL 9.02	[574]

Organic complexants (continued)

Metal ion	Method	I	log β	Ref.
Sm^{3+}	ex	0·1	SmL 6·84; SmL_3 19·50	[454]
Sr^{2+}	ex	0·1	SrL 2·89; SrL_2 3·19	[454]
Th^{4+}	ex	0·1	ThL 10·45; ThL_2 20·4; ThL_3 29·8; ThL_4 38·8	[454]
UO_2^{2+}	pot	0·3	UO_2L 11·25; UO_2L_2 21·0 (50% dioxane)	[575]
Zn^{2+}	pot	0·3	ZnL 9·34; ZnL_2 17·56 (50% dioxane)	[575]

8-Hydroxyquinoline-5-sulphonic acid (Sulphoxine) H_2L

Metal ion	Method	I	log β	Ref.
H^+	pot	0·1	HL 8·35; H_2L 12·19	[576]
Ba^{2+}	sp	0	BaL 2·3	[577]
Ca^{2+}	sp	0	CaL 3·5	[577]
Cd^{2+}	sp	0	CdL 7·7; CdL_2 14·2	[577]
Ce^{3+}	pot	0	CeL 6·05; CeL_2 11·05; CeL_3 14·95	[578]
Co^{2+}	pot	0	CoL 8·1; CoL_2 15·06; CoL_3 20·4	[576]
Cu^{2+}	pot	0	CuL 11·9; CuL_2 21·9	[576]
Er^{3+}	pot	0	ErL 7·16; ErL_2 13·3; ErL_3 18·56	[578]
Fe^{4+}	pot	0·1	FeL 11·6; FeL_2 22·8	[576]
Gd^{3+}	pot	0	GdL 6·64; GdL_2 12·37; GdL_3 17·3	[578]
La^{3+}	pot	0	LaL 5·6; LaL_2 10·1; LaL_3 13·8	[578]
Mg^{2+}	pot	0·1	MgL 4·06; MgL_2 7·63	[576]
Mn^{2+}	pot	0·1	MnL 5·67; MnL_2 10·7	[576]
Nd^{3+}	pot	0	NdL 6·3; NdL_2 11·6; NdL_3 16·0	[578]
Ni^{2+}	pot	0·1	NiL 9·0; NiL_2 16·77; NiL_3 22·9	[576]
Pb^{2+}	pot	0	PbL 8·5; PbL_2 16·1	[577]
Pr^{3+}	pot	0	PrL 6·17; PrL_2 11·37; PrL_3 15·7	[578]
Sr^{2+}	sp	0	SrL 2·75	[577]
Th^{4+}	pot	0·1	ThL 9·56; ThL_2 18·29; ThL_3 25·9; ThL_4 32·0	[576]
UO_2^{2+}	pot	0·1	UO_2L 8·5; UO_2L_2 15·7	[576]
Zn^{2+}	pot	0·1	ZnL 7·64; ZnL_2 14·32	[576]

Iminodiacetic acid (IDA) H_2L

Metal ion	Method	I	log β	Ref.
H^+	pot	0·1	HL 9·38; H_2L 12·03	[579]
Ba^{2+}	pot	0·1	BaL 1·67	[579]
Ca^{2+}	pot	0·1	CaL 2·6	[579]
Cd^{2+}	pot	0·1	CdL 5·35; CdL_2 9·53	[580]
Ce^{3+}	pot	0·1	CeL 6·18; CeL_2 10·71	[581]
Co^{2+}	pot	0·1	CoL 6·95; CoL_2 12·29	[580]
Cu^{2+}	pot	0·1	CuL 10·55; CuL_2 16·2	[580]
Dy^{3+}	pot	0·1	DyL 6·88; DyL_2 12·3	[581]
Er^{3+}	pot	0·1	ErL 7·09; ErL_2 12·68	[581]
Eu^{3+}	pot	0·1	EuL 6·73; EuL_2 12·11	[581]
Fe^{2+}	pot	0·1	FeL 5·8; FeL_2 10·1	[582]

Sec. 4.1] Protonation and complex formation equilibrium constants

Organic complexants (continued)

Metal ion	Method	I	$\log \beta$	Ref.
Gd^{3+}	pot	0·1	GdL 6·68; GdL_2 12·07	[581]
Ho^{3+}	pot	0·1	HoL 6·97; HoL_2 12·47	[581]
In^{3+}	pot	0·3	InL 9·54; InL_2 18·41	[583]
La^{3+}	pot	0·1	LaL 5·88; LaL_2 9·97	[581]
Lu^{3+}	pot	0·1	LuL 7·61; LuL_2 13·73	[581]
Mg^{2+}	pot	0·1	MgL 2·94	[579]
Nd^{3+}	pot	0·1	NdL 6·50; NdL_2 11·39	[581]
Ni^{2+}	pot	0·1	NiL 8·26; NiL_2 14·61	[580]
Pr^{3+}	pot	0·1	PrL 6·44; PrL_2 11·22	[581]
Sr^{2+}	pot	0·1	SrL 2·23	[582]
Sm^{3+}	pot	0·1	SmL 6·64; SmL_2 11·88	[581]
Tb^{3+}	pot	0·1	TbL 6·78; TbL_2 12·24	[581]
Tm^{3+}	pot	0·1	TmL 7·22; TmL_2 12·90	[581]
UO_2^{2+}	pot	0·1	UO_2L 8·93	[584]
Y^{3+}	pot	0·1	YL 6·78; YL_2 12·03	[581]
Yb^{3+}	pot	0·1	YbL 7·42; YbL_2 13·27	[581]
Zn^{2+}	pot	0·1	ZnL 7·03; ZnL_2 12·17	[580]

Lactic acid HL

Metal ion	Method	I	$\log \beta$	Ref.
H^+	pot	0·1	3·75	
Ba^{2+}	pot	1	BaL 0·34; BaL_2 0·42	[569]
Ca^{2+}	pot	1	CaL 0·90; CaL_2 1·24	[569]
Cd^{2+}	cond		CdL 1·69	[585]
Ce^{3+}	pot	2	CeL 2·33; CeL_2 4·1; CeL_3 5·2	[586]
Co^{2+}	pot	1	CoL 1·37; CoL_2 2·32; CoL_3 2·34	[570]
Cu^{2+}	pot	1	CuL 3·02; CuL 4·84	[585]
Er^{3+}	pot	2	ErL 2·77; ErL_2 5·11; ErL_3 6·70	[586]
Eu^{3+}	pot	2	EuL 2·53; EuL_2 4·6; EuL_3 5·88	[586]
Fe^{3+}		0	FeL 7	[446]
Gd^{3+}	pot	2	GdL 2·53; GdL_2 4·63; GdL_3 5·91	[586]
Ho^{3+}	pot	2	HoL 2·71; HoL_2 4·97; HoL_3 6·55	[586]
Mg^{2+}	pot	1	MgL 0·73; MgL_2 1·30	[569]
Nd^{3+}	pot	2	NdL 2·47; NdL_2 4·37; NdL_3 5·60	[586]
Ni^{2+}	pot	1	NiL 1·59; NiL_2 2·67; NiL_3 2·70	[570]
Pb^{2+}	pot	1	PbL 1·98; PbL_2 2·98	[570]
Sm^{3+}	pot	2	SmL 2·56; SmL_2 4·58; SmL_3 5·90	[586]
Sr^{2+}	pot	1	SrL 0·53; SrL_2 0·69	[569]
Tb^{3+}	pot	2	TbL 2·61; TbL_2 4·73; TbL_3 6·01	[586]
UO_2^{2+}	pot	1	UO_2L 2·76; UO_2L_2 4·43; UO_2L_3 5·77	[570]
Y^{3+}	pot	2	YL 2·53; YL_2 4·70; YL_3 6·12	[586]
Yb^{3+}	pot	2	YbL 2·85; YbL_2 5·27; YbL_3 7·96	[586]
Zn^{2+}	pot	1	ZnL 1·61; ZnL_2 2·85; ZnL_3 2·88	[570]

Organic complexants (continued)

Metal ion	Method	I	$\log \beta$	Ref.
			Maleic acid H_2L	
H^+	pot	0.1	HL 5.9; HL_2 7.7	[539]
Ba^{2+}	cond	0	BaL 2.26	[539]
Be^{2+}	pot	0.15	BeL 4.33; BeL_2 6.46	[587]
Ca^{2+}	cond	0	CaL 2.43	[539]
Cd^{2+}	pol	0.2	CdL 2.2; CdL_2 3.6; CdL_3 3.8	[588]
Cu^{2+}	pol	0.2	CuL 3.4; CuL_2 4.9; CuL_3 6.2	[588]
In^{3+}	pol	0.2	InL 5.0; InL_2 7.1; InL_3 6.2	[588]
Ni^{2+}	pot	0.1	NiL 2.0	[589]
Pb^{2+}	pol	0.2	PbL 3.0; PbL_2 4.5; PbL_3 5.4	[588]
Zn^{2+}	pot	0.1	ZnL 2.0	[589]
			Malic acid H_2L	
H^+	pot	0.1	HL 4.72; H_2L 8.00	[590]
Ba^{2+}	pot	0.1	BaHL 6.09; BaH_2L 8.60	[591]
Ca^{2+}	pot	0.1	CaHL 6.67; $CaHL_2$ 8.99	[591]
Co^{2+}	pot	0.1	CoHL 7.67; CoH_2L 9.57	[591]
Cu^{2+}	pot	0.1	CuL 3.59; CuHL 8.13; CuH_2L 9.93	[591]
Fe^{2+}	pot	0.1	FeL 2.5	[592]
Fe^{3+}	pot	0.1	FeL 7.1	[592]
Mg^{2+}	pot	0.1	MgHL 6.4; MgH_2L 8.83	[591]
Ni^{2+}	pot	0.1	NiHL 7.88; NiH_2L 9.76	[591]
Zn^{2+}	pot	0.1	ZnHL 7.64; ZnH_2L 9.59	[591]
			Malonic acid H_2L	
H^+	pot	0.04	HL 5.66; H_2L 8.51	[592]
Al^{3+}	pot	0.2	AlL 5.24; AlL_2 9.40	[593]
Ba^{2+}	pot	0.1	BaL 1.34; BaHL 5.93	[591]
Be^{2+}	pot	0.2	BeL 5.15; BaL_2 8.48	[593]
Ca^{2+}	pot	0.1	CaL 1.85; CaHL 6.12	[591]
Ce^{3+}	pot	0.1	CeL 3.83; CeL_2 6.17	[478]
Cd^{2+}	pot	0.1	CdL 2.51; CdHL 6.37	[591]
Co^{2+}	pot	0.1	CoL 2.98; CoHL 7.53	[591]
Cr^{3+}	pot	0.1	CrL 7.06; CrL_2 12.85; CrL_3 16.15	[594]
Cu^{2+}	pot	0.1	CuL 5.55; CuHL 8.08	[591]
Dy^{3+}	pot	0.1	DyL 4.47; DyL_2 7.17	[478]
Er^{3+}	pot	0.1	ErL 4.42; ErL_2 7.04	[478]
Eu^{3+}	pot	0.1	EuL 4.30; EuL_2 6.99	[478]

Sec. 4.1] Protonation and complex formation equilibrium constants 351

Organic complexants (continued)

Metal ion	Method	I	$\log \beta$	Ref.
Fe^{3+}	pol	0.5	FeL_3 15.65	[595]
Gd^{3+}	pot	0.1	GdL 4.32; GdL_2 6.97	[478]
Ho^{3+}	pot	0.1	HoL 4.39; HoL_2 6.97	[478]
La^{3+}	pot	0.1	LaL 3.69; LaL_2 5.90	[478]
Lu^{3+}	pot	0.1	LuL 4.45; LuL_2 7.13	[478]
Mg^{2+}	pot	0.1	MgL 1.95; MgHL 6.15	[591]
Nd^{3+}	pot	0.1	NdL 3.95; NdL_2 6.41	[478]
Ni^{2+}	pot	0.1	NiL 3.30; NiHL 6.73	[591]
Pb^{2+}	pot	0.1	PbL 3.1	[589]
Pr^{3+}	pot	0.1	PrL 3.91; PrL_2 6.30	[478]
Sm^{3+}	pot	0.1	SmL 4.19; SmL_2 6.84	[478]
Tb^{3+}	pot	0.1	TbL 4.44; TbL_2 7.15	[478]
Tm^{3+}	pot	0.1	TmL 4.42; TmL_2 7.01	[478]
UO_2^{2+}	pot	0.2	UO_2L 4.88; UO_2L_2 8.63	[593]
Y^{3+}	pot	0.1	YL 4.40; YL_2 7.04	[478]
Yb^{3+}	pot	0.1	YbL 4.53; YbL_2 7.27	[478]
Zn^{2+}	pot	0.1	ZnL 2.97; ZnHL 6.56	[591]

Mandelic acid (2-phenyl-2-hydroxyacetic acid) HL

Metal ion	Method	I	$\log \beta$	Ref.
H^+	pot	0.1	HL 3.19	[596]
Be^{2+}	i	0.1	BeL 1.64	[597]
Ce^{3+}	pot	0.1	CeL 2.34; CeL_2 4.14	[596]
Eu^{3+}	ex	0.1	EuL 2.70; EuL_2 4.90	[598]
Fe^{3+}	sp		FeL 3.71	[599]
La^{3+}	pot	0.1	LaL 2.24; LaL_2 3.94	[596]
Nd^{3+}	pot	0.1	NdL 2.49; NdL_2 4.39	[596]
Pr^{3+}	pot	0.1	PrL 2.43; PrL_2 4.27	[596]
Sm^{3+}	pot	0.1	SmL 2.56; SmL_2 4.32	[596]
Th^{4+}	i	0.2	ThL 2.94; ThL_2 4.98; ThL_3 5.91	[600]
Zn^{2+}	oth	2	ZnL 1.48; ZnL_2 2.41; ZnL_3 3.59	[601]

D-Mannitol L

Metal ion	Method	I	$\log \beta$	Ref.
$As(OH)_4^-$	pot	0.1	$As(OH)_2(H_{-2}L)^-$ 0.85	[602]
$B(OH)_4^-$	pot	0.1	$B(OH)_2(H_{-2}L)^-$ 4.0	
			$B(H_{-2}L)_2^-$ 4.88	[603]
$HGeO_3^-$	pot	0.1	$HGeO(H_{-2}L)_2^-$ 4.53	[602]

Tables of equilibrium constants [Ch. 4

Organic complexants (continued)

Metalphthalein (See Phthalein complexone)

Methyl-thymol Blue H_6L

Metal ion	Method	I	$\log \beta$	Ref.
H^+	pot	0.2	HL 13.4; H_2L 24.6; H_3L 32.0; H_4L 35.8; H_5L 39.1; H_6L 42.1	[604] [605]
Fe^{3+}	sp	0.1	FeH_2L 43.3; FeH_6L_2 85.7	[606]
La^{3+}	sp	0.2	La_2L_2 35.8; $La_2(OH)_2L_2$ 23.2	[607]
Y^{3+}	sp	0.2	$Y_2H_2L_2$ 50.4; Y_2HL_2 42.4; Y_2L_2 **32.9**	[605]

Murexide (Purpuric acid) H_3L

Metal ion	Method	I	$\log \beta$	Ref.
H^+	sp	0.1	HL 10.9; H_2L 20.1; H_3L 20.1	[608]
Ca^{2+}	sp	0.1	CaL 5.0; CaHL 14.5; CaH_2L 22.7	[608]
Ce^{3+}	sp	0.1	CeH_2L 23.75	[609]
Cd^{2+}	sp	0.1	CdH_2L 24.3	[608]
Co^{2+}	sp	0.1	CoH_2L 22.56	[609]
Cu^{2+}	sp	0.1	CuH_2L 25.1	[608]
Dy^{3+}	sp	0.1	DyH_2L 23.88	[609]
Er^{3+}	sp	0.1	ErH_2L 23.58	[609]
Eu^{3+}	sp	0.1	EuH_2L 24.27	[609]
Gd^{3+}	sp	0.1	GdH_2L 24.18	[609]
Ho^{3+}	sp	0.1	HoH_2L 23.81	[609]
In^{3+}	sp	0.1	InH_2L 24.71	[609]
La^{3+}	sp	0.1	LaH_2L 23.53	[609]

Sec. 4.1] Protonation and complex formation equilibrium constants 353

Organic complexants (continued)

Metal ion	Method	I	log β	Ref.
Lu^{3+}	sp	0·1	LuH_2L 23·55	[609]
Nd^{3+}	sp	0·1	NdH_2L 24·14	[609]
Ni^{2+}	sp	0·1	NiH_2L 23·46	[609]
Pr^{3+}	sp	0·1	PrH_2L 23·88	[609]
Sc^{3+}	sp	0·1	ScH_2L 24·60	[609]
Sm^{3+}	sp	0·1	SmH_2L 24·30	[609]
Tb^{3+}	sp	0·1	TbH_2L 24·05	[609]
Tm^{3+}	sp	0·1	TmH_2L 23·46	[609]
Y^{3+}	sp	0·1	YH_2L 23·46	[609]
Yb^{3+}	sp	0·1	YbH_2L 23·51	[609]
Zn^{2+}	sp	0·1	ZnH_2L 23·2	[608]

Nitrilotriacetic acid (NTA) H_3L

Metal ion	Method	I	log β	Ref.
H^+	pot	0·1	HL 9·73; H_2L 12·22; H_3L 14·11	[610]
Ag^+	ex	0·1	AgL 5·16	[611]
Al^{3+}	ex	0·1	AlL 9·5	[611]
Ba^{2+}	pot	0·1	BaL 4·72	[612]
Be^{2+}	ex	0·1	BeL 7·11	[611]
Ca^{2+}	pot	0·1	CaL 6·33	[612]
Cd^{2+}	pol	0·1	CdL 9·8	[610]
Ce^{3+}	pot	0·1	CeL 10·8	[613]
Co^{2+}	pot	0·1	CoL 10·4	[610]
Cu^{2+}	pot	0·1	CuL 13·1	[613]
Dy^{3+}	pot	0·1	DyL 11·74; DyL_2 21·15	[613]
Er^{3+}	pot	0·1	ErL 12·03; ErL_2 21·29	[613]
Eu^{3+}	pot	0·1	EuL 11·52; EuL_2 20·70	[613]
Fe^{2+}	pot	0·1	FeL 8·8; Fe(OH)L 12·2	[614]
Fe^{3+}	pot	0·1	FeL 15·87; FeL_2 24·3; Fe(OH)L 25·8	[614]
Ga^{3+}	ex	0·1	GaL_2 25·81	[611]
Gd^{3+}	pot	0·1	GdL 11·54; GdL_2 20·80	[613]
Hg^{2+}	pot	0·1	HgL 14·6	[615]
Ho^{3+}	pot	0·1	HoL 11·90; HoL_2 21·25	[613]
In^{3+}	ex	0·1	InL_2 24·4	[611]
La^{3+}	pot	0·1	LaL 10·36; LaL_2 17·60	[613]
Lu^{3+}	pot	0·1	LuL 12·49; LuL_2 21·91	[613]
Mg^{2+}	pot	0·1	MgL 5·36	[612]
Mn^{2+}	pot	0·1	MnL 8·5	[616]
Nd^{3+}	pot	0·1	NdL 11·26; NdL_2 19·73	[613]
Ni^{2+}	pol	0·1	NiL 11·5	[610]
Pb^{2+}	pol	0·1	PbL 11·39	[610]
Pr^{3+}	pot	0·1	PrL 11·07; PrL_2 19·25	[613]
Sm^{3+}	pot	0·1	SmL 11·53; SmL_2 20·53	[613]
Sr^{2+}	pot	0·1	SrL 4·91	[612]

Organic complexants (continued)

Metal ion	Method	I	$\log \beta$	Ref.
Tb^{3+}	pot	0.1	TbL 11.59; TbL$_2$ 20.97	[613]
TiO^{2+}	ex	0.1	TiOL 12.3	[611]
Tm^{3+}	pot	0.1	TmL 12.22; TmL$_2$ 21.45	[613]
UO_2^{2+}	ex	0.1	UO$_2$L 9.56	[611]
Y^{3+}	pot	0.1	YL 11.48; YL$_2$ 20.43	[613]
Yb^{3+}	pot	0.1	YbL 12.4; YbL$_2$ 21.69	[613]
Zn^{2+}	pol	0.1	ZnL 10.66	[610]
Zr^4	sp	0.1	ZrL 20.8	[617]

Nitroso-R acid H$_3$L

Metal ion	Method	I	$\log \beta$	Ref.
H^+	pot	0.1	HL 6.9	[618]
Cd^{2+}	pot	0.1	CdL 3.4; CdL$_2$ 6.0	[618]
Cu^{2+}	pot	0.1	CuL 7.7; CuL$_2$ 15.0	[618]
La^{3+}	pot	0.1	LaL 4.37; LaL$_2$ 7.83; LaL$_3$ 11.24	[618]
Mn^{2+}	pot	0.1	MnL 2.7	[618]
Ni^{2+}	pot	0.1	NiL 6.9; NiL$_2$ 12.5; NiL$_3$ 17.3	[618]
Pb^{2+}	pot	0.1	PbL 4.64; PbL$_2$ 7.37	[618]
Y^{3+}	pot	0.1	YL 4.48; YL$_2$ 7.83; YL$_3$ 11.29	[618]
Zn^{2+}	pot	0.1	ZnL 4.5; ZnL$_2$ 7.1	[618]

Oxalic acid H$_2$L

Metal ion	Method	I	$\log \beta$	Ref.
H^+	pot	0.1	HL 3.8; H$_2$L 5.2	[619]
Al^{3+}	pot	0	AlL$_2$ 13; AlL$_3$ 16.3	[620]
Ag^+	pot		AgL 2.41	[621]
Be^{2+}	pot	0.15	BeL 4.08; BeL$_2$ 5.91	[622]
Ca^{2+}	pot	0.1	CaL 3.0	[623]
Cd^{2+}	pot	0	CdL 4.0; CdL$_2$ 5.77	[624]
Ce^{3+}	sol	0	CeL 6.52; CeL$_2$ 10.48; CeL$_3$ 11.3	[625]
Co^{2+}	i	0.16	CoL 3.72; CoL$_2$ 6.03; CoHL 5.46; CoH$_2$L$_2$ 10.51	[626]
Cu^{2+}	pot	0.1	CuL 4.8; CuL$_2$ 8.4; CuHL 6.3	[619]
Eu^{3+}	ex	1	EuL 4.77; EuL$_2$ 8.72; EuL$_3$ 11.4	[627]
Fe^{2+}	pol	0.5	FeL$_2$ 4.52; FeL$_3$ 5.22	[595]
Fe^{3+}	pot	0.5	FeL 7.53; FeL$_2$ 13.64; FeL$_3$ 18.49	[628]
Gd^{3+}	ex	0.5	GdL 4.78; GdL$_2$ 8.68	[629]
Hg_2^{2+}	pot	2.5	Hg$_2$L 6.98; Hg$_2$(OH)L 13.04	[630]
In^{3+}	i		InHL 6.88	[631]
Lu^{3+}	ex	0.1	LuL 5.11; LuL$_2$ 9.2; LuL$_3$ 12.79	[627]

Sec. 4.1] Protonation and complex formation equilibrium constants

Organic complexants (continued)

Metal ion	Method	I	$\log \beta$	Ref.
Mg^{2+}	pot	0.5	MgL 2.4	[632]
Mn^{2+}	sol	0	MnL 3.82; MnL_2 5.25	[633]
Mn^{3+}	kin	2	MnL 9.98; MnL_2 16.57; MnL_3 19.42	[634]
Ni^{2+}		1	NiL 4.1; NiL_2 7.2; NiL_3 8.5	[635]
Nd^{3+}		0	NdL 7.21; NdL_2 11.51	[625]
Pb^{2+}	sol	0	PbL_2 6.54	[636]
Sc^{3+}	ex	0.1	ScL_3 16.28	[611]
TiO^{2+}	sp		TiOL 6.60; $TiOL_2$ 9.90	[637]
Tl^{3+}	ex	0.1	TlL_3 16.9	[627]
UO_2^{2+}		0.5	UO_2HL 6.65; $UO_2H_2L_2$ 9.5	[638]
VO^{2+}			VOL_2 12.5	[639]
Zn^{2+}	i	0.1	ZnL 3.88; ZnL_2 6.40; ZnHL 5.5; ZnH_2L_2 10.72	[626]

Oxine
(See 8-hydroxyquinoline)

PAN HL

H^+	pot		HL 12.2; H_2L 14.1	[640]
Co^{2+}	sp	0.05	CoL 12.15; CoL_2 24.16	[641]
Cu^{2+}	sp		CuL 16	[640]
Eu^{3+}	ex	0.05	EuL 12.39; EuL_2 23.80; EuL_3 34.23; EuL_4 43.68	[641]
Ho^{3+}	ex	0.05	HoL 12.76; HoL_2 24.36; HoL_3 34.80; HoL_4 44.08	[641]
Mn^{2+}	pot		MnL 8.5; MnL_2 16.4	[642]
Ni^{2+}	pot		NiL 12.7; NiL_2 25.3	[642]
Zn^{2+}	pot		ZnL 11.2; ZnL_2 21.7	[642]

PAR H_2L

H^+	sp		HL 11.9; H_2L 17.5; H_3L 20.6	[643]
Al^{3+}	sp	0.1	AlL 11.5	[644]

23*

Organic complexants (continued)

Metal ion	Method	I	$\log \beta$	Ref.
Bi^{3+}	sp		BiHL 30·1	[643]
Cd^{2+}	sp		CdL_2 21·6; CdHL 23·4	[643]
Co^{2+}	pot		CoL 10·0; CoL_2 17·1	[645]
Cu^{2+}	sp		CuL_2 38·2; CuHL 29·4	[643]
Dy^{3+}	sp	0·1	DyL 10·6; DyHL 23·4	[644]
Er^{3+}	sp	0·1	ErL 10·1; ErHL 23·2	[644]
Ga^{3+}	sp	0·1	GaL_2 30·3; GaHL 26·8	[643]
In^{3+}			InL 9·6; InL_2 19·2	[646]
Mn^{2+}	pot		MnL 9·7; MnL_2 18·9	[642]
Nd^{3+}	sp	0·1	NdL 9·8; NdHL 23·30	[644]
Ni^{2+}	pot		NiL 13·2; NiL_2 26·0	[642]
Pb^{2+}	pot		PbL_2 26·6; PbHL 24·8	[643]
Pr^{3+}	sp	0·1	PrL 9·3; PrHL 22·7	[644]
Sm^{3+}	sp	0·1	SmL 10·1; SmHL 23·6	[644]
Tl^{3+}			TlL 9·8; TlL_2 19·6	[646]
UO_2^{2+}	pot		UO_2L 12·5; UO_2L_2 20·9	[645]
Y^{3+}	sp	0·1	YL 9·1; YHL 22·4	[644]
Yb^{3+}	sp	0·1	YbL 10·2; YbHL 23·3	[644]
Zn^{2+}	sp	0·1	ZnL_2 25·3; ZnHL 24·5	[643]

Pentaethylenehexamine (Pentene) L

Metal ion	Method	I	$\log \beta$	Ref.
H^+	pot	0·1	HL 10·28; H_2L 20·06; H_3L 29·28; H_4L 37·92	[647]
Cd^{2+}	pot	0·1	CdL 16·8; CdHL 23·3	[647]
Co^{2+}	pot	0·1	CoL 15·75; CoHL 22·7	[647]
Cu^{2+}	pot	0·1	CuL 22·44; CuHL 30·6; CuH_2L 33·7; CuH_3L 37·4	[647]
Fe^{2+}	pot	0·1	FeL 11·2; FeHL 19·9	[647]
Hg^{2+}	pot	0·1	HgL 29·59; HgHL 38·13; HgH_2L 43·6	[647]
Mn^{2+}	pot	0·1	MnL 9·37	[647]
Ni^{2+}	pot	0·1	NiL 19·30; NiHL 26·07; NiH_2L 30·5	[647]
Zn^{2+}	pot	0·1	ZnL 16·24; ZnHL 24·4	[647]

1,10-Phenanthroline L

Metal ion	Method	I	$\log \beta$	Ref.
H^+	pot	0·1	HL 4·95	[434]
Ag^+	pot	0·1	AgL 5·02; AgL_2 12·07	[648]
Ca^{2+}	pot	0·1	CaL 0·7	[434]
Cd^{2+}	pot	0·1	CdL 5·78; CdL_2 10·82; CdL_3 14·92	[434]
Co^{2+}	pot	0·1	CoL 7·25; CoL_2 13·95; CoL_3 19·90	[434]

Sec. 4.1]   Protonation and complex formation equilibrium constants   357

Organic complexants (continued)

Metal ion	Method	I	$\log \beta$	Ref.
Cu^{2+}	pot	0.1	CuL 9.25; CuL_2 16.0; CuL_3 21.35	[434]
Fe^{2+}	pot	0.1	FeL 5.9; FeL_2 11.1; FeL_3 21.3	[649]
Fe^{3+}	pot	0.1	FeL_3 14.1	[650]
Hg^{2+}	pot	0.1	HgL_2 19.65; HgL_3 23.35	[434]
Mg^{2+}	pot	0.1	MgL 1.2	[434]
Mn^{2+}	pot	0.1	MnL 4.13; MnL_3 7.61; MnL_2 10.31	[434]
Ni^{2+}	pot	0.1	NiL 8.8; NiL_2 17.1; NiL_3 24.8	[434]
Pb^{2+}	pot	0.1	PbL 5.1; PbL_2 7.5; PbL_3 9	[15]
VO^{2+}	pot	0.1	VOL 5.47; VOL_2 9.69	[651]
Zn^{2+}	pot	0.1	ZnL 5.65; ZnL_2 12.35; ZnL_3 17.55	[434]

Phenol HL

Metal ion	Method	I	$\log \beta$	Ref.
H^+	pot	0.1	HL 9.62	[652]
Fe^{3+}	sp	0.03	FeL 8.11	[653]
La^{3+}	sp	0.1	LaL 1.51	[654]
UO_2^{2+}	pot	0.1	UO_2L 5.8	[652]
Y^{3+}	sp	0.1	YL 2.40	[654]

Phthalein complexone (Metalphthalein) H_6L

HOOC—CH_2\
            \N—CH_2 CH_2—N/ CH_2—COOH\
HOOC—CH_2/   \CH_2—COOH

                HO   OH

            H_3C CH_3
                        C—O
                        |
                        C=O

Metal ion	Method	I	$\log \beta$	Ref.
H^+	pot	0.1	HL 12.01; H_2L 23.36; H_3L 31.19; H_4L 38.16; H_5L 41.06; H_6L 43.26	[655]
Ba^{2+}	pot	0.1	BaL 6.2; BaHL 16.81; BaH_2L 25.6; BaH_3L 32.5; Ba_2L 11.4	[655]
Ca^{2+}	pot	0.1	CaL 7.8; CaHL 18.9; CaH_2L 26.56; CaH_3L 33.5; Ca_2L 12.8	[655]
Mg^{2+}	pot	0.1	MgL 8.9; MgHL 19.5; MgH_3L 26.96; MgH_3L 33.4; Mg_2L 11.9	[655]
Zn^{2+}	pot	0.1	ZnL 15.1; ZnHL 25.8; ZnH_2L 33.56; ZnH_3L 37.2; Zn_2L 24.9; Zn_2HL 30.8	[655]

Organic complexants (continued)

Metal ion	Method	I	$\log \beta$	Ref.

Phthalic acid H_2L

Metal ion	Method	I	$\log \beta$	Ref.
H^+	pot	0.1	HL 4.92; H_2L 7.68	[589]
Ba^{2+}	cond	0	BaL 2.33	[539]
Ca^{2+}	cond	0	CaL 2.43	[539]
Cd^{2+}	pot	0.1	CdL 2.5	[589]
Cr^{3+}	pot	0.1	CrL 5.52; CrL_2 10.0; CrL_3 12.48	[656]
Cu^{2+}	pot	0.1	CuL 3.1	[589]
Ni^{2+}	pot	0.1	NiL 2.1	[589]
Zn^{2+}	pot	0.1	ZnL 2.2	[589]

Picolinic acid (Pyridine-2-carboxylic acid) HL

Metal ion	Method	I	$\log \beta$	Ref.
H^+	pot	0.1	HL 5.23; H_2L 6.25	[657]
Ag^+	pot	0.1	AgL 3.4; AgL_2 5.9	[657]
Ba^{2+}	pot	0.1	BaL 1.65	[657]
Ca^{2+}	pot	0.1	CaL 1.81	[657]
Cd^{2+}	pot	0.1	CdL 4.55; CdL_2 8.16; CdL_3 10.76	[657]
Co^{2+}	pot	0.1	CoL 5.74; CoL_2 10.44; CoL_3 14.09	[657]
Cr^{3+}	sp	0.5	CrL_2 10.22	[658]
Cu^{2+}	pot	0.1	CuL 7.95; CuL_2 14.95	[657]
Dy^{3+}	pot	0.1	DyL 4.22; DyL_2 7.76; DyL_3 10.8	[659]
Er^{3+}	pot	0.1	ErL 4.28; ErL_2 7.86; ErL_3 10.9	[659]
Eu^{3+}	pot	0.1	EuL 4.07; EuL_2 7.48; EuL_3 10.6	[659]
Fe^{2+}	pot	0.1	FeL 4.90; FeL_2 9.0; FeL_3 12.3	[657]
Fe^{3+}	pot	0.1	FeL_2 12.8; $Fe(OH)L_2$ 23.84; $Fe_2(OH)_2L_4$ 50.76	[657]
Gd^{3+}	pot	0.1	GdL 4.03; GdL_2 7.34; GdL_3 10.5	[659]
Hg^{2+}	pot	0.1	HgL 7.7; HgL_2 15.4	[657]
La^{3+}	pot	0.1	LaL 3.54; LaL_2 6.28; LaL_3 8.9	[659]
Mg^{2+}	pot	0.1	MgL 2.2	[657]
Mn^{2+}	pot	0.1	MnL 3.57; MnL_2 6.32; MnL_3 8.1	[657]
Nd^{3+}	pot	0.1	NdL 3.88; NdL_2 6.92; NdL_3 10.0	[659]
Ni^{2+}	pot	0.1	NiL 6.8; NiL_2 12.58; NiL_3 17.22	[657]
Pb^{2+}	pot	0.1	PbL 4.58; PbL_2 7.92	[657]
Sr^{2+}	pot	0.1	SrL 1.7	[657]
Zn^{2+}	pot	0.1	ZnL 5.3; ZnL_2 7.62; ZnL_3 9.92	[657]

Picric acid (Trinitrophenol) HL

Metal ion	Method	I	$\log \beta$	Ref.
H^+	sp		HL 2.3	[660]
Al^{3+}	sp		AlL 1.05; AlL_3 3.12	[660]
Ca^{2+}	sp		CaL_2 2.48	[660]

[Sec. 4.1] **Protonation and complex formation equilibrium constants**

Organic complexants (continued)

Metal ion	Method	I	log β	Ref.
Ce^{3+}	sp		CeL 1·05; CeL$_3$ 3·09	[660]
Cu^{2+}	sp		CuL$_2$ 2·7	[660]
Fe^{3+}	sp		FeL 1·8; FeL$_3$ 3·10	[660]
Mg^{2+}	sp		MgL$_2$ 2·43	[660]
Ni^{2+}	sp		NiL$_2$ 2·89	[660]
Zn^{2+}	sp		ZnL$_2$ 2·92	[660]

Pyridine L

Metal ion	Method	I	log β	Ref.
H^+	pot	0·5	HL 5·21	[224]
Ag^+	pot	0·5	AgL 2·01; AgL$_2$ 4·16	[224]
Cd^{2+}	pot	0·5	CdL 1·27; CdL$_2$ 2·07	[224]
Co^{2+}	pot	0·5	CoL 1·14; CoL$_2$ 1·54	[224]
Cu^{2+}	pot	0·1	CuL 2·41; CuL$_2$ 4·29; CuL$_3$ 5·43; CuL$_4$ 6·03	[224]
Fe^{2+}	pot	0·5	FeL 0·7; FeL$_4$ 6·7	[224]
Hg^{2+}	pot	0·5	HgL 5·1; HgL$_2$ 10·0; HgL$_3$ 10·4	[224]
Mn^{2+}	pot	0·5	MnL 0·14	[224]
Ni^{2+}	pot	0·5	NiL 1·78; NiL$_2$ 2·83; NiL$_3$ 3·14	[224]
Zn^{2+}	pot	0·5	ZnL 0·95; ZnL$_2$ 1·45	[224]

Pyridine-2-aldoxime HL

Metal ion	Method	I	log β	Ref.
H^+	pot	0·3	HL 10·0; H$_2$L 13·4	[661]
Co^{2+}	pot	0·3	CoL 8·6; CoL$_2$ 17·2	[661]
Cu^{2+}	pot	0·3	CuL 8·9; CuL$_2$ 14·55	[662]
Fe^{2+}	sp	0·3	FeL$_3$ 24·85	[663]
Mn^{2+}	pot	0·3	MnL 5·2; MnL$_2$ 9·1	[661]
Ni^{2+}	pot	0·3	NiL 9·4; NiL$_2$ 16·5; NiL$_3$ 22·0	[661]
Zn^{2+}	pot	0·3	ZnL 5·8; ZnL$_2$ 11·1	[661]

Pyridine-2-carboxylic acid
(See Picolinic acid)

Pyridine-2,6-dicarboxylic acid
(Dipicolinic acid) H$_2$L

Metal ion	Method	I	log β	Ref.
H^+	pot	0·1	HL 4·67; H$_2$L 6·91	[664]
Al^{3+}	pot	0·5	AlL 4·87; AlL$_2$ 8·32	[665]
Ba^{2+}	pot	0·1	BaL 3·43	[664]
Ca^{2+}	pot	0·1	CaL 4·60; CaL$_2$ 7·20	[664]
Cu^{2+}	pot	0·1	CuL$_2$ 17·1; CuH$_2$L$_2$ 18·76	[666]
Fe^{3+}	pot	0·1	FeL$_2$ 16·74	[667]
Mg^{2+}	pot	0·1	MgL 2·32	[664]
Sr^{2+}	pot	0·1	SrL 3·80; SrL$_2$ 5·50	[664]

Organic complexants (continued)

Metal ion	Method	I	$\log \beta$	Ref.
			1-(2-Pyridylazo)-2-naphthol (See PAN)	
			4-(2-Pyridylazo)-resorcinol (See PAR)	
			Pyrocatechol H_2L	
H^+	pot	1	HL 13·05; H_2L 22·28	[668]
Al^{3+}	pot	0·2	AlL 16·56; AlL_2 32·20; AlL_3 45·85	[669]
$B(OH)_4^-$	pot	0·1	$K[B(OH)_4^- + H_2L \rightarrow B(OH)_2L]$ 3·92; $K[B(OH)_4^- + 2H_2L \rightarrow BL_2^-]$ 4·26	[603]
Be^{2+}	pot	0·1	BeL 13·52; BeL_2 23·35; BeHL 18·05; $BeHL_2$ 29·55	[670]
Cd^{2+}	pot	0·1	CdL 7·70	[671]
Co^{2+}	pot	1	CoL 8·32; CoL_2 14·74	[668]
Cu^{2+}	pot	1	CuL 13·60; CuL_2 24·93	[668]
Fe^{2+}	pot	1	FeL 7·95; FeL_2 13·49; FeHL 16·57	[668]
Mg^{2+}	pot	0·1	MgL 5·24	[671]
Mn^{2+}	pot	1	MnL 7·47; MnL_2 12·75; MnHL 15·87	[668]
Ni^{2+}	pot	1	NiL 8·77	[668]
Zn^{2+}	pot	1	ZnL 9·54; ZnL_2 17·51	[668]

Pyrocatechol Violet H_4L

H^+	pot		HL 11·7; H_2L 21·5; H_3L 29·3	[672]
Bi^{3+}			BiL 27·1; Bi_2L 32·3	[672]
Th^{4+}			ThL 23·4; Th_2L 27·8	[672]

Pyruvic acid
(2-oxopropanoic acid) HL

H^+	pot	0·5	HL 2·35	[673]
Mn^{2+}	pot	0·5	MnL 1·26	[674]

Organic complexants (continued)

Metal ion	Method	I	$\log \beta$	Ref.
Ni^{2+}	pot	0·5	NiL 1·12; NiL_2 0·46	[673]
Zn^{2+}	pot	0·5	ZnL 1·26; ZnL_2 1·98	[673]

Quinoline-2-carboxylic acid
(Quinaldinic acid) HL

Metal ion	Method	I	$\log \beta$	Ref.
H^+	pot	0·1	HL 4·92	[675]
Ba^{2+}	pot	0	BaL 1·20	[675]
Ca^{2+}	pot	0	CaL 1·42; CaL_2 4·41	[675]
Cd^{2+}	pot	0	CdL 4·12; CdL_2 10·95	[675]
Co^{2+}	pot	0	CoL 4·49; CoL_2 12·72	[675]
Cu^{2+}	pot	0	CuL 5·91	[675]
Fe^{2+}	pot	0	FeL 3·92; FeL_2 11·59	[675]
Mg^{2+}	pot	0	MgL 1·37; MgL_2 3·92	[675]
Mn^{2+}	pot	0	MnL 2·96; MnL_2 8·88	[675]
Ni^{2+}	pot	0	NiL 4·95; NiL_2 13·60	[675]
Pb^{2+}	pot	0	PbL 3·95; PbL_2 10·97	[675]
Sr^{2+}	pot	0	SrL 1·24	[675]
Zn^{2+}	sp	0	ZnL 4·17	[675]

Rubeanic acid
(See Dithio-oxamide)
Salicylaldoxime H_2L

Metal ion	Method	I	$\log \beta$	Ref.
H^+	pot	0	HL 12·1; H_2L 21·29; H_3L 22·66	[676]
Ba^{2+}	pot	0	$BaHL$ 12·64; BaH_2L_2 27·9	[676]
Ca^{2+}	pot	0	$CaHL$ 13·03; CaH_2L_2 27·9	[676]
Co^{2+}	pot	0	CoH_2L_2 32·3	[677]
Cu^{2+}	sp	0	CuH_2L_2 28·4	[677]
Fe^{2+}	pot	0·1	$FeHL$ 21·48; FeH_2L_2 31·55 (75% dioxan)	[661]
Mg^{2+}	pot	0	$MgHL$ 12·74; MgH_2L_2 26·5	[676]
Mn^{2+}	pot	0·1	$MnHL$ 17·9; MnH_2L_2 30·3 (75% dioxan)	[661]
Ni^{2+}	sp		NiH_2L_2 27·97	[677]
Sr^{2+}	pot	0	SrH_2L_2 27·97	[676]
Zn^{2+}	pot	0·1	$ZnHL$ 18·4; ZnH_2L_2 31·4 (75% dioxan)	[661]

Salicylic acid H_2L

Metal ion	Method	I	$\log \beta$	Ref.
H^+	pot	0·1	HL 13·6; H_2L 16·6	[678]
Al^{3+}	sp	0	AlL 14·1	[679]
Be^{2+}	pot	0·1	BeL 12·37; BeL_2 22·0	[443]
Ca^{2+}	i	0·16	$CaHL$ 13·8	[447]
Ce^{2+}	pot	0·15	CdL 5·55	[678]
Ce^{3+}	pot	0·1	$CeHL$ 16·26	[680]

Organic complexants (continued)

Metal ion	Method	I	lóg β	Ref.
Co^{2+}	pot	0.15	CoL 6.72; CoL_2 11.42	[678]
Cu^{2+}	pot	0.15	CuL 10.6; CuL_2 18.45	[678]
Fe^{2+}	pot	0.15	FeL 6.55; FeL_2 11.25	[678]
Fe^{3+}	sp	0.25	FeL 16.48; FeL_2 28.16; FeL_3 36.84	[680]
La^{3+}	pot	0.1	LaHL 16.24	[422]
Mn^{2+}	pot	0.15	MnL 5.9; MnL_2 9.8	[678]
Nd^{3+}	pot	0.1	NdHL 16.30	[422]
Ni^{2+}	pot	0.15	NiL 6.95; NiL_2 11.75	[678]
Th^{4+}	ex	0.1	ThL 4.25; ThL_2 7.55; ThL_3 10.0; ThL_4 11.55	[681]
TiO^{2+}	sp	0.1	TiOL 15.66; $TiOL_2$ 24.36	[682]
UO_2^{2+}	ex	0.1	UO_2HL 15.8; UO_2H_2L 13; UO_2H_3L 9.1; $UO_2(OH)L$ 12.1	[681]
VO^{2+}	pot	0.1	VOL 13.38	[683]
Zn^{2+}	pot	0.15	ZnL 6.85	[678]

Solochrome Violet R H_3L

H^+	sp	0	HL 13.04; H_2L 20.07	[684]
Al^{3+}	sp	0	AlL 18.4; AlL_2 31.60	[684]
Ca^{2+}	sp	0	CaL 6.6; CaL_2 9.6	[684]
Cr^{3+}	sp	0	CrL_2 17.25 (75°C)	[684]
Cu^{2+}	sp	0	CuL 21.8	[684]
Mg^{2+}	sp	0	MgL 8.6; MgL_2 13.6	[684]
Ni^{2+}	sp	0	NiL 15.9; NiL_2 26.35	[684]
Pb^{2+}	sp	0	PbL 12.5; PbL_2 17.8	[684]
Zn^{2+}	sp	0	ZnL 13.5; ZnL_2 20.9	[684]

D-Sorbitol L

$B(OH)_4^-$	pot	0.1	$B(H_{-2}L)_2^-$ 5.65	[602]
$HGeO_3^-$	pot	0.1	$HGeO(H_{-2}L)_2^-$ 5.09	[602]

Succinic acid H_2L

H^+	pot	0.1	HL 5.28; H_2L 9.28	[591]
Be^{2+}	pot	0.15	BeL 4.69; BeL_2 6.43	[587]
Ca^{2+}	pot	0.1	CaL 1.20; CaHL 5.82	[591]

Sec. 4.1] Protonation and complex formation equilibrium constants

Organic complexants (continued)

Metal ion	Method	I	log β	Ref.
Co^{2+}	pot	0·1	CoL 1·70; CoHL 6·27	[591]
Cr^{3+}	pot	0·1	CrL 6·42; CrL_2 10·99; CrL_3 13·85	[594]
Cu^{2+}	pot	0·1	CuL 2·93; CuHL 6·98	[591]
La^{3+}	i	0·15	LaHL 6·76; LaH_2L_2 13·24	[685]
Lu^{3+}	i	0·15	LuHL 7·04; LuH_2L_2 13·62	[685]
Sm^{3+}	i	0·15	SmHL 7·28; SmH_2L_2 13·96	[685]
Zn^{2+}	pot	0·1	ZnL 1·76; ZnHL 6·24	[591]

5-Sulphosalicylic acid H_3L

Metal ion	Method	I	log β	Ref.
H^+	pot	0·1	HL 11·6; H_2L 14·2	
Al^{3+}	pot	0·1	AlL 13·2; AlL_2 22·8; AlL_3 28·9	[686]
Be^{2+}	pot	0·1	BeL 11·7; BeL_2 20·8	[686]
Cd^{2+}	pot	0·15	CdL 4·65	[678]
Ce^{3+}	pot	0·1	CeL 6·83; CeL_2 12·40; CeHL 13·53	[687]
Co^{2+}	pot	0·1	CoL 6·13; CoL_2 9·82	[686]
Cr^{3+}	pot	0·1	CrL 9·56	[686]
Cu^{2+}	pot	0·15	CuL 9·5; CuL_2 16·5	[678]
Er^{3+}	pot	0·1	ErL 8·15; ErL_2 14·45; ErHL 13·72	[687]
Eu^{3+}	pot	0·1	EuL 7·87; EuL_2 13·90; EuHL 13·86	[687]
Fe^{2+}	pot	0·15	FeL 5·9; FeL_2 9·9	[578]
Fe^{3+}	sp	0·25	FeL 15·0; FeL_2 25·8; FeL_3 32·6	[680]
Gd^{3+}	pot	0·1	GdL 7·58; GdL_2 13·65; GdHL 13·80	[687]
Lu^{3+}	pot	0·1	LuL 8·43; LuL_2 15·46; LuHL 14·07	[687]
Mn^{2+}	pot	0·1	MnL 5·24; MnL_2 8·24	[686]
Ni^{2+}	pot	0·1	NiL 6·4; NiL_2 10·2	[686]
Pr^{3+}	pot	0·1	PrL 7·08; PrL_2 12·69; PrHL 13·69	[687]
Sm^{3+}	pot	0·1	SmL 7·65; SmL_2 13·58; SmHL 13·83	[687]
UO_2^{2+}	pot	0·1	UO_2L 11·14; UO_2L_2 19·2	[686]
Zn^{2+}	pot	0·15	ZnL 6·05; ZnL_2 10·65	[678]

Tartaric acid H_2L

Metal ion	Method	I	log β	Ref.
H^+	pot	0·1	HL 4·1; H_2L 7·0	
Al^{3+}	oth	0·1	AlL 6·35; AlHL 7·93; AlH_2L_2 14·71; Al(OH)L 18·5	[688]
Ba^{2+}	pot	0·2	BaL 1·62; BaHL 5·0	[689]
Bi^{3+}	ex	0·1	BiL_2 11·3	[500]
Ca^{2+}	pot	0·2	CaL 1·8; CaHL 5·2	[689]
Cd^{2+}	pot	0·5	CdL 2·8	[690]
Ce^{3+}	pot		CeL 3·84; CeL_2 6·72; Ce_2L 5·80	[691]
Co^{2+}		0·5	CoL 2·1	[692]
Cu^{2+}	pot	1	CuL 3·2; CuL_2 5·1; CuL_3 5·8; CuL_4 6·2	[361]

Organic complexants (continued)

Metal ion	Method	I	$\log \beta$	Ref.
Fe^{3+}	ex	0·1	FeL_2 11·86	[500]
Ga^{3+}	ex	0·1	GaL_2 9·76	[500]
In^{3+}	ex	0·1	InL 4·48	[500]
La^{3+}	pot		LaL 3·68; LaL_2 6·37; La_2L 5·32	[691]
Mg^{2+}	pot	0·2	MgL 1·36; $MgHL$ 5·0	[689]
Pb^{2+}		0·5	PbL 3·8	[690]
Sc^{3+}	ex	0·1	ScL_2 12·5	[500]
Sr^{2+}	pot	0·2	SrL 1·65; $SrHL$ 5·0	[689]
TiO^{2+}	ex	0·1	$TiOL_2$ 9·7	[500]
Zn^{2+}	pot	0·2	ZnL 2·68; $ZnHL$ 5·5	[689]

Tetraethylenepentamine
(Tetren) L

Metal ion	Method	I	$\log \beta$	Ref.
H^+	pot	0·1	HL 9·78; H_2L 19·16; H_3L 27·30; H_4L 32·13; H_5L 35·28	[693]
Cd^{2+}	pot	0·1	CdL 14·0	[694]
Co^{2+}	pot		CoL 15·07	[695]
Cu^{2+}	pot	0·15	CuL 24·25	[693]
Fe^{2+}	pot		FeL 11·40	[695]
Hg^{2+}	pot	0·1	HgL 27·7	[694]
Mn^{2+}	pot	0·15	MnL 7·62	[695]
Ni^{2+}	pot	0·1	NiL 17·51; $NiHL$ 22·44	[693]
Pb^{2+}	pot	0·1	PbL 10·5	[694]
Zn^{2+}	pot	0·1	ZnL 15·4	[694]

2-Thenoyltrifluoroacetone
(TTA)   HL

Metal ion	Method	I	$\log \beta$	Ref.
H^+	ex		HL 6·23	[696]
Cu^{2+}	pot	0·1	CuL 6·55; CuL_2 13	[696]
Ni^{2+}	pot	0·1	NiL_2 10·0	[696]
Sc^{3+}	ex	0·1	ScL 7·1	[697]

Thiourea L

Metal ion	Method	I	$\log \beta$	Ref.
H^+	pot	0·01	HL 2·03	[698]
Ag^+	pot	0	AgL_3 13·05	[699]
Bi^{3+}		0·1	BiL_6 11·9	[700]
Cd^{2+}	pol	0·1	CdL 1·38; CdL_2 1·71; CdL_3 1·60; CdL_4 3·55	[698]
Cu^{2+}	pol	0·1	CuL_4 15·4	[701]

Sec. 4.1] Protonation and complex formation equilibrium constants 365

Organic complexants (continued)

Metal ion	Method	I	$\log \beta$	Ref.
Hg^{2+}		0.1	HgL_2 22.1; HgL_3 24.7; HgL_4 26.8	[702]
Pb^{2+}	pol	0.1	PbL 0.6; PbL_2 1.04; PbL_3 0.98; PbL_4 2.04	[698]

Thioglycollic acid H_2L

Metal ion	Method	I	$\log \beta$	Ref.
H^+	pot	0.1	HL 10.2; H_2L 13.6	[703]
Ce^{3+}	pot	0.1	CeHL 12.2; CeH_2L_2 23.44	[704]
Co^{2+}	pot	0.1	CoL 5.84; CoL_2 12.15	[703]
Er^{3+}	pot	0.1	ErHL 12.14; ErH_2L_2 23.66	[704]
Eu^{3+}	pot	0.1	EuHL 12.27; EuH_2L_2 23.81	[704]
Fe^{2+}	sol	0	FeL_2 10.92; Fe(OH)L 12.38	[703]
Hg^{2+}	pot	1	HgL_2 43.82	[705]
La^{3+}	pot	0.1	LaHL 12.18; LaH_2L_2 23.38	[704]
Mn^{2+}	pot	0.1	MnL 4.38; MnL_2 7.56	[703]
Ni^{2+}	pot	0.1	NiL 6.98; NiL_2 13.53	[703]
Pb^{2+}	pot		PbL 8.5	[706]
Zn^{2+}	pot	0.1	ZnL 7.86; ZnL_2 15.04	[703]

Thiosalicylic acid H_2L

Metal ion	Method	I	$\log \beta$	Ref.
H^+	pot	0.1	HL 9.96; H_2L 14.88 (50% dioxan)	[707]
Co^{2+}	pot	0.1	CoL 6.03; CoL_2 10.47 (45% ethanol; 30°C)	[708]
Fe^{2+}	pot	0.1	FeL 5.45; FeL_2 9.86 (45% ethanol; 30°C)	[708]
Mn^{2+}	pot	0.1	MnL 5.04; MnL_2 9.09 (45% ethanol; 30°C)	[708]
Ni^{2+}	pot	0.1	NiL 7.08; NiL_2 11.54 (45% ethanol; 30°C)	[708]
Zn^{2+}	pot	0.1	ZnL 8.45; ZnL_2 14.4 (45% ethanol; 30°C)	[708]

Threonine HL

Metal ion	Method	I	$\log \beta$	Ref.
H^+	pot	0.1	HL 9.16; H_2L 11.45	[709]
Be^{2+}	pot	0	BeL_2 11.9	[710]
Cd^{2+}	pot	0	CdL_2 7.2	[710]
Co^{2+}	pot	0.1	CoL 4.58	[709]
Cu^{2+}	pot	0.1	CuL 8.34; CuL_2 15.32	[709]
Fe^{3+}	pot	1	FeL 8.6	[411]
Hg^{2+}	pot	0	HgL_2 17.5	[710]
$Ni^2\varepsilon^+$	pot	0.1	NiL 5.66; NiL_2 10.20	[709]
Zn^{2+}	pot	0.1	ZnL 4.87	[709]

Tiron
(4.5-Dihydroxybenzene-1,3-disulphonic acid) H_4L

Metal ion	Method	I	$\log \beta$	Ref.
H^+	pot	0.1	HL 12.7; H_2L 20.4	[711]
Al^{3+}	pot	0	AlL 19.02; AlL_2 31.10; AlL_3 33.5	[712]
Ba^{2+}	pot	0.1	BaL 4.1; BaHL 14.7	[711]
Be^{2+}	pot	0.1	BeL 12.88; BeL_2 22.25; BeHL 16.90; $BeHL_2$ 27.88	[443]

Organic complexants (continued)

Metal ion	Method	I	$\log \beta$	Ref.
Ca^{2+}	pot	0·1	CaL 5·8; CaHL 14·9	[711]
Cd^{2+}	pot	1	CdL 7·69; CdL_2 13·3	[712]
Co^{2+}	pot	0·1	CoL 9·49; CoHL 15·8	[712]
Cu^{2+}	pot	0·1	CuL 14·5; CuHL 18·2	[712]
Fe^{3+}	pot	0·1	FeL 20·7; FeL_2 35·9; FeL_3 46·9; FeHL 22·7	[711]
In^{3+}	sp	0·1	InL 3·79	[713]
La^{3+}	pot	0·1	LaL 12·9; La(OH)L 18·6	[714]
Mg^{2+}	pot	0·1	MgL 6·86; MgHL 14·7	[711]
Mn^{2+}	pot	0·1	MnL 8·6	[714]
Ni^{2+}	pot	0·1	NiL 9·96; NiHL 15·7	[711]
Pb^{2+}	pot	1	PbL 11·95; PbL_2 18·3	[712]
Sr^{2+}	pot	0·1	SrL 4·55; SrHL 14·58	[711]
TiO^{2+}	sp	0·1	$TiOH_2L_2$ 40·5; TiL_3 58·0	[715]
UO_2^{2+}	sp	0·1	UO_2HL 19·2	[716]
VO^{2+}	pot	0·1	VOL 16·74	[683]
Zn^{2+}	pot	0·1	ZnL 10·41; ZnHL 16·0	[711]

Trichloroacetic acid HL

H^+	pot	0·1	HL 0·5	[717]
Th^{4+}	ex	0·5	ThL 1·62; ThL_2 2·8	[717]

Triethanolamine L

H^+	pot	0·1	HL 7·9	[718]
Ag^+	pot	0·5	AgL 2·3; AgL_2 3·64	[718]
Cd^{2+}	pol	1	CdL 2·3; CdL_2 5·0; $CdL_2(OH)$ 8; $CdL_2(OH)_2$ 11; $CdL(OH)_3$ 11·7; $CdL_2(OH)_3$ 13·1; $CdL_2(PO_4)_2$ 9·7; $CdL(CO_3)$ 5·2; $CdL_2(CO_3)$ 6·2; $CdL(CO_3)_2$ 6·5	[719]
Co^{2+}	pot	0·5	CoL 1·73	[718]
Cu^{2+}	pot	0·5	CuL 4·23; Cu(OH)L 12·5	[718]
Fe^{3+}	pot	0·1	$Fe(OH)_4L$ 41·2	[15]
Hg^{2+}	pot	0·5	HgL 6·9; HgL_2 20·08	[718]
Ni^{2+}	pot		NiL 2·95; $Ni_2L_2(OH)_2$ 18·2	[719]
Zn^{2+}	pot	0·5	ZnL 2·0	[718]

Triethylenetetramine L (Trien)

H^+	pot	0·1	HL 9·92; H_2L 19·12; H_3L 25·79; H_4L 29·11	[720]
Ag^+	pot	0·1	AgL 7·7; AgHL 15·72	[720]
Cd^{2+}	pot	0·1	CdL 10·75; CdHL 17·0	[720]
Co^{2+}	pot	0·1	CoL 11·0; CoHL 16·7	[720]
Cr^{3+}	pot	0·1	CrL 7·71	[721]

Sec. 4.1] Protonation and complex formation equilibrium constants

Organic complexants (continued)

Metal ion	Method	I	log β	Ref.
Cu^{2+}	pot	0·1	CuL 20·4; CuHL 23·9	[720]
Fe^{2+}	pot	0·1	FeL 7·8	[720]
Fe^{3+}	k	0	FeL 21·94	[722]
Hg^{2+}	pot	0·5	HgL 25·26; HgHL 30·8	[720]
Mn^{2+}	pot	0·1	MnL 4·9	[720]
Ni^{2+}	pot	0·1	NiL 14·0; NiHL 18·8	[720]
Pb^{2+}	pot	0·1	Pb 10·4	[723]
Zn^{2+}	pot	0·1	ZnL 12·1; ZnHL 17·2	[720]

Triethylenetetraminehexa-acetic acid (TTHA) H_6L

Metal ion	Method	I	log β	Ref.
H^+	pot	0·1	HL 10·19; H_2L 19·59; H_3L 25·75; H_4L 29·91; H_5L 32·86; H_6L 35·28	[724]
Al^{3+}	pot	0·1	AlL 19·7; AlHL 25·55; Al_2L 28·6; $Al_2(OH)_2L$ 44·5	[725]
Ba^{2+}	pot	0·1	BaL 8·22; BaHL 15·74; BaH_2L 21·29; Ba_2L 11·63	[726]
Ca^{2+}	pot	0·1	CaL 9·89; CaHL 18·4; CaH_2L 23·36; Ca_2L 14·21; Ca_3L 17·2	[724]
Cd^{2+}	pot	0·1	CdL 18·65; CdHL 26·97; CdH_2L 30·12; Cd_2L 26·8	[725]
Co^{2+}	pot	0·1	CoL 17·1; CoHL 25·2; Co_2L 28·8	[725]
Cu^{2+}	pot	0·1	CuL 19·2; CuHL 27·2; Cu_2L 32·6	[725]
Er^{3+}	pot	0·1	ErL 23·2; ErHL 27·9; Er_2L 26·9	[725]
Fe^{3+}	pot	0·1	FeL 26·8; FeHL 34·4; Fe_2L 40·5	[725]
Hg^{2+}	pot	0·1	HgL 26·8; HgHL 33·1; Hg_2L 39·1	[725]
La^{3+}	pot	0·1	LaL 22·2; LaHL 25·5; La_2L 25·6	[725]
Mg^{2+}	pot	0·1	MgL 8·10; MgHL 17·4; Mg_2L 14·0	[725]
Mn^{2+}	pot	0·1	MnL 14·6; MnHL 23·4; Mn_2L 21·2	[725]
Ni^{2+}	pot	0·1	NiL 18·1; NiHL 26·1; Ni_2L 32·4	[725]
Pb^{2+}	pot	0·1	PbL 17·1; PbHL 25·3; Pb_2L 28·1	[725]
Sr^{2+}	pot	0·1	SrL 9·26; SrHL 16·9; SrH_2L 21·2	[726]
Th^{4+}	pot	0·1	ThL 31·9; ThHL 35·0	[725]
Zn^{2+}	pot	0·1	ZnL 16·65; ZnHL 24·8; Zn_2L 28·7	[725]

Uramil-N,N-diacetic acid H_3L

Metal ion	Method	I	log β	Ref.
H^+	pot	0·1	HL 9·6; H_2L 12·27; H_3L 13·97	[727]
Ba^{2+}	pot	0·1	BaL 6·13; BaL_2 9·83	[727]

Organic complexants (continued)

Metal ion	Method	I	$\log \beta$	Ref.
Be^{2+}	pot	0.1	BeL 10.36; BeHL 13.0	[727]
Ca^{2+}	pot	0.1	CaL 8.31; CaL_2 13.58	[727]
K^+	pot	0.1	KL 1.23	[727]
Li^+	pot	0.1	LiL 4.9	[727]
Mg^{2+}	pot	0.1	MgL 8.19; MgL_2 11.81	[727]
Na^+	pot	0.1	NaL 2.72	[727]
Pb^{2+}	pot	0.1	PbL 12	[727]
Sr^{2+}	pot	0.1	SrL 6.93; SrL_2 11.0	[727]
Tl^+	pot	0.1	TlL 5.99	[727]

Xylenol Orange H_6L

H^+		0.2	HL 12.58; H_2L 23.04; H_3L 29.44; H_4L 32.67; H_5L 35.25; H_6L 36.40; H_7L 37.16; H_8L 36.07; HgL 34.33	[728]
Bi^{3+}	sp	0.2	Bi_2L_2 75.6	[729]
Cd^{2+}	sp	0.3	CdL 16.36	[730]
Fe^{3+}	sp	0.2	Fe_2L 39.80	[731]
Gd^{3+}	sp	0.2	Gd_2L_2 43.1	[732]
Sc^{3+}	sp	0.2	Sc_2L_2 61.2	[729]
Sm^{3+}	sp	0.2	Sm_2L_2 47.0	[732]
UO_2^{2+}	sp	0.2	$(UO_2)_2L_2$ 38.57	[731]
VO_2^+	sp	0.2	$(VO_2)_2L_2$ 63.1	[729]
Yb^{3+}	sp	0.2	Yb_2L_2 45.7	[732]

4.2. Precipitate Formation Constants

The following table contains the precipitate formation constants of metal salts and complexes which are only slightly soluble in water, defined by Eqs. (3.2) and (3.8) respectively, for room temperature. I = ionic strength; v = variable; Ox = oxalate, Oxin = oxinate; An = anthranilate; DMG = dimethylglyoximate; DDTC = diethyldithiocarbamate; Cup = cupferronate.

Constants

Precipitate	I	$\log K_s$	$\log K_{si}$	Ref.
Ag_3AsO_4	v	22.0		[733]
$AgBr$	0.1	12.3		[734]
$AgBrO_3$	0	4.24		[735]
$AgCN$	v	15.8		[736]
$AgCNO$	0	6.64		[737]
Ag_2CO_3	v	11.19		[738]
$AgCl$	0	9.75		[739]
Ag_2CrO_4	0	11.61		[740]
$Ag_4[Fe(CN)_6]$	0	19.2		[741]
AgI	0	16.08		[742]
$AgIO_3$	0	7.52		[743]
Ag_2MoO_4	0	11.55		[744]
$AgOH$	0	7.84		[745]
Ag_3PO_4	v	17.86		[746]
Ag_2S	v	49.7		[747]
$AgSCN$	0	11.97		[748]
Ag_2SO_4	0	4.83		[749]
$Ag_2(HVO_4)$	v	13.7		[750]
$Ag(H_2VO_4)$	v	6.3		[750]
Ag_2WO_4	v	9.28		[751]
$Ag_2(Ox)$	0	11		[15]
$AlAsO_4$	v	15.8		[733]
$Al(OH)_3$	0	33.5		[752]
$AlPO_4$	v	18.24		[753]
$Au(OH)_3$	0.45	45.26	$\log K_{s4}$ 3.28	[754]
			$\log K_{s5}$ 2.64	[755]
$BaCO_3$	0	8.31		[756]
$BaCrO_4$	0	9.93		[757]
BaF_2	0	5.76		[758]
$Ba(IO_3)_2$	0	8.82		[759]
$Ba_3(PO_4)_2$	0	29.3		[760]
$Ba(HPO_4)$	0	7.56		[760]

Constants (continued)

Precipitate	I	$\log K_s$	$\log K_{s1}$	Ref.
$BaSO_4$	0	9.96		[761]
$Ba_2S_2O_3$	0	4.79		[762]
$BaSeO_4$	0	7.46		[763]
$Ba(Ox)$	0.1	6.0		[611]
$Ba(Oxin)_2$	0.1	7.7		[15]
$Be(OH)_2$	0	20.8	$\log K_{s3}$ 2.9 $\log K_{s4}$ 3.1	[764]
BiI_3	2	18.09		[189]
$Bi(OH)_2Cl$	0	30.75		[15]
$BiPO_4$	v	22.9		[753]
Bi_2S_3	0	97		[15]
$Bi_2(Ox)_3$	0.1	35.4		[611]
$Ca_3(AsO_4)_2$	v	18.17		[733]
$CaCO_3$	0	8.14		[756]
CaF_2	0	10.5		[758]
$Ca(IO_3)_2$		6.15		[759]
$CaMoO_4$	0	8.0		[744]
$Ca(OH)_2$	0	5.03		[765]
$Ca_3(PO_4)_2$	0	26.0		[766]
$CaHPO_4$	0	7.0		[766]
$CaSO_3$	0	6.5		[767]
$CaSO_4$	0	5.04		[768]
$CaWO_4$	v	8.06		[769]
$Ca(Ox)$	0	8.64		[15]
$Ca(Oxin)_2$	0.1	10.4		[15]
$Cd_3(AsO_4)_2$	v	32.66		[733]
$CdCO_3$	0	11.28		[11]
$Cd_2[Fe(CN)_6]$	v	16.49		[770]
$Cd(OH)_2$	0	14.61		[745]
CdS	1	25.76		[771]
$Cd(An)_2$	0	8.39		[421]
$Cd(DDTC)_2$	0.1	22.0		[772]
$Cd(Ox)$	0	7.82		[15]
CeF_3	v	15.1		[773]
$Ce(OH)_3$	v	20.2		[745]
$Ce_2(Ox)_3$	0	25.4		[15]
$Ce(IO_3)_4$	0	9.50		[774]
$Ce(OH)_4$	v	50.4		[775]
$CoCO_3$	0	12.84		[756]
$Co_2[Fe(CN)_6]$	v	14.74		[770]

Precipitate formation constants

Constants (continued)

Precipitate	I	$\log K_s$	$\log K_{sl}$	Ref.
$Co(OH)_2$ blue	0	14.2		[764]
$Co(OH)_2$ pink	0	14.8		[764]
$CoS\ \alpha$	0	20.4		[15]
$CoS\ \beta$	0	24.7		[15]
$Co(An)_2$	0	10.97		[421]
$Co(Oxin)_2$	0.1	24.2		[15]
$Co(OH)_3$	0	44.5		[776]
$Cr(OH)_3$	v	30		[745]
$CrPO_4$ green	v	22.6		[753]
$CrPO_4$ violet	v	17.0		[753]
$CuBr$	v	7.38	$\log K_{s2}\ 2.34$	[777]
$CuCN$	0	19.49		[778]
$CuCl$	0	6.73		[11]
CuI	0.1	11.7	$\log K_{s3}\ 2.58$ $\log K_{s4}\ 2.23$	[779]
$CuOH$	0	14.7		[764]
$CuSCN$	v	12.7		[780]
$CuCO_3$	0	9.63		[756]
$CuCrO_4$	0	5.44		[781]
$Cu_2[Fe(CN)_6]$	v	17		[741]
$Cu(IO_3)_2$	0	7.13		[782]
$Cu(OH)_2$	0	18.2		[745]
$Cu_2P_2O_7$	v	15.08		[783]
CuS	0	47.6		[15]
$Cu(An)_2$	0	14.18		[421]
$Cu(DDTC)_2$	0.1	29.6		[772]
$Cu(Cup)_2$		16.03		[14]
$Cu(Ox)$	0	7.54		[15]
$Cu(Oxin)_2$	0.1	29.1		[15]
$FeCO_3$	0	10.50		[756]
$Fe(OH)_2$	v	14.01		[745]
FeS	0	17.2		[15]
$FeAsO_4$	v	20.24		[733]
$Fe_4[Fe(CN)_6]_3$	v	40.5		[770]
$Fe(OH)_3$		38.6		[784]
$FePO_4$	v	21.9		[753]
$Fe_4(P_2O_7)_3$	v	22.5		[785]
$Fe(Cup)_3$		25		[14]
$Fe(Oxin)_3$	0.1	43.5		[786]

Constants (continued)

Precipitate	I	$\log K_s$	$\log K_{sl}$	Ref.
$Ga_4[Fe(CN)_6]_3$	v	33·82		[770]
$Ga(OH)_3$	v	35·4		[745]
$Ga(Oxin)_3$	v	32·06		[787]
Hg_2Br_2	0·5	21·29		[788]
Hg_2CO_3	0	16·05		[789]
Hg_2Cl_2	0	17·27		[790]
$Hg_2(CN)_2$	0	39·3		[789]
Hg_2CrO_4	0	8·70		[789]
Hg_2I_2	0	28·35		[11]
$Hg_2(OH)_2$	0	23·7		[789]
$Hg_2(HPO_4)$	0	12·4		[791]
Hg_2SO_4	0	6·13		[15]
$Hg_2(SCN)_2$	v	19·8		[792]
Hg_2WO_4	0·1	16·96		[769]
$Hg_2(CH_3COO)_2$	0	14·7		[15]
$Hg(OH)_2$	v	25·4		[792]
HgS	1	51	$\log K_2$ 0·57	[793]
$Hg(DDTC)_2$	0·1	43·5		[475]
$In_4[Fe(CN)_6]_3$	v	43·72		[770]
$In(OH)_3$	v	33·2		[745]
In_2S_3	0	73·2		[794]
$In(DDTC)_3$	0·1	25·0		[795]
$In(Oxin)_3$	v	31·34		[787]
$La(IO_3)_3$	0	11·2		[796]
$La(OH)_3$	0	18·7		[745]
$La_2(Ox)_3$	0·1	25·0		[611]
$Mg_3(AsO_4)_2$	v	19·68		[733]
$MgCO_3$	0	5·0		[797]
MgF_2	0	8·2		[758]
$Mg(OH)_2$	0	10·74		[798]
$MgNH_4PO_4$	v	12·6		[759]
Mg(Ox)	v	4·07		[15]
$Mg(Oxin)_2$	0·1	14·8		[15]
$MnCO_3$	0	9·30		[756]
$Mn_2[Fe(CN)_6]$	v	12·1		[770]
$Mn(OH)_2$	v	12·9		[745]
MnS	0	8		[800]
$Mn(Oxin)_2$	0	21·7		[15]

Constants (continued)

Precipitate	I	$\log K_s$	$\log K_{sl}$	Ref.
$NiCO_3$	0	6·87		[756]
$Ni_2[Fe(CN)_6]$	v	14·89		[770]
$Ni(OH)_2$	v	14·5		[745]
NiS	0	24		
$Ni(An)_2$	0	11·72		[421]
$Ni(DMG)_2$	v	23·66	$\log K_{s2}$ 5·68	[118]
$Ni(Oxin)_2$	0·1	25·5		[15]
$Pb_3(AsO_4)_2$	v	35·39		[733]
$PbBr_2$	v	4·56		[801]
$Pb(BrO_3)_2$	0	5·10		[802]
$PbCO_3$	0	13·14		[756]
$PbCl_2$	0	4·67		[803]
$PbCrO_4$	v	12·55		[804]
PbF_2	0	7·43		[758]
$Pb_2[Fe(CN)_6]$	v	15·5		[741]
PbI_2	v	8·01		[803]
$Pb(IO_3)_2$	v	12·5		[805]
$PbMoO_4$	0	13·0		[744]
$Pb(OH)_2$	0	15·2		[764]
$Pb_3(PO_4)_2$	0	43·53		[806]
$Pb(HPO_4)$	0	11·36		[806]
PbS	0	27·5		
$PbSO_4$	0	8·0		
$PbSeO_4$	0	6·84		[807]
$Pb(An)_2$	0	9·81		[421]
$Pb(DDTC)_2$	0·1	21·7		[727]
$Pb(Ox)$	0	10·5		[15]
$Pd(OH)_2$	0	31		[11]
$Sb(OH)_3$	0	41·4		[808]
Sb_2S_3	0	92·7		[809]
$Sc(OH)_3$	v	30·1		[745]
$SrCO_3$	0	9·03		[756]
$SrCrO_4$	0	4·65		[810]
SrF_2	0	8·5		[758]
$Sr(IO_3)_2$	0	6·5		
$Sr_3(PO_4)_2$	0	27·8		[760]
$Sr(HPO_4)$	0	7·06		[760]
$SrSO_4$	v	6·2		[811]
$Sr(Ox)$	0·1	6·4		[611]
$Sr(Oxin)_2$	0·1	8·7		[15]

Constants (continued)

Precipitate	I	$\log K_s$	$\log K_{si}$	Ref.
$Sn(OH)_2$	0	26.2		[764]
SnS	0	25.0		[15]
$Sn(OH)_4$	0	56		[15]
ThF_4	v	25.3		[812]
$Th(IO_3)_4$	0.5	14.62	$\log K_{s6}$ 3.6; $\log K_{s6}$ 1.66	[813]
$Th(OH)_4$	v	44.9		[745]
$Th(Ox)_2$	0	21.4		[814]
$TiO(OH)_2$	0	29		[11]
$TlBr$	0	5.41		[815]
$TlBrO_3$	0	3.41		[735]
$TlCl$	0	3.77		[816]
Tl_2CrO_4	0	12.01		[817]
TlI	0	7.19		[818]
$TlIO_3$	0	5.51		[796]
Tl_2S	0	20.3		[15]
$TlSCN$	0	3.8		[819]
$Tl(OH)_3$	v	34.1		[745]
$Tl(Oxin)_3$		32.4		[787]
UF_4	v	21.24		[812]
$UO_2(OH)_2$	v	23.7		[820]
$(UO_2)_2[Fe(CN)_6]$	v	13.15		[770]
$UO_2(Ox)_2$	1	8.66		[821]
$Zn_3(AsO_4)_2$	v	27.89		[833]
$ZnCO_3$	0	10.78		[756]
$Zn_2[Fe(CN)_6]$	v	16.8		[741]
$Zn(OH)_2$	0	15		
$Zn_3(PO_4)_2$	0	32.04		[753]
ZnS	1	24.37		[822]
$Zn(An)_2$	0	9.75		[421]
$Zn(DDTC)_2$	0.1	16.9		[772]
$Zn(Ox)$	0	8.89		[15]
$Zn(Oxin)_2$	0.1	23.7		[15]
$ZrO(OH)_2$	v	48.2		[745]
$ZrO(H_2PO_4)_2$	v	17.64		[823]

4.3. Redox Equilibrium Constants

Inorganic redox systems

The equilibrium constants given in the table are defined by Eq. (1.56) and are valid for 25°C (and ionic strength $I = 1$). (aq) = aqueous solution; (s) = solid; (Hg) = amalgam; (g) = gas; (l) = liquid.

Redox system	$\log K_r$	E^0 Volts	Ref.
$Ag^+ + e \rightleftharpoons Ag_{(s)}$	13·51	0·80	[703]
$Al^{3+} + 3e \rightleftharpoons Al_{(s)}$	−84·3	−1·66	[11]
$AsO_4^{3-} + 2e + 2H^+ \rightleftharpoons AsO_3^{3-}$	8·1	0·24	[11]
$Au^{3+} + 3e \rightleftharpoons Au_{(s)}$	76	1·50	[11]
$Au^{3+} + 2e \rightleftharpoons Au^+$	(47·8)	(1·41)	[11]
$AuCl_4^- + 2e \rightleftharpoons AuCl_2^- + 2Cl^-$	31·6	0·93	[45]
$AuCl_4^- + 3e \rightleftharpoons Au_{(s)} + 4Cl^-$	50·4	0·99	[45]
$AuCl_2^- + e \rightleftharpoons Au_{(s)} + 2Cl^-$	18·8	1·11	[45]
$1/2\ Br_{2(aq)} + e \rightleftharpoons Br^-$	18·35	1·085	[824]
$BrO_3^- + 5e + 6H^+ \rightleftharpoons 1/2 Br_{2(aq)} + 3H_2O$	125·6	1·48	[11]
$BrO_3^- + 6e + 6H^+ \rightleftharpoons Br^- + 3H_2O$	146·4	1·44	[11]
$BrO^- + 2e + 2H^+ \rightleftharpoons Br^- + H_2O$	53·7	1·58	[11]
$Ca^{2+} + 2e \rightleftharpoons Ca_{(s)}$	−96·9	−2·87	[11]
$Cd^{2+} + 2e \rightleftharpoons Cd_{(s)}$	−13·5	−0·40	[11]
$Cd^{2+} + 2e \rightleftharpoons Cd_{(Hg)}$	−11·9	−0·35	[825]
$Cd(CN)_4^{2-} + 2e \rightleftharpoons Cd_{(s)} + 4CN^-$	36·9	1·09	[11]
$Ce^{4+} + e \rightleftharpoons Ce^{3+}$ (1 M $HClO_4$)	29·47	1·74	[826]
$1/2\ Cl_{2(g)} + e \rightleftharpoons Cl^-$	22·97	1·36	[11]
$1/2\ Cl_{2(aq)} + e \rightleftharpoons Cl^-$	23·58	1·39	[11]
$ClO^- + 2e + 2H^+ \rightleftharpoons Cl^- + H_2O$	58·20	1·72	[11]
$ClO_3^- + 5e + 6H^+ \rightleftharpoons 1/2\ Cl_{2(g)} + 3H_2O$	124·2	1·47	[11]
$Co^{3+} + e \rightleftharpoons Co^{2+}$	31	1·84	[827]
$Cr^{3+} + e \rightleftharpoons Cr^{2+}$	−6·9	−0·41	[11]
$1/2\ Cr_2O_7^{2-} + 3e + 7H^+ \rightleftharpoons Cr^{3+} + 3.5H_2O$	67·6	1·33	[11]
$Cu^{2+} + e \rightleftharpoons Cu^+$	2·68	0·16	[825]
$Eu^{3+} + e \rightleftharpoons Eu^{2+}$	−7·3	−0·43	[828]
$Fe^{3+} + e \rightleftharpoons Fe^{2+}$	12·97	0·77	[10]
$Fe(CN)_6^{3-} + e \rightleftharpoons Fe(CN)_6^{4-}$	6·1	0·36	[11]
$Ga^{3+} + 3e \rightleftharpoons Ga_{(s)}$	−28·4	−0·56	[829]
$GeO_2 + 2e + 4H^+ \rightleftharpoons Ge^{2+} + 2H_2O$	∼10	∼0·3	[11]
$H^+ + e \rightleftharpoons 1/2\ H_{2(g)}$	±0·00	±0·00	
$2Hg^{2+} + 2e \rightleftharpoons Hg_2^{2+}$	30·68	0·907	[830]
$Hg_2^{2+} + 2e \rightleftharpoons 2Hg_{(l)}$	26·79	0·79	[830]
$1/2\ I_{2(aq)} + e \rightleftharpoons I^-$	10·52	0·62	[831]
$I_3^- + 2e \rightleftharpoons 3I^-$	18·13	0·536	[11]
$IO^- + 2e + 2H^+ \rightleftharpoons I^- + H_2O$	44·4	1·31	[11]
$IO_3^- + 6e + 6H^+ \rightleftharpoons I^- + 3H_2O$	110·1	1·08	[11]

Inorganic redox systems (continued)

Redox system	log K_r	E^0 Volts	Ref.
$In^{3+} + 2e \rightleftharpoons In^+$	-13.7	-0.40	[832]
$IrO_{2(s)} + 4e + 4H^+ \rightleftharpoons Ir_{(s)} + 2H_2O$	62.6	0.93	[11]
$IrCl_6^{2-} + 4e \rightleftharpoons Ir_{(s)} + 6Cl^-$	56.4	0.835	[11]
$Mg^{2+} + 2e \rightleftharpoons Mg_{(s)}$	-80.3	-2.375	[833]
$Mn^{2+} + 2e \rightleftharpoons Mn_{(s)}$	-40.0	-1.182	[834]
$Mn^{3+} + e \rightleftharpoons Mn^{2+}$	25.15	1.488	[835]
$MnO_{2(s)} + 2e + 4H^+ \rightleftharpoons Mn^{2+} + 2H_2O$	41.60	1.23	[836]
$MnO_4^{2-} + 2e + 4H^+ \rightleftharpoons MnO_{2(s)} + 2H_2O$	76.6	2.26	[11]
$MnO_4^- + e \rightleftharpoons MnO_4^{2-}$	9.53	0.564	[11]
$MnO_4^- + 3e + 4H^+ \rightleftharpoons MnO_{2(s)} + 2H_2O$	85.8	1.695	[11]
$MnO_4^- + 5e + 8H^+ \rightleftharpoons Mn^{2+} + 4H_2O$	127.4	1.51	[11]
$MoO_4^{2-} + 6e + 8H^+ \rightleftharpoons Mo_{(s)} + 4H_2O$	6	0.06	[11]
$Mo^{6+} + e \rightleftharpoons Mo^{5+}$	9.2	0.53	[11]
$Nb^{3+} + 3e \rightleftharpoons Nb_{(s)}$	-56	-1.1	[11]
$NbO^{3+} + 2e + 2H^+ \rightleftharpoons Nb^{3+} + H_2O$	-11.88	-0.34	[837]
$NO_3^- + 2e + 2H^+ \rightleftharpoons NO_2^- + H_2O$	28.3	0.835	[11]
$NO_3^- + 3e + 4H^+ \rightleftharpoons NO_{(g)} + 2H_2O$	48.5	0.96	[11]
$NO_2^- + e + 2H^+ \rightleftharpoons NO_{(g)} + H_2O$	20.3	1.20	[11]
$N_2 + 2e + 2H^+ + 2H_2O \rightleftharpoons 2NH_2OH$	-65.5	-1.94	[11]
$N_2 + 4e + 4H^+ \rightleftharpoons N_2H_4$	-23.4	-0.345	[11]
$Ni^{2+} + 2e \rightleftharpoons Ni_{(s)}$	-8.45	-0.25	[11]
$Ni(OH)_{3(s)} + e + 3H^+ \rightleftharpoons Ni^{2+} + 3H_2O$	35.2	2.08	[838]
$1/2\ O_{2(g)} + 2e + 2H^+ \rightleftharpoons H_2O$	41.55	1.23	[11]
$HO_2^- + 2e + 3H^+ \rightleftharpoons 2H_2O$	71.7	2.12	[11]
$H_2O_2 + 2e + 2H^+ \rightleftharpoons 2H_2O$	60.0	1.77	[11]
$O_{3(g)} + 2e + 2H^+ \rightleftharpoons O_{2(g)} + H_2O$	70.2	2.07	[11]
$OsO_{4(s)} + 8e + 8H^+ \rightleftharpoons Os_{(s)} + 4H_2O$	114	0.85	[11]
$P_{(s)} + 3e + 3H^+ \rightleftharpoons PH_{3(g)}$	3.2	0.06	[11]
$PO_4^{3-} + 2e + 3H^+ \rightleftharpoons HPO_3^{2-} + H_2O$	4.1	0.12	[11]
$HPO_3^{2-} + 2e + 3H^+ \rightleftharpoons H_2PO_2^- + H_2O$	-10.9	-0.32	[11]
$Pb^{2+} + 2e \rightleftharpoons Pb_{(s)}$	-4.23	-0.125	[839]
$PbO_{2(s)} + 2e + 4H^+ \rightleftharpoons Pb^{2+} + 2H_2O$	49.19	1.455	[11]
$Pb^{4+} + 2e \rightleftharpoons Pb^{2+}$	(57.5)	(1.7)	[11]
$Pd^{2+} + 2e \rightleftharpoons Pd_{(s)}$	33.4	0.987	[11]
$PdCl_6^{2-} + 2e \rightleftharpoons PdCl_4^{2-} + 2Cl^-$	43.5	1.29	[840]
$PtCl_4^{2-} + 2e \rightleftharpoons Pt_{(s)} + 4CL^-$	24.5	0.726	[11]
$PtCl_6^{2-} + 2e \rightleftharpoons PtCl_4^{2-} + 2Cl^-$	23.0	0.68	[11]
$ReO_4^- + 3e + 4H^+ \rightleftharpoons ReO_{2(s)} + 2H_2O$	25.9	0.51	[11]
$ReO_{2(s)} + 4e + 4H^+ \rightleftharpoons Re_{(s)} + 2H_2O$	17.6	0.26	[841]
$Rh^{3+} + e \rightleftharpoons Rh^{2+}$	20	1.2	[11]
$RhO^{2+} + e + 2H^+ \rightleftharpoons Rh^{3+} + H_2O$	24	1.4	[11]
$RhCl_6^{2-} + e \rightleftharpoons RhCl_6^{3-}$	20	1.2	[11]
$Ru^{4+} + e \rightleftharpoons Ru^{3+}$	(15)	(0.9)	[842]
$RuO_{2(s)} + 4e + 4H^+ \rightleftharpoons Ru_{(s)} + 2H_2O$	53.3	0.79	[11]

Inorganic redox systems (continued)

Redox system	$\log K_r$	E^0 Volts	Ref.
$RuO_4 + e \rightleftharpoons RuO_4^-$	16·9	1·0	[829]
$S_{(s)} + 2e \rightleftharpoons S^{2-}$	−16·1	−0·48	[843]
$2S_{(s)} + 2e \rightleftharpoons S_2^{2-}$	−14·48	−0·43	[843]
$SO_3^{2-} + 4e + 6H^+ \rightleftharpoons S_{(s)} + 3H_2O$	39·6	0·58	[11]
$SO_4^{2-} + 2e + 2H^+ \rightleftharpoons SO_3^{2-} + H_2O$	−3·3	−0·10	[11]
$2SO_3^{2-} + 4e + 6H^+ \rightleftharpoons S_2O_3^{2-} + 3H_2O$	45·4	0·67	[11]
$S_4O_6^{2-} + 2e \rightleftharpoons 2S_2O_3^{2-}$	2·7	0·08	[11]
$2SO_3^{2-} + 2e + 4H^+ \rightleftharpoons S_2O_4^{2-} + 2H_2O$	18	0·53	[11]
$2SO_4^{2-} + 2e + 4H^+ \rightleftharpoons S_2O_6^{2-} + 2H_2O$	−7·46	−0·22	[11]
$S_2O_8^{2-} + 2e \rightleftharpoons 2SO_4^{2-}$	67·9	2·01	[11]
$SbO^+ + 3e + 2H^+ \rightleftharpoons Sb_{(s)} + H_2O$	10·76	0·21	[844]
$Sb^{5+} + 2e \rightleftharpoons Sb^{3+}$	(26)	(0·8)	[845]
$Se_{(s)} + 2e \rightleftharpoons Se^{2+}{}_{(s)}$	−31·2	−0·92	[11]
$SeO_3^{2-} + 4e + 6H^+ \rightleftharpoons Se_{(s)} + 3H_2O$	59·2	0·87	[11]
$SeO_4^{2+} + 2e + 2H^+ \rightleftharpoons SeO_3^{2-} + H_2O$	29·8	0·88	[11]
$SiO_3^{2-} + 4e + 6H^+ \rightleftharpoons Si_{(s)} + 3H_2O$	−31	−0·46	[11]
$Sn^{2+} + 2e \rightleftharpoons Sn_{(s)}$	−4·6	−0·136	[11]
$Sn^{4+} + 2e \rightleftharpoons Sn^{2+}$	(5·1)	(0·15)	[11]
$Te_{(s)} + 2e \rightleftharpoons Te^{2-}$	−38·6	−1··14	[11]
$TeO_3^{2-} + 4e + 6H^+ \rightleftharpoons Te_{(s)} + 3H_2O$	4·55	0·67	[11]
$TeCl_6^{2-} + 4e \rightleftharpoons Te_{(s)} + 6Cl^-$	43·7	0·65	[846]
$TeO_4^{2-} + 2e + 2H^+ \rightleftharpoons TeO_3^{2-} + H_2O$	(41·5)	(1·22)	[11]
$Ti^{2+} + 2e \rightleftharpoons Ti_{(s)}$	−55	−1·63	[11]
$Ti^{3+} + e \rightleftharpoons Ti^{2+}$	−6·2	−0·37	[11]
$Ti(OH)^{3+} + e + H^+ \rightleftharpoons Ti^{3+} + H_2O$	−0·9	−0·05	[847]
$Tl^+ + e \rightleftharpoons Tl_{(s)}$	−5·68	−0·336	[825]
$Tl^+ + e \rightleftharpoons Tl_{(Hg)}$	−6·61	−0·39	[848]
$Tl^{3+} + 2e \rightleftharpoons Tl^+$	42·6	1·26	[825]
$U^{3+} + 3e \rightleftharpoons U_{(s)}$	−91	−1·80	[11]
$U^{4+} + e \rightleftharpoons U^{3+}$	−10·3	−0·61	[11]
$UO_2^{2+} + 2e + 4H^+ \rightleftharpoons U^{4+} + 2H_2O$	11·3	0·33	[11]
$V^{2+} + 2e \rightleftharpoons V_{(s)}$	−40·1	−1·18	[11]
$V^{3+} + e \rightleftharpoons V^{2+}$	−4·9	−0·28	[849]
$VO^{2+} + e + 2H^+ \rightleftharpoons V^{3+} + H_2O$	5·7	0·34	[850]
$VO_2^+ + e + 2H^+ \rightleftharpoons VO^{2+} + H_2O$	16·9	1·00	[851]
$WO_4^{2-} + 6e + 8H^+ \rightleftharpoons W_{(s)} + 4H_2O$	5	0·05	[11]
$W(CN)_8^{3-} + e \rightleftharpoons W(CN)_8^{4-}$	7·7	0·45	[852]
$Yb^{3+} + e \rightleftharpoons Yb^{2+}$	−9·8	−0·58	[853]
$Zn^{2+} + 2e \rightleftharpoons Zn_{(s)}$	−25·79	−0·76	[825]
$Zn(CN)_4^{2-} + 2e \rightleftharpoons Zn_{(s)} + 4CN^-$	−42·6	−1·26	[11]

Organic redox systems

The constants, as defined by Eq. (1.56) are given for 25°C. The logarithms of the protonation constants of the oxidized and reduced forms are given in parentheses after the reaction equations.

System	$\log K_r$	E^0 Volts	Ref.
$CH_2O_{(aq)} + 2e + 2H^+ \rightleftharpoons CH_3OH_{(aq)}$	6·4	0·19	[11]
$HCOO^- + 2e + 3H^+ \rightleftharpoons CH_2O + H_2O$ ($\log K_{01} = 3·7$)	5·6	0·165	[11]
$CO_{2(g)} + 2e + H^+ \rightleftharpoons HCOO^-$ ($\log K_{r1} = 3·7$)	−10·4	−0·30	[11]
$CH_3CHO_{(aq)} + 2e + 2H^+ \rightleftharpoons C_2H_5OH_{(aq)}$	6·4	0·19	[11]
$CH_3CO^- + 2e + 3H^+ \rightleftharpoons CH_3CHO + H_2O$ ($\log K_{01} = 4·7$)	0·7	0·02	[11]
$2CO_{2(g)} + 2e \rightleftharpoons C_2O_4^{2-}$ ($\log K_{r1} = 3·8$; $\log K_{r2} = 1·4$)	−21·8	−0·64	[11]
9,10-Anthraquinone disulphonate/ 9,10-Anthrahydroquinone disulphonate $AQDS^{2-} + 2e \rightleftharpoons AHQDS^{4-}$ ($\log K_{01}$, $\log K_{02} < 0$, $\log K_{r1} = 10·52$; $\log K_{r2} = 8·1$; $\log K_{r3}$, $\log K_{r4} < 0$)	−10·9	−0·32	[854]
Bindschedler's Green: $(CH_3)_2N^+ \cdot C_6H_4 \cdot NC_6H_4N(CH_3)_2 + 2e + + H^+ \rightleftharpoons (CH_3)_2NC_6H_4NH \cdot C_6H_4 \cdot N(CH_3)_2$ ($\log K_{01} = 3·27$; $\log K_{r1} = 6·46$; $\log K_{r2} = 5·1$)	14·7	0·434	[855]
Dehydroascorbic acid/Ascorbinate: $C_6H_6O_6 + 2e \rightleftharpoons C_6H_6O_6^{2-}$ ($\log K_{r1} = 11·56$; $\log K_{r2} = 4·17$)	−2·5	−0·07	[856]
2·6-Dichlorophenol-indophenol/ Leucodichlorophenol-indophenol: $DCPIP^- + 2e + H^+ \rightleftharpoons LDCPIP^{2-}$ ($\log K_{01} = 5·7$; $\log K_{r1} = 10·13$; $\log K_{r2} = 7·0$)	11·2	0·33	[857]
2,6-Dichloroquinone/Dichlorohydroquinonate $DCQ + 2e \rightleftharpoons DCHQ^{2-}$ ($\log K_{r1} = 10$; $\log K_{r2} = 7·3$)	6·8	0·20	[858]
Indigo-disulphonate/Leucoindigo-disulphonate: $IDS^{2-} + 2e + H^+ \rightleftharpoons LIDS^{3-}$ ($\log K_{r1} = 7·44$)	2·52	0·074	[859]
Methylene Blue/Leucomethylene Blue: $MB^+ + 2e + H^+ \rightleftharpoons LMB$ ($\log K_{r1} = 5·85$; $\log K_{r2} = 4·52$)	7·6	0·22	[860]

Sec. 4.4]   Extraction constants   379

Organic redox systems (continued)

System	log K_r	E^0 Volts	Ref.
o-Quinone/Pyrocatecholate: $C_6H_4O_2 + 2e \rightleftharpoons C_6H_4O_2^{2-}$ (log $K_{r1} = 13.05$; log $K_{r2} = 9.23$)	4.4	0.13	
p-Quinone/Hydroquinonate: $C_6H_4O_2 + 2e \rightleftharpoons C_6H_4O_2^{2-}$ (log $K_{r1} = 11.4$; log $K_{r2} = 9.85$)	2.4	0.07	[858]
Tetramethylquinone/Tetramethyl hydroquinonate: TMQ + 2e $\rightleftharpoons$ TMHQ^{2-} (log $K_{r1} = 12.7$; log $K_{r2} = 11.25$)	−7	−0.21	[858]
Variamine Blue/Leucovariamine Blue: VB + 2e + 2H$^+$ $\rightleftharpoons$ LVB (log $K_{01} = 6.6$; log $K_{r1} = 5.9$)	24.8	0.73	[139]

4.4. Extraction Constants

The log K_e values are the logarithms of the mass-action fractions given in terms of concentrations according to Eq. (3.141). All the values are referred to room temperature and an ionic strength of $I = 0.1$; n, if not indicated otherwise, is equal to the charge number of the metal ion.

The general formula giving the ligand to proton ratio of the compound, and the name of the organic solvent are given following the name of the complex-forming compound. These are followed by the protonation constant of the ligand, and distribution coefficient of the complex-forming reagent molecule [see Eq. (3.139)], the reagent concentration used in the determination (C_{HL}) and the reference.

Acetylacetone, HL, benzene

log $K = 8.8$;   log $d_{HL} = 0.77$; $C_{HL} = 0.1 M$ [861]

Ion	log K_e	Ion	log K_e	Ion	log K_e
Al^{3+}	−6.48	Fe^{3+}	−1.39	Pd^{2+}	<+2
Be^{2+}	−2.79	Ga^{3+}	−5.51	Sc^{3+}	−5.82
Cu^{2+}	−3.93	In^{3+}	−7.20	Th^{4+}	−12.16
		Pb^{2+}	−10.15		

Benzoylacetone, HL, benzene

$\log K = 8.7$; $\log d_{HL} = 3.14$; $C_{HL} = 0.1 M$ [861]

Ion	$\log K_e$	Ion	$\log K_e$	Ion	$\log K_e$
Ag^+	− 7·81	Eu^{3+}	−18·9	Pb^{2+}	− 9·61
Al^{3+}	− 7·60	Fe^{3+}	− 0·50	Pd^{2+}	+ 1·2
Be^{2+}	− 3·88	Ga^{3+}	− 6·34	Sc^{3+}	− 5·99
Ca^{2+}	−18·28	In^{3+}	− 9·30	Sr^{2+}	−20
Cd^{2+}	−14·11	La^{3+}	−20·46	Th^{4+}	− 7·68
Co^{2+}	−11·11	Mg^{2+}	−16·65	UO_2^{2+}	− 4·68
Cu^{2+}	− 4·17	Mn^{2+}	−14·63	Y^{3+}	−16·95
		Ni^{2+}	−12·12	Zn^{2+}	−10·79

Cupferron (nitrosophenylhydroxylamine), HL, chloroform

$\log K = 4.16$; $d_{HL} = 2.18$; $C_{HL} = 0.05$ and $0.005 M$ [862]

Ion	$\log K_e$	Ion	$\log K_e$	Ion	$\log K_e$
Al^{3+}	− 3·50	Fe^{3+}	9·8	Pb^{2+}	−1·53
Be^{2+}	− 1·54	Ga^{3+}	4·92	Sc^{3+}	3·34
Bi^{3+}	5·08	Hg^{2+}	0·91	Th^{4+}	4·44
Co^{2+}	− 3·65	In^{3+}	2·42	Tl^{3+}	3
Cu^{2+}	2·66	La^{3+}	−6·22	Y^{3+}	−4·74

Dibenzoylmethane, HL, benzene

$\log K = 9.35$; $\log d_{HL} = 5.35$; $C_{HL}^0 = 0.1 M$ [861]

Ion	$\log K_e$	Ion	$\log K_e$	Ion	$\log K_e$
Ag^+	− 8·78	Fe^{3+}	− 1·93	Pb^{2+}	− 9·45
Al^{3+}	− 8·92	Ga^{3+}	− 5·76	Sc^{3+}	− 6·04
Be^{2+}	− 3·46	In^{3+}	− 7·61	Sr^{2+}	−20·9
Ca^{2+}	−18·00	La^{3+}	−19·46	Th^{4+}	− 6·39
Cd^{2+}	−13·98	Mg^{2+}	−14·72	UO_2^{2+}	− 4·12
Co^{2+}	−10·78	Mn^{2+}	−13·71	Zn^{2+}	−10·67
Cu^{2+}	− 3·80	Ni^{2+}	−11·02		

Diethyldithiocarbamic acid, HL, carbon tetrachloride

$\log K \simeq 3\cdot8$; $\log d_{HL} = 2\cdot38$; $C_{HL} = 10^{-1}$ and $10^{-2} M$ [863]

Ion	$\log K_e$	Ion	$\log K_e$	Ion	$\log K_e$
Ag^+	11·9	Fe^{2+}	1·20	Pb^{2+}	7·77
Bi^{3+}	16·79	Hg^{2+}	31·94	Pd^{2+}	>32
Cd^{2+}	5·41	In^{3+}	10·34	Tl^+	−0·53
Co^{2+}	2·33	Mn^{2+}	−4·42	Zn^{2+}	2·96
Cu^{2+}	13·70	Ni^{2+}	11·58		

Dithizone (diphenylthiocarbazone), H(HL), carbon tetrachloride and chloroform

$(\log K_1 > 12)$; $\log K_2 = 4\cdot5$; $\log d_{H(HL)}(CCl_4) = 4\cdot0$;

$\log d_{H(HL)}(CHCl_3) = 5\cdot7$; $C_{H(HL)} = 10^{-4} M$ [864, 865]

| Ion | $\log K_e$ | | Ion | $\log K_e$ | |
	CCl_4	$CHCl_3$		CCl_4	$CHCl_3$
Ag^+	7·18	6·0	In^{3+}	4·84	1
Bi^{3+}	9·98	5·4	Ni^{2+}	1·18	−2·5
Cd^{2+}	2·14	0·5	Pb^{2+}	0·44	0·0
Co^{2+}	1·53	−0·5	Pd^{2+}	>27	
Cu^{2+}	10·5	6·5	Sn^{2+}	2	
Ga^{3+}	−1·3	−1·3	Tl^+	−3·3	−3·8
Hg^{2+}	26·8		Zn^{2+}	2·3	0·8

α-Nitroso-β-naphthol, HL, chloroform

$\log K = 7\cdot63$; $\log d_{HL} = 2\cdot97$; $C_{HL} = 0\cdot1 M$ [59, 866]

Ion	$\log K_e$
Th^{4+}	−1·64

Oxine (8-Hydroxyquinoline), HL, chloroform

$\log K_1 = 9.9$; $\log K_2 = 5.0$; $d_{HL} = 2.6$; $C_{HL} = 10^{-1}$ and $10^{-2}M$ [611]

Ion	$\log K_e$	Ion	$\log K_e$	Ion	$\log K_e$
Ag^+	-4.51	Ga^{3+}	3.72	Sc^{3+}	-6.64
Al^{3+}	-5.22	Hg^{2+}	-3	Sm^{3+}	-13.41
Ba^{2+}	-20.9	In^{3+}	0.89	Sr^{2+}	-19.71
Be^{2+}	-9.62	La^{3+}	-16.37	Th^{4+}	-7.18
Bi^{3+}	-1.2	Mg^{2+}	-15.13	TiO^{2+}	0.9
Ca^{2+}	-17.89	Mn^{2+}	-9.32	Tl^{3+}	5.0
Cd^{2+}	-5.29	MoO_2^{2+}	9.88	UO_2^{2+}	-1.60
Co^{2+}	-2.16	Ni^{2+}	-2.18	VO_2^+	1.67
Cu^{2+}	1.77	Pb^{2+}	-8.04	Zn^{2+}	-2.41
Fe^{3+}	4.11	Pd^{2+}	15	ZrO^{2+}	2.71

PAN [1-(2-Pyridylazo)-2-naphthol], HL; carbon tetrachloride and chloroform

$\log K = 11.2$; $\log d_{HL}(CCl_4) = 4$; $\log d_{HL}(CHCl_3) = 5.4$; $C_{HL} = 2 \times 10^{-3}M$ [863a, 863b]

Ion	$\log K_e$		Ion	$\log K_e$	
	CCl_4	$CHCl_3$		CCl_4	$CHCl_3$
Cd^{2+}	-7.4	-7.9	Mn^{2+}	-9.8	-11.0
Cu^{2+}		2.7^*	Ni^{2+}	-3.6	-7.7

* The extracted species not CuL_2 but $CuLCl$.

Ion-exchange equilibrium constants

β-i-Propyltropolone; HL; chloroform

$\log K = 7\cdot04$; $\log d_{HL} = 3\cdot37$; $C_{HL}^0 = 0\cdot1M$ [867]

Ion	$\log K_e$	Ion	$\log K_e$	Ion	$\log K_e$
Ag^+	$-8\cdot7$	In^{3+}	$4\cdot01$	Tb^{3+}	$-7\cdot1$
Ba^{2+}	-17	La^{3+}	$-10\cdot3$	Th^{4+}	$6\cdot2$
Ca^{2+}	$-14\cdot4$	Lu^{3+}	$-4\cdot27$	Tm^{3+}	$-6\cdot1$
Cd^{2+}	$-8\cdot1$	Ni^{2+}	$-7\cdot7$	U^{4+}	$2\cdot63$
Cu^{2+}	$1\cdot7$	Pr^{3+}	$-8\cdot49$	Y^{3+}	$-7\cdot4$
Eu^{3+}	$-6\cdot24$	Sc^{3+}	$1\cdot08$	Yb^{3+}	$-4\cdot89$
Fe^{3+}	$10\cdot0$	Sm^{3+}	$-2\cdot52$	Zn^{2+}	$-6\cdot2$
Ho^{3+}	$-6\cdot13$	Sr^{2+}	$-15\cdot2$		

Thenoyltrifluoroacetone, HL, benzene

$\log K = 6\cdot23$; $\log d_{HL} = 1\cdot6$; $C_{HL} = 0\cdot2M$ [864, 868, 869, 870]

Ion	$\log K_e$	Ion	$\log K_e$	Ion	$\log K_e$
Al^{3+}	$-5\cdot23$	Ga^{3+}	$-7\cdot57$	Sc^{3+}	$-0\cdot77$
Ba^{2+}	$-14\cdot4$	Hf^{4+}	$7\cdot8$	Sm^{3+}	$-7\cdot68$
Be^{2+}	$-3\cdot2$	Ho^{3+}	$-7\cdot25$	Sr^{2+}	$-14\cdot1$
Ca^{2+}	$-12\cdot0$	In^{3+}	$-4\cdot34$	Tb^{3+}	$-7\cdot51$
Ce^{3+}	$-9\cdot43$	La^{3+}	$-10\cdot31$	Th^{4+}	$0\cdot8$
Co^{2+}	$-6\cdot7$	Lu^{3+}	$-6\cdot77$	Tl^+	$-5\cdot2$
Cu^{2+}	$-1\cdot32$	Mn^{2+}	-1	Tm^{3+}	$-6\cdot96$
Dy^{3+}	$-7\cdot03$	Nd^{3+}	$-8\cdot58$	U^{4+}	$5\cdot3$
Er^{3+}	$-7\cdot2$	Pb^{2+}	$-5\cdot2$	UO_2^{2+}	$-2\cdot26$
Eu^{3+}	$-7\cdot66$	Pm^{3+}	$-8\cdot05$	Y^{3+}	$-7\cdot39$
Fe^{3+}	$3\cdot3$	Pr^{3+}	$-8\cdot85$	Yb^{3+}	$-6\cdot72$
				Zr^{4+}	$9\cdot15$

4.5. Ion-exchange Equilibrium Constants

The volume ion-exchange equilibrium constants given in the table, as defined by Eqs. (3.154) and (3.180) have been calculated from data in the literature cited, and are suitable for making exploratory calculations.

Cation-exchange equilibrium constants are given for strongly acidic Dowex 50×8 ($Q = 1.8$ meq/ml, $\sigma = 0.35$ g/ml) resin, whereas anion-exchange equilibrium constants are given for various resin types indicated in the table. The data for the anion-exchange resins are as follows:

Deacidite FF   ($Q = 1.5$, $\sigma = 0.42$)   Dowex 1×8 ($Q = 1.4$, $\sigma = 0.4$)
Lewatit MP 500 ($Q = 1.0$, $\sigma = 0.3$)   Dowex 1×10 ($Q = 1.5$, $\sigma = 0.5$)
Dowex 2×8 ($Q = 1.4$, $\sigma = 0.39$)

Cation-exchange equilibrium constants

Ion pair	K_x	Ref.	Ion pair	K_x	Ref.
$Ag^+ - H^+$	4.5	[871]	$Na^+ - H^+$	1.4	[871]
$Cs^+ - H^+$	4.5	[871]	$NH_4^+ - H^+$	1.9	[871]
$Hg^+ - H^+$	17	[871]	$Rb^+ - H^+$	3.0	[871]
$K^+ - H^+$	2.4	[872]			
$Li^+ - H^+$	0.79	[871]	$Tl^+ - H^+$	5.0	[871]
$Ba^{2+} - H^+$	16	[871]	$Ni^{2+} - H^+$	3.9	[871]
$Be^{2+} - H^+$	3.8	[871]	$Pb^{2+} - H^+$	6.3	[871]
$Ca^{2+} - H^+$	5	[871]	$Pd^{2+} - H^+$	2	[871]
$Cd^{2+} - H^+$	4.1	[871]	$Sr^{2+} - H^+$	7.1	[871]
$Co^{2+} - H^+$	3.3	[871]	$TiO^{2+} - H^+$	3.1	[871]
$Cu^{2+} - H^+$	4	[873]	$Tl(OH)^{2+} - H^+$	20.0	[871]
$Fe^{2+} - H^+$	3	[871]	$UO_2^{2+} - H^+$	2.58	[871]
$Hg^{2+} - H^+$	3.2	[871]	$VO^{2+} - H^+$	2.5	[871]
$Mg^{2+} - H^+$	3.3	[873]	$Zn^{2+} - H^+$	3.6	[871]
$Mn^{2+} - H^+$	3.8	[871]			
$Cd^{2+} - NH_4^+$	0.74	[874]	$Cu(NH_3)_4^{2+} - NH_4^+$	8.7	[874]
$Cd(NH_4)_4^{2+} - NH_4^+$	7.8	[874]	$Ni^{2+} - NH_4^+$	2.0	[874]
$Cd(NH_4)_6^{2+} - NH_4^+$	17	[874]	$Ni(NH_3)_4^{2+} - NH_4^+$	11.1	[874]
$Co^{2+} - NH_4^+$	1.2	[874]	$Ni(NH_3)_6^{2+} - NH_4^+$	13.4	[874]
$Co(NH_3)_4^{2+} - NH_4^+$	10.7	[874]	$Zn^{2+} - NH_4^+$	1.7	[874]
$Cu^{2+} - NH_4^+$	1.6	[874]	$Zn(NH_3)_4^{2+} - NH_4^+$	7.5	[874]
$Al^{3+} - H^+$	10	[871]	$Hf(OH)^{3+} - H^+$	250	[871]
$Bi^{3+} - H^+$	10	[871]	$In^{3+} - H^+$	14	[871]

Ion-exchange equilibrium constants

Cation-exchange equilibrium constants (continued)

Ion pair	K_x	Ref.	Ion pair	K_x	Ref.
$Ce^{3+} - H^+$	26	[871]	$La^{3+} - H^+$	26	[871]
$Cr^{3+} - H^+$	11	[871]	$Sc^{3+} - H^+$	14	[871]
$Er^{3+} - H^+$	18	[871]	$Sm^{3+} - H^+$	20	[871]
$Eu^{3+} - H^+$	19	[871]	$Th(OH)^{3+} - H^+$	120	[871]
$Fe^{3+} - H^+$	10	[871]	$Y^{3+} - H^+$	20	[871]
$Ga^{3+} - H^+$	6	[871]	$Yb^{3+} - H^+$	19	[871]
$Gd^{3+} - H^+$	18	[871]	$Zr(OH)^{3+} - H^+$	600	[871]

Anion-exchange equilibrium constants

Ion pair	Ion-exchange resin	K_x	Ref.
$Br^- - Cl^-$	Dowex 1×8	2.8	[875]
$BrO_3^- - Cl^-$	,,	0.8	[876]
$HCO_3^- - Cl^-$	,,	0.38	[876]
$ClO_3^- - Cl^-$	,,	3	[876]
$ClO_4^- - Cl^-$	,,	24	[876]
$CN^- - Cl^-$	,,	1.6	[876]
$F^- - Cl^-$	,,	0.09	[875]
$I^- - Cl^-$	,,	8.7	[875]
$IO_3^- - Cl^-$	,,	0.20	[876]
$NO_2^- - Cl^-$	,,	1.2	[875]
$NO_3^- - Cl^-$	Deacidite FF	2.45	[877]
$OH^- - Cl^-$	Dowex 1×8	0.04	[876]
$H_2PO_4^- - Cl^-$	,,	0.25	[875]
$SCN - Cl^-$	..	16	[876]
$HSO_3^- - Cl^-$	,,	1.3	[875]
$HSO_4^- - Cl^-$	,,	4.1	[875]
$SO_4^{2-} - Cl^-$	Deacidite FF	0.10	[877]
$SeO_4^{2-} - Cl^-$	,,	0.16	[877]
$AuCl_4^- - Cl^-$	Dowex 1×10	7×10^5	[878]
$CdCl_4^{2-} - Cl^-$	,,	2×10^4	[878]
$FeCl_4^- - Cl^-$	,,	4×10^5	[878]
$GaCl_4^- - Cl^-$	,,	7×10^5	[878]
$HgCl_4^{2-} - Cl^-$	,,	3×10^4	[878]
$TlCl_4^- - Cl^-$	,,	3.6×10^4	[878]
$ZnCl_4^{2-} - Cl^-$	,,	$\sim 10^4$	[878]
$ZnCl_3^- - Cl^-$	,,	$\sim 10^3$	[878]

Anion-exchange equilibrium constants (continued)

Ion pair	Ion-exchange resin	K_x	Ref.
$HCOO^- - Cl^-$	Dowex 1x8	0·22	[875]
$CH_3COO^- - Cl^-$	,,	0·17	[875]
$H_2NCH_2COO^- - Cl^-$	,,	0·10	[875]
Acetylsalicylate$^-$ $-$ Cl$^-$	Lewatit MP 500	1·1	[174]
Benzoate$^-$ $-$ Cl$^-$	,,	0·46	[174]
Dinitrophthalate^{2-} $-$ Cl$^-$	,,	2·9	[174]
Phthalate^{2-} $-$ Cl$^-$	,,	0·48	[174]
Salicylate$^-$ $-$ Cl$^-$	,,	2·75	[174]
Sulphosalicylate^{2-} $-$ Cl$^-$	,,	6·9	[174]
Oxalate^{2-} $-$ acetate$^-$	Dowex 1×8	2·86	[879]
Malate^{2-} $-$ acetate	,,	1·67	[879]
$Br^- - Cl^-$	Dowex 2×8	2·5	[880]
$HCO_3^- - Cl^-$	,,	0·53	[875]
$ClO_4^- - Cl^-$	,,	9	[880]
$CN^- - Cl^-$	,,	1·3	[875]
$F^- - Cl^-$	,,	0·10	[880]
$I^- - Cl^-$	,,	18	[880]
$IO_3^- - Cl^-$	,,	0·21	[880]
$NO_2^- - Cl^-$	,,	1·3	[875]
$NO_3^- - Cl^-$	,,	3·0	[880]
$OH^- - Cl^-$	,,	0·65	[875], [880]
$H_2PO_4^- - Cl^-$	,,	0·34	[875]
$SCN^- - Cl^-$	,,	4·5	[880]
$HSO_3^- - Cl^-$	,,	1·3	[875]
$HSO_4^- - Cl^-$	,,	6·1	[875]
$HCOO^- - Cl^-$	,,	0·22	[875]
$CH_3COO^- - Cl^-$	,,	0·17	[875], [880]
$H_2NCH_2COO^- - Cl^-$	,,	0·10	[875]
$CH_2ClCOO^- - Cl^-$	,,	0·21	[880]
$CHCl_2COO^- - Cl^-$	,,	2	[880]
$CCl_3COO^- - Cl^-$	,,	2·3	[880]

REFERENCES

[1] G. Briegleb *Elektronen-Donator-Acceptor Komplexe*, Springer Verlag, Berlin (1961).
[2] L. J. Andrews, R. M. Keefer *Molecular Acceptor Donor Complexes in Organic Chemistry*, Holden-Day, London (1964).
[3] C. W. Davies *Ion Association*, Butterworths, London (1962).
[4] N. Bjerrum *Z. Anorg. Chem.* **119,** 179 (1921).
[5] I. Leden *Svensk Kem. Tidskr.* **56,** 31 (1944).
[6] L. G. Sillén *Acta Chem. Scand.* **8,** 299, 318 (1954).
[7] S. Hietanen *Acta Chem. Scand.* **8,** 1626 (1954).
[8] R. Näsänen *Acta Chem. Scand.* **6,** 352 (1952).
[9] H. Jørgensen *Redox Maalinger*, Gjellerup, Copenhagen (1945).
[10] W. C. Schumb, M. S. Sherill, S. B. Sweetser *J. Am. Chem. Soc.* **59,** 2630 (1937).
[11] W. M. Latimer *Oxidation Potentials*, 2nd Ed. Prentice-Hall, New York (1952).
[12] L. G. Sillén, A. E. Martell *Stability Constants of Metal-ion Complexes*. The Chemical Society, London (1964); *Supplement No. 1* (1971).
[13] G. Kortüm, W. Vogel, K. Andrussow *Dissociation Constants of Organic Acids in Aqueous Solutions*, Butterworths, London (1961).
[14] W. F. Linke, A. Seidell *Solubilities*, Van Nostrand, New York (1958).
[15] A. Ringbom *Complexation in Analytical Chemistry*, Wiley, New York (1963).
[16] J. Bjerrum *Metal Ammine Formation in Aqeous Solution*, P. Haase and Son, Copenhagen (1941).
[17] R. M. Izatt, W. C. Fernelius, C. G. Haas, B. P. Block *J. Phys. Chem.* **59,** 170 (1955).
[18] C. W. Davies, C. B. Monk *Trans. Faraday Soc.* **50,** 132 (1954).
[19] A. Schufle, H. M. Eiland *J. Am. Chem. Soc.* **76,** 960 (1954).
[20] S. Fronaeus *Acta Chem. Scand.* **7,** 21 (1953).
[21] S. Chaberek, R. C. Courtney, A. E. Martell *J. Am. Chem. Soc.* **74,** 5057 (1952).
[22] B. O. A. Hedström *Arkiv Kemi* **6,** 1 (1954).
[23] C. Bereczki-Biedermann, G. Biedermann, L. G. Sillén *Report to Anal. Section*, IUPAC (July 1953).
[24] G. Schwarzenbach *Die komplexometrische Titration*, Enke Verlag, Stuttgart (1957).
[25] G. Schwarzenbach, R. Gut, G. Anderegg *Helv. Chim. Acta* **7,** 937 (1954).
[26] D. H. Everett, W. F. K. Wynne-Jones *Proc. Roy. Soc.* **169** A, 190 (1938).
[27] H. Deelstra, F. Verbeek *Anal. Chim. Acta* **31,** 251 (1964).

References

[28] J. E. Prue, G. Schwarzenbach *Helv. Chim. Acta* **33**, 985 (1950).
[29] D. Dyrssen, P. Lumme *Proceedings "11. Nordiska Kemistmötet"*, Turku (1962).
[30] M. Clyde, J. Selbin *Theoretical Inorganic Chemistry*, 2nd Ed., Reinhold, New York (1969).
[31] C. K. Jørgensen *Modern Aspects of Ligand-Field Theory*, North-Holland, Amsterdam (1971).
[32] M. T. Beck *Chemistry of Complex Equilibria*, Van Nostrand, London (1969).
[33] H. Irving, R. J. P. Williams *J. Chem. Soc.* 3192 (1953).
[34] R. G. Pearson *J. Am. Chem. Soc.* **85**, 3533 (1963).
[35] R. G. Pearson *J. Chem. Educ.* **45**, 581, 643 (1968).
[36] F. Basolo, R. G. Pearson *Mechanisms of Inorganic Reactions*, Wiley, New York (1958).
[37] G. Schwarzenbach *Helv. Chim. Acta* **35**, 2344 (1952).
[38] K. Denbigh *The Principles of Chemical Equilibrium*, University Press, Cambridge (1966).
[39] H. S. Harned, B. B. Owen *The Physical Chemistry of Electrolyte Solutions*, Reinhold, New York (1958).
[40] T. A. Bohigian, A. E. Martell *U.S. At. Energy Comm. Contract* No. AT (30−1)−1823. *Progress Rept.* (1960).
[41] R. G. Bates, A. G. Canham *J. Res. Natl. Bur. Stds.* **47**, 5 (1951).
[42] C. D. Hodgman *Handbook of Chemistry and Physics*, Chemical Rubber Co., Cleveland (1951).
[43] J. H. Perry *Chemical Engineers' Handbook*, 4th Ed., McGraw Hill, New York (1963).
[44] R. Luther, G. V. Sammet *Z. Elektrochem.* **11**, 293 (1905).
[45] C. Tschappat, E. Robert *Helv. Chim. Acta* **37**, 333 (1954).
[46] I. Gyenes *Titrations in Non-Aqueous Media*, Van Nostrand, New Jersey (1967).
[47] G. Charlot, B. Trémillon *Chemical Reactions in Solvents and Melts*, Pergamon, Oxford (1969).
[48] J. N. Brønsted *Chem. Rev.* **5**, 231 (1928).
[49] A. Albert, E. P. Serjeant *The Determination of Ionization Constants*, 2nd Ed., Chapman and Hall, London (1971).
[50] A. Albert, E. P. Serjeant *Ionization Constants of Acids and Bases*, Methuen, London (1962).
[51] G. Schwarzenbach, H. Ackermann *Helv. Chim. Acta* **31**, 1029 (1948).
[52] I. M. Kolthoff, V. A. Stenger *Volumetric Analysis*, Vol. I, Interscience, New York (1942).
[53] J. Inczédy *Analytical Application of Ion Exchangers*, Pergamon, Oxford (1966).
[54] L. G. van Uitert, C. G. Haas *J. Am. Chem. Soc.* **75**, 451 (1953).
[55] Kok-Peng Ang *J. Phys. Chem.* **62**, 1109 (1958).
[56] B. J. Thamer, A. E. Voigt, *J. Phys. Chem.* **56**, 225 (1952).
[57] M. Kilpatrick *J. Am. Chem. Soc.* **56**, 2048 (1934).
[58] K. N. Bascombe, R. P. Bell *J. Chem. Soc.* 1096 (1959).
[59] D. Dyrssen, E. Johansson *Acta Chem. Scand.* **9**, 763 (1955).
[60] H. L. Schläfer *Komplexbildung in Lösung*, Springer Verlag, Berlin (1961).
[61] F. J. C. Rossotti, H. Rossotti *The Determination of Stability Constants*, McGraw-Hill, New York (1961).

[62] K. B. Yatsimirskii, V. P. Vasilev *Instability Constants of Complex Compounds*, Pergamon, Oxford (1960).
[63] L. N. Klatt, R. L. Rouseff *Anal. Chem.* **42**, 1234 (1970).
[64] K. Momoki, H. Sato, H. Ogawa *Anal. Chem.* **39**, 1072 (1967).
[65] K. O. Watkins, M. M. Jones *J. Inorg. Nucl. Chem.* **16**, 187 (1961).
[66] I. Leden *Z. Phys. Chem.* (Leipzig) **188** A, 160 (1941).
[67] M. Calvin, N. C. Melchior *J. Am. Chem. Soc.* **70**, 3270 (1948).
[68] N. Ingri, L. G. Sillén *Acta Chem. Scand.* **16**, 173 (1962); *Arkiv. Kem.* **23**, 97 (1964).
[69] J. Rydberg *Acta Chem. Scand.* **14**, 157 (1960). **15**, 1723 (1961).
[70] E. A. Unwin, R. G. Beimer, Q. Fernando *Anal. Chim. Acta* **39**, 95 (1967).
[71] L. P. Varga *Anal. Chem.* **41**, 323 (1969).
[72] R. S. Tobias, M. Yasuda *Inorg. Chem.* **2**, 1307 (1963).
[73] V. S. Sharma *Biochim. Biophys. Acta* **148**, 37 (1967).
[74] I. G. Sayce *Talanta* **15**, 1397 (1968).
[75] S. Fronaeus *Acta Chem. Scand.* **4**, 72 (1950).
[76] A. Gergely, I. Nagypál, J. Mojzes *Acta Chim. Acad. Sci. Hung.* **51**, 381 (1967).
[77] L. G. Sillén *Acta Chem. Scand.* **10**, 186 (1956).
[78] A. Ringbom, L. Harju *Anal. Chim. Acta* **59**, 33 (1972).
[79] P. Job *Ann. Chim. Phys.* **9**, 113 (1928). *Compt. Rend.* **180**, 928 (1925).
[80] J. H. Yoe, A. L. Jones *Ind. Eng. Chem, Anal. Ed.* **16**, 111 (1944).
[81] A. E. Harvey, D. L. Manning *J. Am. Chem. Soc.* **72**, 4488 (1950).
[82] T. W. Newton, G. M. Accand *J. Am. Chem. Soc.* **75**, 2449 (1953).
[83] J. Bjerrum *Kgl. Danske Videnskab. Selskab Math. Fys. Medd.* **21**, 4 (1944).
[84] K. B. Yatsimirskii *Zh. Neorgan. Khim.* **10**, 2306 (1956).
[85] E. Asmus *Z. Anal. Chem.* **178**, 104 (1960).
[86] E. W. Wentworth, W. Hirsch, E. Chen. *J. Phys. Chem.* **71**, 218 (1967).
[87] J. J. Kankara *Anal. Chem.* **42**, 1322 (1970).
[88] K. Burger, I. Egyed, F. Ruff, I. Ruff *Magy. Kém. Folyóirat* **71**, 472 (1965).
[89] I. M. Kolthoff, J. J. Lingane *Polarography*, Vols. 1, 2. Interscience, New York (1952).
[90] D. D. DeFord, D. N. Hume *J. Am. Chem. Soc.* **73**, 5321 (1951).
[91] J. J. Lingane *Chem. Rev.* **20**, 1 (1941).
[92] J. Inczédy, L. Erdey *Acta Chim. Acad. Sci. Hung.* **51**, 349 (1967).
[93] J. Heyrovsky *Polarographisches Praktikum*, Springer Verlag, Berlin (1960).
[94] J. Rydberg *Acta Chem. Scand.* **4**, 1503 (1950); *Arkiv Kemi* **8**, 101, 113 (1955); *Rec. Trav. Chim.* **75**, 207 (1955).
[95] D. Dyrssen, L. G. Sillén *Acta Chem. Scand.* **7**, 663 (1953).
[96] J. Starý *Collection Czech. Chem. Commun* **25**, 2630 (1960).
[97] J. Starý *Anal. Chim. Acta* **28**, 132 (1963); *Talanta* **13**, 421 (1961).
[98] J. Rydberg, J. C. Sullivan *Acta Chem. Scand.* **13**, 2057 (1959).
[99] L. P. Varga, W. D. Wakley, L. S. Nicolson, M. L. Madden, J. Patterson *Anal. Chem.* **37**, 1003 (1965).
[100] D. Dyrssen, T. Sekine *Acta Chem. Scand.* **15**, 1399 (1961).
[101] S. Fronaeus *Acta Chem. Scand.* **3**, 859 (1951).
[102] K. A. Kraus, F. Nelson *The Structure of Electrolytic Solutions*, Wiley, New York (1959).
[103] S. Fronaeus *Svaensk Kem. Tidskr.* **65**, 1 (1953).
[104] Y. Marcus, C. D. Coriell *Bull. Res. Council Israel* **A8**, 1 (1959).
[105] T. Lengyel, J. Törkő *Acta Chim. Acad. Sci. Hung.* **54**, 27 (1967).

[106] T. Lengyel *Acta Chim. Acad. Sci. Hung.* **58**, 133 (1968).
[107] P. Gábor-Klatsmányi, J. Inczédy, L. Erdey *Acta Chim. Acad. Sci. Hung.* **57**, 5 (1968).
[108] I. Leden. G. Persson *Acta Chem. Scand.* **15**, 607, 1141 (1961).
[109] J. I. Watters *J. Am. Chem. Soc.* **81**, 1560 (1959).
[110] G. Schwarzenbach, J. Heller *Helv. Chim. Acta* **34**, 576 (1951).
[111] K. Burger, B. Pintér, *Magy. Kém. Folyóirat* **73**, 209 (1967).
[112] A. K. Babko *Talanta* **15**, 721 (1968).
[113] T. Spiro, D. N. Hume *J. Am. Chem. Soc.* **83**, 4305 (1961).
[114] P. Huhn, M. T. Beck *Acta Chim. Acad. Sci. Hung.* **51**, 7 (1967).
[115] F. S. Sadek, C. N. Reilley *Anal. Chem.* **31**, 494 (1959).
[116] D. D. Perrin, I. G. Sayce *J. Chem. Soc.* (A) **82**, (1967).
[117] L. Erdey *Gravimetric Analysis* Vols I—III, Pergamon Oxford (1963).
[118] C. V. Banks, D. W. Barnum *J. Am. Chem. Soc.* **80**, 3579 (1958).
[119] J. Inczédy *Magy. Kém. Lapja* **26**, 75 (1971).
[120] L. H. Jones, R. H. Penneman *J. Chem. Phys.* **22**, 965 (1954).
[121] O. Tomiček *Chemical Indicators* Butterworths, London (1951).
[122] J. Inczédy *J. Chem. Educ.* **47**, 769 (1970).
[123] G. Gran *Analyst* **77**, 661 (1952).
[124] F. Ingman, E. Still *Talanta* **13**, 1431 (1966).
[124a] D. Midgeley, C. McCallum *Talanta* **21**, 723 (1974).
[125] P. E. Sturrock *J. Chem. Educ.* **45**, 258 (1968).
[125a] C. Meites, D. M. Barry *Talanta* **20**, 1173 (1973).
[125b] B. H. Campbell, L. Meites *Talanta* **21**, 117, 393 (1974).
[126] G. Schwarzenbach, W. Biederman *Helv. Chim. Acta* **31**, 459 (1948).
[127] M. Beck, Z. Szabó, *Magy. Kém. Folyóirat* **57**, 143 (1951).
[128] J. R. Lalanne *J. Chem. Educ.* **48**, 266 (1971).
[129] E. Pungor *Anal. Chem.* **39**, No. 13, 28A (1967).
[130] E. Pungor, K. Tóth *Anal. Chim. Acta* **47**, 291 (1969).
[131] H. A. Flaschka *EDTA Titrations. An Introduction to Theory and Practice*, Pergamon Oxford (1959).
[132] A. Ringbom, B. Skrifvars, E. Still *Anal. Chem.* **39**, 1217 (1967).
[133] C. N. Reilley, R. W. Schmid *Anal. Chem.* **30**, 947 (1958).
[134] M. Tanaka, G. Nakagawa *Anal. Chim. Acta* **32**, 123 (1965).
[135] J. Inczédy *Anal. Letters* **2**, 601 (1969); *Periodica Polytechn.* **14**, 131 (1970).
[136] F. Vydra, R. Přibil *Talanta* **8**, 824 (1961); *Collection Czech. Chem. Commun.* **26**, 3081 (1961).
[137] R. Přibil, J. Sykora *Chem. Listy* **45**, 105 (1951).
[138] T. J. McDonald, B. J. Barker, J. A. Caruso *J. Chem. Educ.* **49**, 200 (1972).
[139] É. Bányai, L. Erdey, F. Szabadváry *Acta Chim. Acad. Sci. Hung.* **20**, 307 (1959).
[140] L. Erdey, K. Vigh, I. Buzás *Acta Chim. Acad. Sci. Hung.* **26**, 93 (1961).
[141] I. M. Kolthoff, E. B. Sandell *Textbook of Quantitative Inorganic Analysis*, p. 602, McMillan, New York (1952).
[142] L. Erdey, G. Svehla *Z. Anal. Chem.* **150**, 407 (1956).
[143] M. Kopanica, J. Doležal, J. Zýka *Chelates in Inorganic Polarographic Analysis.* in H. A. Flaschka, A. J. Barnard *Chelates in Analytical Chemistry*, Vol. I, Dekker, New York (1967).
[144] R. Přibil, Z. Roubal, E. Svátek *Collection Czech. Chem. Commun.* **18**, 43 (1953).

References

[145] G. W. Latimer *Talanta* **15**, 1 (1968).
[146] R. Přibil, Z. Zábranský *Chem. Listy* **45**, 427 (1951); *Collection Czech. Chem. Commun.* **16**, 554 (1951).
[147] R. Přibil *Komplexone in der Chemischen Analyse*, p. 171. VEB Deutscher Verlag, Berlin (1961).
[148] D. Monnier, Y. Rusconi, P. Wenger *Helv. Chim. Acta* **29**, 521 (1946).
[149] H. Richter *Z. Anal. Chem.* **124**, 161 (1942).
[150] J. Iwasaki, S. Utsumi, T. Ozawa *Bull. Chem. Soc. Japan* **25**, 226 (1952).
[151] J. Inczédy, G. Nemeshegyi, L. Erdey *Acta Chim. Acad. Sci. Hung.* **43**, 1 (1965).
[152] J. Inczédy, G. Nemeshegyi, L. Erdey *Acta Chim. Acad. Sci. Hung.* **43**, 9 (1965).
[153] A. Galík *Talanta* **14**, 731 (1967); **15**, 771 (1968).
[154] F. Sebesta *J. Radioanal. Chem.* **6**, 41 (1970).
[155] J. Růžička, J. C. Tjell *Anal. Chim. Acta* **49**, 346 (1970).
[156] A. Durst, *Ion Selective Electrodes, Natl. Bureau Standards Spec. Publ.* 314, Washington (1969).
[157] H. M. Irving, R. J. P. Williams *J. Chem. Soc.* 1841 (1949).
[158] R. Bock *Z. Anal. Chem.* **133** 110 (1951).
[159] S. Lacroix *Anal. Chim. Acta* **1**, 260 (1947).
[160] O. Samuelson *Ion-Exchange Separations in Analytical Chemistry*, Wiley, New York (1963).
[160a] W. Rieman III, H. F. Walton *Ion Exchange in Analytical Chemistry*, Pergamon, Oxford (1970).
[161] F. W. E. Strelow *Anal. Chem.* **32**, 1185 (1960).
[162] J. E. Salmon *Notes on Ion exchange Equilibria*, Manuscript (1968).
[163] B. J. Birch, J. P. Redfern, J. E. Salmon *Trans. Faraday Soc.* **63**, 2362 (1967).
[164] J. Inczédy *Magy. Kém. Lapja* **23**, 488 (1968).
[165] J. Inczédy, P. Gábor-Klatsmányi, L. Erdey *Acta Chim. Acad. Sci. Hung.* **61**, 261 (1969).
[166] L. Wünsch *Chem. Listy* **51**, 376 (1957); *Z. Anal. Chem.* **158**, 364 (1957).
[167] J. Inczédy *J. Chromatog.* **50**, 112 (1970).
[168] E. Glueckauf *Ion Exchange and its Application*, p. 34, Soc. Chem. Ind., London (1954).
[169] J. Inczédy *Planning of Ion-exchange Chromatographic Separations*, in 'Kunstharz', Ionenaustauscher, pp. 558—566. Akademie Verlag, Berlin (1970).
[170] G. Gaál, J. Inczédy *Acta Chim. Acad Sci. Hung.* **76**, 113 (1973).
[171] J. Inczédy *Acta Chim. Acad. Sci. Hung.* **69**, 265 (1971).
[172] F. Helfferich *Nature* **189**, 1001 (1961); *Angew. Chem.* **73**, 446 (1961).
[173] J. Inczédy *Magy. Kém. Lapja* **23**, 621 (1968).
[174] J. Inczédy, L. Glósz *Acta Chim. Acad. Sci. Hung.* **62**, 241 (1969).
[175] J. Inczédy *Magy. Kém. Lapja* **24**, 232 (1969).
[175a] C. Liteanu, S. Gocan, *Gradient Liquid Chromatography*, Horwood, Chichester (1975).
[176] J. Inczédy *Planning of Ion-exchange Chromatographic Separations using Complex Equilibria*, Soc. Chem. Ind., London (1969).
[177] E. Zimonyi, M. Péterfalvi *Student Thesis*, Budapesti Műszaki Egyetem (1967).
[178] J. Inczédy, E. Zimonyi *Acta Chim. Acad. Sci. Hung.* **67**, 391 (1971).
[179] A. J. P. Martin, R. L. M. Synge *Biochem. J.* **35**, 358 (1941).

[180] D. I. Ryabtshikov, V. F. Osipova *Dokl. Akad. Nauk SSSR* **96**, 761 (1954); through *Anal. Abstr.* **3**, 100 (1956).
[181] J. Inczédy, P. Gábor-Klatsmányi, L. Erdey *Acta Chim. Acad. Sci. Hung.* **50**, 105 (1966).
[182] M. Grimaldi, A. Liberti *J. Chromatog.* **15**, 510 (1964).
[183] M. Lederer *An Introduction to Paper Electrophoresis and Related Methods*, Elsevier, Amsterdam (1955).
[184] H. McDonald *J. Chem. Educ.* **29**, 428 (1952).
[185] M. K. Hargreaves, E. A. Stevinson, J. Evans *J. Chem. Soc.* 4582 (1965).
[186] I. E. Flis, K. P. Mischenko, T. A. Tumanova *Zh. Neorgan. Khim.* **4**, 277 (1959).
[187] B. B. Owen *J. Am. Chem. Soc.* **57**, 1526 (1935).
[188] E. Berne, I. Leden *Z. Naturforsch.* **8A**, 719 (1953).
[189] S. Ahrland, I. Grenthe *Acta Chem. Scand.* **11**, 1111 (1957).
[190] P. Kiválo, P. Ekari *Suomen Kemistilehti* **30 B**, 116 (1957).
[191] A. Surinarski, A. Grodzicki *Roczniki Chem.* **41**, 1205 (1967).
[192] S. Matsuo *Nippon Kagaku Zasshi* **82**, 1330, 1334 (1961).
[193] M. W. Lister, D. E. Rivington *Can. J. Chem.* **33**, 1603 (1955).
[194] L. G. Sillén *Acta Chem. Scand.* **3**, 539 (1949).
[195] M. T. Beck, F. Gaizer *Acta Chim. Acad. Sci. Hung.* **41**, 423 (1964).
[196] C. E. Vanderzee *J. Am. Chem. Soc.* **74**, 4806 (1952).
[197] F. Ya. Kulba, V. E. Mironov *Zh. Neorgan. Khim.* **3**, 1851 (1958); **5**, 1898 (1964).
[198] H. I. Busev, V. G. Tipstova, T. A. Sokolova *Vestn. Mosk. Univ. Khim.* **6**, 42 (1960).
[199] S. A. Shchukarev, L. S. Lilich, V. A. Latysheva *Zh. Neorgan. Khim.* **1**, 225 (1956).
[200] V. E. Mironov *Radiokhimiya* **4**, 707 (1962).
[201] M. G. Vladimirov, J. A. Kakaovskii *Zh. Prikl. Khim.* **23**, 580 (1950).
[202] N. Bjerrum, A. Kirschner *Kgl. Danske Videnskab. Selskab. Skrifter* No. 1, 5 (1918).
[203] P. Kiválo, R. Luoto *Suomen Kemistilehti* **30 B**, 163 (1957).
[204] D. F. C. Morris, E. L. Short *J. Chem. Soc.* 2672 (1962).
[205] H. Olerup *Thesis*, Lund (1944).
[206] D. F. C. Morris, E. L. Short *J. Chem. Soc.* 5148 (1961).
[207] F. Nelson, K. A. Kraus *J. Am. Chem. Soc.* **76**, 5916 (1954).
[208] K. Burger *Magy. Kém. Folyóirat* **70**, 179 (1964).
[209] D. Grdenić, B. Kamenar *Proc. Chem. Soc.* 304 (1961).
[210] W. C. Waggener, R. W. Stoughton *J. Phys. Chem.* **56**, 1 (1952).
[211] C. J. Nyman, D. K. Roe, R. A. Plane *J. Am. Chem. Soc.* **83**, 323 (1961).
[212] D. Peschanski, S. Valladas-Dubois *Bull. Soc. Chim. France* 1170 (1956).
[213] R. A. Day, R. N. Wilhite, F. D. Hamilton *J. Am. Chem. Soc.* **77**, 3180 (1955).
[214] J. O. Hefley, E. S. Amis *J. Phys. Chem.* **64**, 870 (1960).
[215] D. Dyrssen, M. De Jesus Tavares in *"Solvent Extraction Chemistry"*, Ed. Dryssen, Liljenzin, Rydberg, p. 465. North-Holland, Amsterdam (1967).
[216] A. Skrabal *Z. Elektrochem.* **48**, 314 (1942).
[217] R. M. Izatt, J. J. Christensen, R. T. Pack, R. Bench *Inorg. Chem.* **1**, 828 (1962).
[218] H. T. S. Britton, E. N. Dodd *J. Chem. Soc.* 100 (1935).
[219] G. D. Watt, J. J. Christensen, R. M. Izatt *Inorg. Chem.* **4**, 220 (1965).

[220] G. Anderegg *Helv. Chim. Acta* **40**, 1022 (1957).
[221] H. Freund, C. R. Schneider *J. Am. Chem. Soc.* **81**, 4780 (1959).
[222] I. M. Kolthoff, J. J. Lingane *Polarography*, Interscience, New York (1941).
[223] R. M. Izatt, G. D. Watt, D. Eatough, J. J. Christensen *J. Chem. Soc.* A 1304 (1967).
[224] J. Bjerrum *Chem. Rev.* **46**, 381 (1950).
[225] M. S. Blackie, V. Gold *J. Chem. Soc.* 3932 (1959).
[226] K. Täufel, C. Wagner, H. Dünwald *Z. Elektrochem.* **34**, 115 (1928).
[227] R. Cohen-Adad *Compt. Rend.* **238**, 810 (1954).
[228] A. Lodzinska *Roczniki Chem.* **41**, 1155, 1437 (1967).
[229] A. K. Babko, V. S. Kodenskaya *Zh. Neorgan. Khim.* **5**, 2568 (1960).
[230] W. G. Davies, J. E. Prue *Trans. Faraday Soc.* **51**, 1045 (1955).
[231] D. C. Feay *Thesis*, Berkeley Univ (1954); UCRL 2547.
[232] E. L. King, P. K. Gallagher *J. Phys. Chem.* **63**, 1073 (1959).
[233] L. M. Yates *Thesis*, State College Washington (1955).
[234] A. S. Wilson, M. Taube *J. Am. Chem. Soc.* **74**, 3509 (1952).
[235] S. Ahrland, K. Rosengren *Acta Chem. Scand.* **10**, 727 (1956).
[236] H. V. Dodgen, G. K. Rollefson *J. Am. Chem. Soc.* **71**, 2600 (1949).
[237] A. D. Paul *Thesis* Univ. California, Berkeley (1955); UCRL 2926.
[238] N. Sunden *Svensk Kem. Tidskr.* **66**, 50 (1954).
[239] J. W. Kury, A. D. Paul, L. G. Hepler, R. E. Connick *J. Am. Chem. Soc.* **81**, 4185 (1959).
[240] R. E. Connick, M. S. Tsao *J. Am. Chem. Soc.* **76**, 5311(1954).
[241] K. E. Kleiner, G. I. Gridehina *Zh. Neorgan. Khim.* **4**, 2020 (1959); *Chem. Abstr.* **54**, 11791 (1960).
[242] W. B. Schaap, J. A. Davies, W. H. Nebergall *J. Am. Chem. Soc.* **76**, 5226 (1954).
[243] V. Cagliotti, L. Ciavatta, A. Liberti *J. Inorg. Nucl. Chem.* **15**, 115 (1956).
[244] S. Ahrland, R. Larsson, K. Rosengren *Acta Chem. Scand.* **10**, 705 (1956).
[245] R. E. Connick, W. H. McVey *J. Am. Chem. Soc.* **71**, 31, 82 (1949).
[246] G. I. H. Hanania, D. H. Irvine, W. A. Zaton, P. George *J. Phys. Chem.* **71**, 2022 (1967).
[247] J. C. James *Trans. Faraday Soc.* **45**, 855 (1949).
[248] S. R. Cohen, R. A. Plane *J. Phys. Chem.* **61**, 1096 (1957).
[249] M. H. Panckhurst, K. G. Woolmington *Proc. Roy. Soc.* **244** A, 124 (1958).
[250] J. Jordan, G. J. Ewing *Inorg. Chem.* **1**, 587 (1962).
[251] C. B. Monk *J. Chem. Soc.* 423 (1949).
[252] C. W. Gibby, C. B. Monk *Trans. Faraday Soc.* **48**, 632 (1952).
[253] J. C. James *Thesis*, London (1947).
[254] I. Leden *Acta Chem. Scand.* **10**, 540 (1956).
[255] H. L. Riley, V. Gallafent *J. Chem. Soc.* 514 (1932).
[256] J. Quarfort, L. G. Sillén *Acta Chem. Scand.* **3**, 505 (1949).
[257] E. N. Rengevich, E. A. Shilov *Ukr. Khim. Zh.* **28**, 1080 (1962).
[258] B. G. F. Carleson, H. Irving *J. Chem. Soc.* 4390 (1954).
[259] P. Kiváló, A. Ekman *Soumen Kemistilehti* **29** B, 139 (1956).
[260] H. Halban, J. Brüll *Helv. Chim. Acta* **27**, 1719 (1944).
[261] R. A. Day, R. W. Stoughton *J. Am. Chem. Soc.* **72**, 5662 (1950).
[262] Y. Sasaki, L. G. Sillén *Acta Chem. Scand.* **18**, 1014 (1963).
[263] O. L. Leussing, I. M. Kolthoff *J. Am. Chem. Soc.* **75**, 2476 (1953).
[264] H. Ölander *Z. Phys. Chem.* (Leipzig) **129**, 1 (1927).

[265] I. Szilárd *Acta Chem. Scand.* **17,** 2674 (1963).
[266] C. J. Nyman *J. Am. Chem. Soc.* **77,** 1371 (1955).
[267] K. Savallo, P. Lumme *Suomen Kemistilehti* **40** B, 155 (1967).
[268] R. L. Rebertus *Thesis,* Illinois Univ. (1954). Microfilm 9125.
[269] D. Banerjea, I. P. Single *Z. Anorg. Chem.* **349,** 213 (1967).
[270] H. Schmid, R. Marchgraber, F. Dunkl *Z. Elektrochem.* **43,** 337 (1937).
[271] A. Kossiakov, D. F. Sickman *J. Am. Chem. Soc.* **68,** 442 (1946).
[272] H. Pick, R. Abegg *Z. Anorg. Chem.* **51,** 1 (1906).
[273] D. S. Jain, J. N. Gaur *Acta Chim. Acad. Sci. Hung.* **55,** 2311 (1933).
[274] V. P. Vasilyev, V. N. Vasilyeva, N. A. Klindukhova, A. N. Parfenova *Izv. Russhikh Uchebn. Zavedenii Khim. i Khim. Tekhnol.* **6,** 339 (1963).
[275] E. E. Kapantsyan, B. I. Nabivanets *Ukr. Khim. Zh.* **33,** 961 (1967).
[276] G. R. Choppin, W. F. Strazik *Inorg. Chem.* **4,** 1250 (1965).
[277] D. Peschanski, J. M. Fruchart *Compt. Rend.* **260,** 3073 (1965).
[278] B. M. L. Bansal, S. K. Patil, H. D. Sharma *J. Inorg. Nucl. Chem.* **26,** 993 (1964).
[279] R. Hugel *Bull. Soc. Chim. France* **971,** 2017 (1965).
[280] A. P. Samodelov *Radiokhimiya* **6,** 568 (1964).
[281] P. H. Tedesco, V. B. de Rumi, J. A. Gonzalez Quintana *J. Inorg. Nucl. Chem.* **30,** 987 (1968).
[282] F. Ya. Kulba, Yu. B. Yakovlev, V. E. Mironov *Zh. Neorgan. Khim.* **10,** 1624 (1965).
[283] H. L. Johnston, F. Čůta, A. B. Garrett *J. Am. Chem. Soc.* **55,** 2311 (1933).
[284] C. Brosset, G. Biedermann, L. G. Sillén *Acta Chem. Scand.* **8,** 1917 (1954).
[285] F. G. R. Gimblett, C. B. Monk *Trans. Faraday Soc.* **50,** 965 (1954).
[286] B. Carrel, A. Olin *Acta Chem. Scand.* **15,** 1875 (1961).
[287] A. Olin *Svensk Kem. Tidskr.* **73,** 482 (1961).
[288] C. W. Davies, B. E. Hoyle *J. Chem. Soc.* 233 (1951).
[289] T. Moeller *J. Phys. Chem.* **50,** 242 (1946).
[290] H. G. Offner, D. A. Skoog *Anal. Chem.* **38,** 1520 (1966).
[291] P. R. Danesi *Acta Chem. Scand.* **21,** 143 (1967).
[292] J. A. Bolzan, J. J. Podesta, A. J. Arvia *Anales Asoc. Quim. Arg.* **51,** 43 (1963).
[293] T. J. Conochiolli, G. H. Nancollas, N. Sutin *Inorg. Chem.* **5,** 1 (1966).
[294] N. Bjerrum *Thesis,* Copenhagen (1908).
[295] K. J. Pedersen *Kgl. Danske Videnskab. Selskab, Mat. Fys. Medd.* **20,** 7 (1943).
[296] B. O. A. Hedström *Arkiv Kemi* **5,** 457 (1953).
[297] A. S. Wilson, H. Taube *J. Am. Chem. Soc.* **74,** 3509 (1952).
[298] W. Forsling, S. Hietanen, L. G. Sillén *Acta Chem. Scand.* **6,** 901 (1952).
[299] S. Hietanen, L. G. Sillén *Acta Chem. Scand.* **6,** 747 (1952).
[300] G. Biedermann, N. C. Li, J. Yu *Acta Chem. Scand.* **15,** 555 (1961).
[301] G. Biedermann, L. Ciavatta *Acta Chem. Scand.* **15,** 1347 (1961).
[302] D. I. Stock, C. W. Davies *Trans. Faraday Soc.* **44,** 856 (1948).
[303] G. Biedermann, M. Kilpatrick, L. Pokras, L. G. Sillén *Acta Chem. Scand.* **10,** 1327 (1956).
[304] R. S. Tobias *Acta Chem. Scand.* **12,** 198 (1958).
[305] S. Hietanen, L. G. Sillén *Acta Chem. Scand.* **13,** 533 (1959).
[306] R. L. Pecsok, A. N. Fletscher *Inorg. Chem.* **1,** 156 (1962).
[307] J. Beukenkamp, K. D. Herrington *J. Am. Chem. Soc.* **82,** 261 (1960).

References

[308] R. P. Bell, M. H. Panckhurst *Rec. Trav. Chim.* **75**, 725 (1956).
[309] G. Biedermann *Rec. Trav. Chim.* **75**, 716 (1956).
[310] S. Hietanen *Acta Chem. Scand.* **10**, 1531 (1956).
[311] A. Ahrland, S. Hietanen, L. G. Sillén *Acta Chem. Scand.* **8**, 1907 (1954).
[312] F. J. C. Rossotti, H. S. Rossotti *Acta Chem. Scand*, **9**, 1177 (1955).
[313] G. Schorsch *Bull. Soc. Chim. France* 1449, 1456 (1964).
[314] A. S. Solovkin *Zh. Neorgan. Khim.* **2**, 611 (1957).
[315] R. A. Joyner *Z. Anorg. Chem.* **77**, 103 (1912).
[316] J. H. Baxendale, C. F. Wells *Trans. Faraday Soc.* **53**, 800 (1957).
[317] M. G. Evans, N. Uri, P. George *Trans. Faraday Soc.* **45**, 230 (1949).
[318] K. E. Kleiner *Zh. Obshch. Khim.* **22**, 17 (1952).
[319] A. K. Babko, A. I. Volkova, S. L. Lisichenok *Zh. Neorgan. Khim.* **11**, 478 (1966).
[320] C. Morton *Quart. J. Pharm. Pharmacol.* **3**, 438 (1930).
[321] N. Bjerrum, A. Unmack *Kgl. Dansake Videnskab. Selskab. Mat. Fys. Medd.* **9**, 1, 126 (1929).
[322] R. M. Smith, R. A. Alberty *J. Am. Chem. Soc.* **78**, 2376 (1956).
[323] H. Siegel, K. Becker, D. B. McCormick *Biochim. Biophys. Acta* **148**, 655 (1967).
[324] O. E. Lanford, S. J. Kiehl *J. Am. Chem. Soc.* **64**, 291 (1942).
[325] H. Gnepf, O. Glübeli, G. Schwarzenbach *Helv. Chim. Acta* **45**, 1171 (1962).
[326] G. Schwarzenbach, J. Zurc *Monatsh.* **81**, 202 (1950).
[327] J. I. Watters, S. M. Lambert *J. Am. Chem. Soc.* **81**, 3201 (1959).
[328] J. A. Walhoff, J. T. G. Overbeck *Rec. Trav. Chim.* **78**, 759 (1959).
[329] P. I. Yakshova *Tr. Voronezhsk Univ.* **42**, No. 2, 63 (1956).
[330] T. Yamane, N. Davidson *J. Am. Chem. Soc.* **81**, 4438 (1959).
[331] B. C. Haldar *Current Sci. (India)* **19**, 244 (1950); *Chem. Zentr.* 2856 (1951).
[332] P. Senise, P. Delahay *J. Am. Chem. Soc.* **74**, 6128 (1952).
[333] G. Anderegg *Helv. Chim. Acta* **48**, 1712 (1965).
[334] C. F. Wells, M. A. Salam *J. Chem. Soc.* A 308 (1968).
[335] L. P. Andrusenko, I. A. Aheka *Zh. Neorgan. Khim.* **13**, 347, 2645 (1968).
[336] W. M. NcNabb, J. F. Hazel, R. A. Baxter *J. Inorg. Nucl. Chem.* **30**, 1585 (1968).
[337] J. I. Watters, S. Matsumoto *J. Am. Chem. Soc.* **86**, 3961 (1964).
[338] J. I. Watters, R. Machen *J. Inorg. Nucl. Chem.* **30**, 2163 (1968).
[339] J. I. Watters, S. Matsumoto *Inorg. Chem.* **5**, 361 (1966).
[340] J. I. Watters, S. Matsumoto *J. Inorg. Nucl. Chem.* **29**, 2955 (1967).
[341] H. Züst *Thesis*, Zürich (1958).
[342] J. Knox *Z. Elektrochem.* **12**, 477 (1906).
34 3] G. S. Cave, D. N. Hume, *J. Am. Chem. Soc.* **75**, 2893 (1953).
[344] A. M. Golub, J. A. Babko, N. A. Levitskaya, *Ukr. Khim. Zh.* **25**, 50 (1959).
[345] P. Senise, E. F. de Almeida Neves *J. Am. Chem. Soc.* **83**, 4146 (1961).
[346] P. Senise, M. Perrier *J. Am. Chem. Soc.* **80**, 4194 (1958).
[347] J. D. Fridman, D. S. Sarbaev *Zh. Neorgan. Khim.* **4**, 1849 (1959).
[348] N. Tanaka, T. Takamura *J. Inorg. Nucl. Chem.* **9**, 15 (1959).
[349] M. Möller *Thesis*, Copenhagen (1937).
[350] A. K. Babko, K. E. Kleiner *Zh. Analit. Khim.* **1**, 106 (1946).
351] C. J. Nyman, G. S. Alberto *Anal. Chem.* **32**, 207 (1960).
352] K. B. Yatsimirskii, V. D. Kovableva *Zh. Neorgan. Khim.* **3**, 339 (1938).
353] G. W. Leonard, M. E. Smith, D. N. Hume *J. Phys. Chem.* **60**, 1493 (1956).

[354] R. E. Frank, D. N. Hume *J. Am. Chem. Soc.* **75**, 1736 (1953).
[355] R. P. Bell, J. B. H. George *Trans. Faraday Soc.* **49**, 619 (1953).
[356] I. Leden *Acta Chem. Scand.* **6**, 971 (1952).
[357] S. Fronaeus *Svensk Kem. Tidskr.* **64**, 317 (1952).
[358] T. J. Hardwick, E. Robertson *Can. J. Chem.* **29**, 828 (1951).
[359] R. W. Money, C. W. Davies *Trans. Faraday Soc.* **28**, 609 (1932).
[360] N. Tanaka, K. Ogino-Ebata, G. Sato *Bull. Chem. Soc. Japan* **31**, 366 (1966).
[361] S. Fronaeus *Thesis*, Lund (1948).
[362] T. Schine *J. Inorg. Nucl. Chem.* **26**, 1463 (1964).
[363] M. W. Lister, D. E. Rivington *Can. J. Chem.* **77**, 5501 (1955).
[364] N. Sunden *Svensk Kem. Tidskr.* **66**, 345 (1954).
[365] R. W. Chlebek, M. W. Lister *Can. J. Chem.* **44**, 437 (1966).
[366] T. Sekine *Acta Chem. Scand.* **19**, 1469 (1965).
[367] T. Sekine *J. Inorg. Nucl. Chem.* **26**, 1463 (1964).
[368] V. S. K. Nair, G. H. Nancollas *J. Chem. Soc.* 3706 (1958).
[369] S. K. Kor, G. S. Verma *J. Chem. Phys.* **29**, 9 (1958).
[370] A. Tateda *Bull. Chem. Soc. Japan* **38**, 165 (1965).
[371] E. W. Davies, C. B. Monk *Trans. Faraday Soc.* **53**, 442 (1957).
[372] E. L. Zebrovki, H. W. Alter, F. K. Neuman *J. Am. Chem. Soc.* **73**, 5646 (1951).
[373] M. G. Panova, N. E. Berezhneva, V. I. Levin *Radiokhimiya* **2**, 208 (1960).
[374] B. B. Owen, R. W. Gurry *J. Am. Chem. Soc.* **60**, 3074 (1938).
[375] F. M. Page *J. Chem. Soc.* 1719 (1953).
[376] H. Chateau, J. Pouradier *Sci. Ind. Phot.* **24**, 129 (1953).
[377] T. O. Denney, C. B. Monk *Trans. Faraday Soc.* **47**, 992 (1951).
[378] F. G. R. Gimblett, C. B. Monk *Trans. Faraday Soc.* **51**, 793 (1955).
[379] V. F. Toropova, J. A. Sivotina, T. J. Lisova *Uch. Zap. Kazansk. Gos. Univ.* **115**, No. 3, 43 (1955).
[380] F. M. Page *Trans. Faraday Soc.* **50**, 120 (1954).
[381] V. F. Toropova *Zh. Obshch. Khim.* **24**, 423 (1954).
[382] K. B. Yatsimirkii *Zh. Fiz. Khim.* **25**, 475 (1951).
[383] M. S. Novakovskii, T. M. Smayeva *Ukr. Kim. Zh.* **20**, 615 (1954).
[384] H. Hagisava *Bull. Inst. Phys. Chem. Res. Tokoy* **20**, 384 (1941).
[385] H. Hagisava *Bull. Inst. Phys. Chem. Res. Tokyo* **18**, 648 (1939).
[386] A. V. Pamfilov, A. L. Agafonova *Zh. Fiz. Khim.* **24**, 1147 (1950).
[387] H. Bilinski, N. Ingri *Acta Chem. Scand.* **21**, 2503 (1967).
[388] W. J. Weber, W. Stumm *J. Inorg. Chem.* **27**, 237 (1965).
[389] J. E. Early, D. Fortnum, A. Wojcicki, J. O. Edwards *J. Am. Chem. Soc.* **81**, 1295 (1959).
[390] F. H. MacDougall, M. Allen *J. Phys. Chem.* **46**, 730, 738 (1942).
[391] G. N. Nancollas *J. Chem. Soc.* 744 (1956).
[392] R. S. Kolat, J. E. Powell *Inorg. Chem.* **1**, 293 (1962).
[393] S. K. Siddhanta, S. N. Banerjee *J. Indian Chem. Soc.* **35**, 323, 343, 419 (1958).
[394] L. Sommer, K. Pliska *Collection Czech. Chem. Commun.* **26**, 2754 (1961).
[395] D. W. Archer, C. B. Monk *J. Chem. Soc.* 3117 (1964).
[396] J. F. Spencer, R. Abegg *Z. Anorg. Chem.* **44**, 379 (1905).
[397] J. Starý *Collection Czech. Chem. Commun.* **25**, 2630 (1960).
[398] C. H. Cartlege *J. Am. Chem. Soc.* **73**, 4416 (1951).
[399] R. M. Izatt, W. C. Fernelius, B. P. Block *J. Phys. Chem.* **59**, 80. 235 (1955).

References

[400] J. Grenthe, W. C. Fernelius *J. Am. Chem. Soc.* **82,** 6258 (1960).
[401] A. Krishen, H. Freiser *Anal. Chem.* **31,** 923 (1959).
[402] H. A. Droll, B. P. Block, W. C. Fernelius *J. Phys. Chem.* **61,** 1000 (1957).
[403] J. Rydberg *Svensk Kem. Tidskr.* **67,** 499 (1955).
[404] J. Inczédy, J. Maróthy *Acta Chim. Acad. Sci. Hung.* (In press).
[405] M. M. Taqui Khan, A. E. Martell *J. Am. Chem. Soc.* **88,** 668 (1966).
[406] L. F. Nims, P. K. Smith *J. Biol. Chem.* **101,** 401 (1933).
[407] C. B. Monk *Trans. Faraday Soc.* **47,** 285, 292, 297, 1233 (1951).
[408] C. W. Davies, G. M. Waind *J. Chem. Soc.* 301 (1950).
[409] J. H. Smith. A. M. Cruickshank, F. T. Donoghue, J. F. Pysz *Inorg. Chem.* **1,** 148 (1950).
[410] A. Albert *J. Biochem.* **47,** 531 (1950).
[411] D. D. Perrin *J. Chem. Soc.* 3125 (1958).
[412] L. E. Maley, F. P. Mellor *Nature* **165,** 453 (1950).
[413] C. B. Monk, C. A. Colman-Porter *J. Chem. Soc.* 4363 (1952).
[414] D. L. Leussing, E. M. Hanna *J. Am. Chem. Soc.* **88,** 643 (1966).
[415] D. J. Alner, R. C. Lansbury, A. G. Smith *J. Chem. Soc.* A 417 (1968).
[416] G. N. Rao, R. S. Subrahmanya *Proc. Indian Acad. Sci.* **60,** 165, 185 (1964).
[417] V. S. Sharma, H. B. Mathur, P. S. Kilkarni *Indian J. Chem.* **3,** 146, 475 (1965).
[418] M. Bartušek, J. Zelinka *Collection Czech. Chem. Commun.* **32,** 992 (1967).
[419] H. E. Zittel, T. M. Florence *Anal. Chem.* **39,** 320 (1967).
[420] T. H. Wirth, N. Davidson *J. Am. Chem. Soc.* **86,** 4314, 4318 (1964).
[421] P. O. Lumme *Suomen Kemistilehti* 30 B, 176 (1957); 31 B, 232, 250 (1958); 33 B, 69, 85 (1960).
[422] M. Cefola, A. S. Tompa, A. V. Celiano, P. S. Gentile *Inorg. Chem.* **1,** 290 (1962).
[423] B. Buděšínský *Collection Czech. Chem. Commun.* **28,** 2902 (1963).
[424] M. M. Taqui Khan *Thesis,* Clark Univ. (1962).
[425] L. Sommer *Collection Czech. Chem. Commun.* **28,** 449 (1963).
[426] D. G. Vartek, R. S. Shetiya *J. Inorg. Nucl. Chem.* **29,** 1261 (1967).
[427] D. S. Jain, J. N. Gaur *J. Electroanal. Chem.* **11,** 310 (1966).
[428] H. Erlenmeyer, R. Griesser, B. Prijs, H. Siegel *Helv. Chim. Acta* **51,** 339 (1968).
[429] D. S. Jain, J. N. Gaur *Indian J. Chem.* **2,** 503 (1964).
[430] S. Ramamoorthy, M. Santappa *Bull. Chem. Soc. Japan* **7,** 1330 (1968).
[431] R. Griesser, B. Prijs, H. Siegel *Inorg. Nucl. Chem. Letters* **4,** 443 (1968).
[432] H. Irving, D. H. Mellor *J. Chem. Soc.* 5222 (1962).
[433] C. Luca *Bull. Soc. Chim. France* 2556 (1967).
[434] G. Anderegg *Helv. Chim. Acta* **46,** 2397 (1963).
[435] F. Lindstrom, H. Diehl *Anal. Chem.* **32,** 1123 (1960).
[436] D. K. Cabbiness, E. S. Amis *Bull. Chem. Soc. Japan* **40,** 435 (1967).
[437] L. P. Varga, F. C. Veatch *Anal. Chem.* **39,** 1101 (1967).
[438] A. Semb, F. J. Langmyhr *Anal. Chim. Acta* **35,** 286 (1966).
[439] L. Sommer, V. Kuban *Collection Czech. Chem. Commun.* **32,** 4355 (1967).
[440] F. J. Langmyhr, K. S. Klausen *Anal. Chim. Acta* **29,** 149 (1963).
[441] L. Sommer *Collection Czech. Chem. Commun.* **28,** 2393 (1963).
[442] S. N. Dubey, R. C. Mehrotra *J. Indian Chem. Soc.* **42,** 685 (1965).
[443] M. Bartušek, J. Zelinka *Collection Czech. Chem. Commun.* **32,** 992 (1967).
[444] G. A. L'Heureux, A. E. Martell *J. Inorg Nucl. Chem.* **28,** 481 (1966).

[445] A. Banerjee, A. K. Dey *J. Inorg. Nucl. Chem.* **30,** 995 (1968).
[446] C. Bertin-Natsch *Ann. Chim.* **7,** 481 (1952).
[447] J. Schubert *J. Am. Chem. Soc.* **76,** 3442 (1954).
[448] I. Feldman, T. Y. Toribara, J. R. Havill, W. F. Neuman *J. Am. Chem. Soc.* **77,** 878 (1955).
[449] C. W. Davies, B. E. Hoyle *J. Chem. Soc.* 1038 (1955).
[450] N. C. Li, A. Lindenbaum, J. M. White *J. Inorg. Nucl. Chem.* **12,** 122 (1959).
[451] R. C. Warner, J. Weber *J. Am. Chem. Soc.* **75,** 5086 (1953).
[452] R. E. Hamm, S. M. Shull, D. M. Grant *J. Am. Chem. Soc.* **76,** 2111 (1954).
[453] J. Schubert *J. Phys. Chem.* **56,** 113 (1952).
[454] D. Dyrssen *Svensk Kem. Tidskr.* **65,** 43 (1953).
[455] A. E. Klygin, N. S. Kolyada *Russ. J. Inorg. Chem.* **6,** 107, 216 (1961).
[456] G. R. Lenz, A. E. Martell *Biochemistry* **3,** 745 (1964).
[457] A. Albert *J. Biochem.* **50,** 690 (1952).
[458] W. Stricks, I. M. Kolthoff *J. Am. Chem. Soc.* **75,** 5673 (1953).
[459] T. Tanaka, I. M. Kolthoff, W. Stricks *J. Am. Chem. Soc.* **77,** 1996 (1955).
[460] M. T. Beck, A. Gergely *Acta Chim. Acad. Sci. Hung.* **50,** 155 (1966).
[461] J. Bond, T. J. Jones *Trans. Faraday Soc.* **55,** 1310 (1959).
[462] M. T. Beck, B. Csiszár *Acta Chim. Acad. Sci. Hung.* **32,** 1 (1962).
[463] J. H. Holloway, C. N. Reilley *Anal. Chem.* **32,** 249 (1960).
[464] P. Urech *Thesis*, Zürich (1962).
[465] R. Nasanen, P. Merilainen, O. Butkewitsch *Suomen Kemistilehti* B **35,** 219 (1962).
[466] R. Nasanen, P. Merilainen, S. Lukkari *Acta Chem. Scand.* **17,** 2384 (1962).
[467] D. K. Roe, D. B. Masson, C. J. Nyman *Anal. Chem.* **33,** 1464 (1961).
[468] R. Nasanen, E. Merilainen, E. Heinanen *Suomen Kemistilehti* **35** B, 15 (1962).
[469] G. Schwarzenbach, H. Ackermann, B. Maissen, G. Anderegg *Helv. Chim. Acta* **35,** 2337 (1952).
[470] J. Poulsen, J. Berrum *Acta Chem. Scand.* **9,** 1407 (1955).
[471] J. F. Fischer, J. L. Hall *Anal. Chem.* **39,** 1550 (1967).
[472] P. K. Migal', K. I. Ploae *Russ. J. Inorg. Chem.* **10,** 1368 (1965).
[473] P. K. Migal', G. F. Serova *Russ. J. Inorg. Chem.* **9,** 972 (1964).
[474] R. Zahradnik, P. Zuman *Collection Chem. Czech. Commun.* **24,** 1132 (1959).
[475] W. Kemula, A. Hulanicki, W. Nawrot *Roczniki Chem.* **36,** 1717 (1962); **38,** 1065 (1964).
[476] A. Hulanicki *Thesis*, Warsaw (1967).
[477] H. Thun, F. Verbeek *J. Inorg. Nucl. Chem.* **27,** 1813 (1965).
[478] J. E. Powell, J. L. Farrell, W. F. S. Neillie, R. Russell *J. Inorg. Nucl. Chem.* **30,** 2223 (1968).
[479] H. J. Jonassen *J. Phys. Chem.* **56,** 16 (1952).
[480] C. J. Nyman, D. K. Roe, D. B. Mason *J. Am. Chem. Soc.* **77,** 4191 (1955).
[481] E. Wänninen *Acta Acad. Aboensis, Math. Phys.* **21,** 17 (1960).
[482] H. Wickberg, A. Ringbom *Suomen Kemistilehti* **41** B, 177 (1968).
[483] T. Moeller, S. Chu *J. Inorg. Nucl. Chem.* **28,** 153 (1966).
[484] E. Bottari, G. Anderegg *Helv. Chim. Acta* **50,** 2349 (1967).
[485] G. Anderegg, P. Nägeli, F. Müller, G. Schwarzenbach *Helv. Chim. Acta* **42,** 827 (1959).
[486] T. Moeller, L. C. Thompson *J. Inorg. Nucl. Chem.* **24,** 499 (1962).
[487] J. H. Holloway, C. Reilley *Anal. Chem.* **32,** 249 (1960).

[488] E. J. Durham, D. P. Ryskiewicz *J. Am. Chem. Soc.* **80,** 4813 (1958).
[489] G. Anderegg, E. Bottari *Helv. Chim. Acta* **50,** 2341 (1967).
[490] O. Navratil, J. Kotas *Collection Czech. Chem. Commun.* **30,** 2736 (1965).
[491] C. V. Banks, S. Anderson *Inorg. Chem.* **2,** 112 (1963).
[492] D. Dyrssen, F. Krašovec, L. G. Sillén *Acta Chem. Scand.* **13,** 50 (1959).
[493] K. Burger, D. Dyrssen *Acta Chem. Scand.* **17,** 1489 (1963).
[494] R. P. Yaffe, A. F. Voight *J. Am. Chem. Soc.* **74,** 2941, 3163 (1952).
[495] R. W. Geiger, E. S. Sandell *Anal. Chim. Acta* **8,** 197 (1953).
[496] J. F. Duncan, F. G. Thomas *J. Chem. Soc.* 2814 (1960).
[497] E. Wänninen, A. Ringbom *Anal. Chim. Acta* **12,** 308 (1955).
[498] M. T. Beck, S. Görög *Acta Chim. Acad. Sci. Hung.* **22,** 159 (1960).
[499] A. Ringbom, E. Linko *Anal. Chim. Acta* **9,** 80 (1953).
[500] J. Starý *Anal. Chim. Acta* **28,** 132 (1963).
[501] M. T. Beck, A. Gergely *Acta Chim. Acad. Sci. Hung.* **50,** 155 (1966).
[502] P. Szarvas, E. Brücher *Acta Chim. Acad. Sci. Hung.* **50,** 279 (1966).
[503] R. Dyke, W. C. E. Higginson *J. Chem. Soc.* 1998 (1960).
[504] G. Schwarzenbach, J. Sandeva *Helv. Chim. Acta* **36,** 1089 (1960).
[505] T. R. Bath, M. Krishnamurthy *J. Inorg. Nucl. Chem.* **25,** 1147 (1963).
[506] K. Saito, H. Terrey *J. Chem. Soc.* 4701 (1956).
[507] I. Sajó *Acta Chim. Acad. Sci. Hung.* **16,** 115 (1958).
[508] F. Nelson, R. A. Day, K. A. Kraus *J. Inorg. Nucl. Chem.* **15,** 140 (1960).
[509] T. D. Smith *J. Chem. Soc.* 2555 (1961).
[510] P. Urech *Thesis*, Zürich (1962).
[511] R. L. Pecsok, E. F. Maverick *J. Am. Chem. Soc.* **76,** 358 (1954).
[512] A. I. Busev, V. G. Tipdiva, T. A. Sokolova *Zh. Neorgan. Khim.* **5,** 2749 (1960).
[513] J. Stand *Collection Czech. Chem. Commun.* **25,** 2630 (1960).
[514] A. Ringbom, S. Siitonen, B. Skrifvars *Acta Chem. Scand.* **11,** 551 (1957).
[515] G. Anderegg *Helv. Chim. Acta* **47,** 1801 (1964).
[516] H. Wikberg, A. Ringbom *Suomen Kemistilehti* **41** B, 177 (1968).
[517] T. R Bhat, R. R. Das, J. Shankar *Indian J. Chem.* **5,** 324 (1967).
[518] T. Nozaki, K. Koshiba *Nippon Kagaku Zasshi* **88,** 1287 (1967).
[519] K. H. Schrøder *Acta Chem. Scand.* **17,** 1509 (1963).
[520] M. Tanaka, S. Funabashi, K. Shirai *Inorg. Chem.* **7,** 573 (1968).
[521] G. Schwarzenbach, W. Biedermann *Helv. Chim. Acta* **31,** 678 (1948).
[522] G. P. Hildebrand, C. N. Reilley *Anal. Chem.* **29,** 258 (1957).
[523] C. N. Reilley, R. W. Schmid *Anal. Chem.* **31,** 887 (1959).
[524] M. Kodama, C. Sasaki *Bull. Chem. Soc. Japan* **41,** 127 (1968).
[525] M. Kodama *Bull. Chem. Soc. Japan* **40,** 2575 (1967).
[526] M. Kodama, H. Ebine *Bull. Chem. Soc. Japan* **40,** 1857 (1967).
[527] L. Sommer, V. Kuban *Collection Czech. Chem. Commun.* **32,** 4355 (1967).
[528] F. J. Langmyhr, T. Stumpe *Anal. Chim. Acta* **32,** 535 (1965).
[529] G. A. Carlson, J. P. McReynolds, F. H. Verhoek *J. Am. Chem. Soc.* **67,** 1334 (1945).
[530] M. Ohta, H. Matsukawa, T. Tsuchiya *Bull. Chem. Soc. Japan* **37,** 692 (1964).
[531] J. Bjerrum, E. J. Nelson *Acta Chem. Scand.* **2,** 307 (1948).
[532] J. I. Watters, J. G. Mason *J. Am. Chem. Soc.* **78,** 285 (1956).
[533] F. J. Langmyhr, A. R. Storm *Acta Chem. Scand.* **15,** 1461 (1961).
[534] P. Lingaiah, J. Mohan Rao, U. V. Seshaiah *Current Sci. (India)* **36,** 197 (1967).
[535] H. Tsubota *Bull. Chem. Soc. Japan* **35,** 640 (1962).

[536] G. N. Nancollas *J. Chem. Soc.* 744 (1956).
[537] J. N. Gaur, D. S. Jain *Australian J. Chem.* **18**, 1687 (1965).
[538] H. M. Hershenson, R. J. Brooks, M. E. Murphy *J. Am. Chem. Soc.* **79**, 2046 (1957).
[539] N. E. Topp, C. W. Davies *J. Chem. Soc.* 87 (1940).
[540] J. M. Peacock, J. C. James *J. Chem. Soc.* 2233 (1951).
[541] J. Schubert, A. Lindenbaum *J. Am. Chem. Soc.* **74**, 3529 (1952).
[542] K. Burger, E. Papp Molnár *Acta Chim. Acad. Sci. Hung.* **53**, 111 (1967).
[543] R. F. Lumb, A. E. Martell *J. Phys. Chem.* **57**, 690 (1953).
[544] R. L. Rebertus *Thesis*, Univ. Illinois (1954).
[545] G. N. Rao, R. S. Subrahmanya *Proc. Indian Acad. Sci.* **60**, 165, 185 (1964).
[546] I. Feldman, L. Koval, *Inorg. Chem.* **2**, 145 (1963).
[547] A. Gergely, I. Nagypál, J. Mojzes *Acta Chim. Acad. Sci. Hung.* **51**, 381 (1967).
[548] A. Gergely, I. Nagypál *Acta Univ. Debrecen* 113 (1965).
[549] G. Anderegg *Helv. Chim. Acta* **44**, 1673 (1961).
[550] A. Albert *J. Biochem.* **54**, 646 (1953).
[551] H. V. Flood, V. Lorzs *Tidskr. Kjemi Bergvesen Met.* **5**, 83 (1945).
[552] C. A. Colman-Porter, C. B. Monk *J. Chem. Soc.* 4363 (1952).
[553] L. F. Nims *J. Am. Chem. Soc.* **58**, 987 (1936).
[554] P. B. Davies, C. B. Monk *Trans. Faraday Soc.* **50**, 128 (1954).
[555] T. Lengyel *Acta Chim. Acad. Sci. Hung.* **58**, 313 (1968).
[556] J. Törkő *Magy. Kém. Folyóirat* **74**, 590 (1968).
[557] T. Moeller, R. Ferrus *J. Inorg. Nucl. Chem.* **20**, 261 (1961).
[558] H. Wickberg, A. Ringbom *Suomen Kemistilehti* **41** B, 177 (1968).
[559] D. A. Aikens, F. J. Bahbah *Anal. Chem.* **39**, 646 (1967).
[560] S. Chaberek, A. E. Martell *J. Am. Chem. Soc.* **77**, 1477 (1955).
[561] J. Powell, J. L. Mackey *Inorg. Chem.* **1**, 418 (1962).
[562] R. Skochdopole, S. Chaberek *J. Inorg. Nucl. Chem.* **11**, 222 (1959).
[563] T. Moeller, Shu-Kung Chu *J. Inorg. Nucl. Chem.* **28**, 153 (1966).
[564] F. G. Pawelka *Z. Elektrochem.* **30**, 180 (1924).
[565] A. C. Andrews, J. K. Romary *J. Chem. Soc.* 405 (1964).
[566] A. C. Andrews, D. M. Zebolsky *J. Chem. Soc.* 742 (1965).
[567] D. D. Perrin, V. S. Sharma *J. Chem. Soc.* A 724 (1967).
[568] V. F. Toropova, Yu. M. Azizov *Russ. J. Inorg. Chem.* **11**, 288 (1966).
[569] F. Verbeek, H. Thun *Anal. Chim. Acta* **33**, 378 (1965).
[570] H. Thun, W. Guns, F. Verbeek *Anal. Chim. Acta* **37**, 332 (1967).
[571] T. Lengyel *Acta Chim. Acad. Sci. Hung.* **57**, 291 (1968).
[572] A. Albert *Biochem J.* **54**, 646 (1953).
[573] A. Albert, A. Hampton *J. Chem. Soc.* 505 (1954).
[574] R. Näsänen *Suomen Kemistilehti* **26** B, 2, 11 (1953).
[575] H. Irving, H. S. Rossotti *J. Chem. Soc.* 2910 (1954).
[576] C. F. Richard, R. L. Gustafson, A. E. Martell *J. Am. Chem. Soc.* **81**, 1033 (1959).
[577] R. Näsänen, E. Uisitalo *Acta Chem. Scand.* **8**, 112 (1954).
[578] B. F. Freasier, A. G. Oberg, W. W. Wendlandt *J. Phys. Chem.* **62**, 700 (1958).
[579] G. Schwarzenbach, E. Kampitsch, R. Stener *Helv. Chim. Acta* **28**, 113 (1945).
[580] S. Chaberek, A. E. Martell *J. Am. Chem. Soc.* **74**, 5052 (1952).
[581] L. C. Thompson *Inorg. Chem.* **1**, 490 (1962).

References

[582] G. Anderegg *Helv. Chim. Acta* **47**, 1801 (1964).
[583] S. Misumi, M. Aihara *Bull. Chem. Soc. Japan* **39**, 2677 (1966).
[584] K. S. Rajan, A. E. Martell *J. Inorg. Nucl. Chem.* **26**, 789 (1964).
[585] W. P. Evans, C. B. Monk *J. Chem. Soc.* 550 (1954).
[586] G. R. Choppin, J. A. Chopoorian *J. Inorg. Nucl. Chem.* **22**, 97 (1961).
[587] H. J. de Bruin, D. Kaitis, R. B. Temple *Australian J. Chem.* **15**, 457 (1962).
[588] T. Nozaki, Z. Mise, K. Higaki *Nippon Kagaku Zasshi* **88**, 1168 (1967).
[589] M. Yasuda, K. Yamasaki, H. Ohtaki *Bull. Chem. Soc. Japan* **33**, 1067 (1960).
[590] C. F. Timberlake, *J. Chem. Soc.* 5078 (1964).
[591] E. Campi *Ann. Chim.* (Rome) **53**, 96 (1963).
[592] D. I. Stock, C. W. Davies *J. Chem. Soc.* 1371 (1949).
[593] V. T. Athavale, N. Mahadevan, P. K. Mathur, R. M. Sathe *J. Inorg. Nucl. Chem.* **29**, 1947 (1967).
[594] H. Muro, T. Tsuchiya *Bull. Chem. Soc. Japan* **39**, 1589 (1966).
[595] W. B. Schaap, H. A. Laitinen, J. C. Bailar *J. Am. Chem. Soc.* **76**, 5868 (1954).
[596] J. E. Powell, W. F. S. Neillie *J. Inorg. Nucl. Chem.* **29**, 2371 (1967).
[597] T. A. Belyavskaya, I. F. Kolosova *Russ. J. Inorg. Chem.* **10**, 236 (1965).
[598] P. G. Manning *Can. J. Chem.* **45**, 1643 (1967).
[599] S. P. Bhardwaj, G. V. Bakore *J. Indian Chem. Soc.* **38**, 967 (1961).
[600] I. Geletseanu, A. V. Lapitskii *Dokl. Akad. Nauk SSSR* **144**, 573 (1962), **147**, 1372 (1962).
[601] R. Larsson, B. Folkeson *Acta Chem. Scand.* **19**, 53 (1965).
[602] P. J. Antikainen, V. M. K. Rossi *Suomen Kemistilehti* **32** B, 182, 185 (1959).
[603] P. J. Antikainen, I. P. Pitkanen *Suomen Kemistilehti* **41** B, 65 (1968).
[604] J. Körbl, B. Kakáč *Collection Czech. Chem. Commun.* **23**, 889 (1958).
[605] G. S. Tereshin, A. R. Rubinstein, I. V. Tananaev *J. Anal. Chem. USSR* **20**, 1138 (1965).
[606] B. Kardalov, D. Kantcheva, P. Nenov *Talanta* **15**, 525 (1968).
[607] B. Buděšinský, E. Antonescu *Collection Czech. Chem. Commun* **28**, 3264 (1963).
[608] G. Schwarzenbach, H. Gysling *Helv. Chim. Acta* **32**, 1314 (1949).
[609] G. Geier *Ber. Bunsenges. Phys. Chem.* **69**, 617 (1965).
[610] G. Schwarzenbach, R. Gut *Helv. Chim. Acta* **34**, 1589 (1956).
[611] J. Starý *Anal. Chim. Acta* **28**, 132 (1963).
[612] T. A. Bohigian, A. E. Martell *Progr. Rept. US. At. Energy Comm.* No. AT (30–1)–1823 (1960).
[613] T. Moeller, R. Ferrus *Inorg. Chem.* **1**, 55 (1962).
[614] G. Schwarzenbach, J. Heller *Helv. Chim. Acta* **34**, 1889 (1951).
[615] N. A. Skorik, V. N. Kunmok, V. V. Serebrennikov *Russ. Inorg. Chem.* **12**, 1429, 1788 (1967).
[616] V. L. Hughes, A. E. Martell *J. Am. Chem. Soc.* **78**, 1319 (1956).
[617] B. J. Intorre, A. E. Martell *Inorg. Chem.* **3**, 81 (1964).
[618] O. Makitie *Suomen Kemistilehti* **39B**, 171 (1966), **40B**, 27 (1967).
[619] A. McAuley, G. H. Nancollas *Trans. Faraday Soc.* **56**, 1165 (1960).
[620] S. Lacroix *Ann. Chim. (Paris)* **4**, 5 (1949).
[621] S. H. Cohen, R. J. Iwomota, J. Kleinberg *J. Am. Chem. Soc.* **82**, 1844 (1960).
[622] H. J. de Bruin, D. Kaitis, R. B. Temple *Australian J. Chem.* **15**, 457 (1962).
[623] E. Gelles, R. M. Hay *J. Chem. Soc.* 3684 (1940). 3673 (1958).
[624] W. C. Vosburgh, J. F. Beckman *J. Am. Chem. Soc.* **62**, 1028 (1940).

[625] C. E. Crouthamel, D. S. Martin *J. Am. Chem. Soc.* **73,** 569 (1951).
[626] J. L. Schubert, E. L. Lind, W. M. Westfall, R. Plager, N. C. Li *J. Am. Chem. Soc.* **80,** 4799 (1958).
[627] T. Sekine *J. Inorg. Nucl. Chem.* **26,** 1463 (1964), *Acta Chem. Scand.* **19,** 1476 (1965).
[628] M. Deneux, R. Meilleur, R. L. Benoit *Can. J. Chem.* **46,** 1383 (1968).
[629] S. J. Lyle, S. J. Naqui *J. Inorg. Nucl. Chem.* **29,** 2441 (1967).
[630] Y. Yamane, N. Davidson *J. Am. Chem. Soc.* **82,** 2123 (1960).
[631] J. M. White, P. Tang, N. C. Li *J. Inorg. Nucl. Chem.* **14,** 255 (1960).
[632] J. Raaflaub *Helv. Chim. Acta* **43,** 629 (1960).
[633] R. W. Money, C. W. Davies *J. Chem. Soc.* 400 **(1934).**
[634] H. Taube *J. Am. Chem. Soc.* **70,** 3928 (1948).
[635] J. I. Watters, R. de Witt *J. Am. Chem. Soc.* **82,** 1333 (1960).
[636] I. M. Kolthoff, R. W. Perlich, D. Weiblen *J. Phys. Chem.* **46,** 561 (1942).
[637] A. K. Babko, L. I. Dubovenko *Zh. Neorgan Khim.* **4,** 372 (1959).
[638] N. C. Li, W. M. Westfall, A. Lindenbaum, J. M. White, J. Schubert *J. Am. Chem. Soc.* **79,** 5864 (1957).
[639] V. L. Zolotovin *Zh. Neorgan Khim.* **12,** 2713 (1959).
[640] B. F. Pease, M. B. Williams *Anal. Chem.* **31,** 1044 (1959).
[641] O. Navratil *Collection Czech. Chem. Commun.* **29,** 2490 (1964), **31,** 2492 (1966), **32,** 2004 (1967).
[642] A. Corsini, I. Mai-ling Yih, O. Fernando, H. Freiser *Anal. Chem.* **34,** 1090 (1962).
[643] M. Hniličková, L. Sommer *Collection Czech. Chem. Commun.* **26,** 2189 (1961); *Z. Anal. Chem.* **193,** 171 (1963).
[644] L. Sommer, H. Novotná *Talanta* **14,** 457 (1967).
[645] W. J. Geary, G. Nickless, F. H. Pollard *Anal. Chim. Acta* **27,** 71 (1962).
[646] C. D. Dwivedi, K. N. Munski, A. K. De *J. Inorg. Nucl. Chem.* **28,** 245 (1966).
[647] G. Schwarzenbach, P. Moser *Helv. Chim. Acta* **36,** 581 (1953).
[648] J. M. Dale, C. V. Banks *Inorg. Chem.* **2,** 591 (1963).
[649] H. Irving, D. H. Mellor *J. Chem. Soc.* 5222 (1962).
[650] T. S. Lee, I. M. Kolthoff, D. L. Leussing *J. Am. Chem. Soc.* **70,** 2348, 3596 (1948).
[651] R. Truillo, F. Brito *Anales Real Soc. Espan. Fis. Qim.* **53,** 249 (1957).
[652] M. Bartušek, J. Růžičková *Collection Czech. Chem. Commun.* **31,** 207 (1966).
[653] W. A. E. McBryde *Can. J. Chem.* **46,** 2385 (1968).
[654] C. Postmus, L. B. Magnusson, C. A. Craig *Inorg. Chem.* **5,** 1154 (1966).
[655] G. Anderegg, H. Flaschka, R. Sallmann, G. Schwarzenbach *Helv. Chim. Acta* **37,** 113 (1954).
[656] K. Higashi, K. Hor, R. Tsuchiya *Bull. Chem. Soc. Japan* **40,** 2569 (1967).
[657] G. Anderegg *Helv. Chim. Acta* **43,** 414, 1530 (1960).
[658] R. C. Mercier, M. R. Paris *Compt. Rend.* 349, 598 (1966).
[659] L. C. Thompson *Inorg. Chem.* **3,** 1319 (1964).
[660] M. Bobtelsky, S. Kertes *Bull. Soc. Chim. France* 328 (1955).
[661] K. Burger, I. Egyed *J. Inorg. Nucl. Chem.* **27,** 2361 (1965), **28,** 139 (1966).
[662] B. Kirson *Bull. Soc. Chim. France* 1030 (1962).
[663] G. I. H. Hanania, D. H. Irvine *J. Chem. Soc.* 2745 (1962).
[664] W. E. Bennett, D. O. Sklorin *J. Inorg. Nucl. Chem.* **28,** 591 (1966).
[665] A. Napoli *Talanta* **15,** 189 (1968).

References

[666] J. Francherre, C. Petitfaux, B. Charlier *Bull. Soc. Chim. France* 1091 (1967).
[667] C. F. Timberlike *J. Chem. Soc.* 1229 (1964).
[668] C. A. Tyson, A. E. Martell *J. Am. Chem. Soc.* **90**, 3379 (1968).
[669] S. N. Dubey, R. C. Mehrotra *J. Inorg. Nucl. Chem.* **26**, 1543 (1964).
[670] M. Bartušek, J. Zelinka *Collection Czech. Chem. Commun.* **32**, 992 (1967).
[671] V. T. Athavale, L. H. Prebhu, D. G. Vartak *J. Inorg. Nucl. Chem.* **28**, 123 (1966).
[672] O. Ryba, J. Cífka, M. Malát, S. Suk *Collection Czech. Chem. Commun.* **21**, 349 (1956); **23**, 71 (1958).
[673] D. L. Leussing, E. M. Hanna *J. Am. Chem. Soc.* **88**, 693 (1966).
[674] D. L. Leussing, D. C. Schultz *J. Am. Chem. Soc.* **86**, 4846 (1964).
[675] P. Lumme *Ann. Acad. Sci. Fennicae* **68** A, II, 7 (1955).
[676] P. O. Lumme *Suomen Kemistilehti* **30** B, 194 (1957); **31** B, 253 (1958).
[677] M. Bobtelsky, E. Jungreis *J. Inorg. Nucl. Chem.* **3**, 38 (1956).
[678] D. D. Perrin *Nature* **182**, 741 (1958).
[679] B. Das, S. Aditya *J. Indian Chem. Soc.* **36**, 473 (1959).
[680] J. C. Colleter *Ann. Chim. (Paris)* **5**, 415 (1960).
[681] B. Hok-Bernstrom *Acta Chem. Scand.* **10**, 163, 174 (1956).
[682] A. K. Babko, A. I. Volkova, T. E. Getman *Russ. J. Inorg. Chem.* **7**, 145, 1121 (1962).
[683] G. E. Mont, A. E. Martell *J. Am. Chem. Soc.* **88**, 1387 (1966).
[684] E. Coates, B. Rigg *Trans. Faraday Soc.* **58**, 88, 2058 (1962).
[685] C. M. Ke, P. C. Kong, M. S. Chen, N. C. Li *J. Inorg. Nucl. Chem.* **30**, 961 (1968).
[686] V. C. Banks, R. S. Singh *J. Inorg. Nucl. Chem.* **15**, 125 (1960).
[687] C. R. Kanekar, N. V. Thakur, S. M. Jogdeo *Bull. Chem Soc. Japan* **41** 759 (1968).
[688] V. Frei *Collection Czech. Chem. Commun.* **32**, 1815 (1967).
[689] R. K. Cannan, A. Kibrick *J. Am. Chem. Soc.* **60**, 2314 (1938).
[690] S. Suzuki *Sci. Rept. Res. Inst. Tohoku Univ.* **4**, 176 (1952).
[691] R. Pastorek, F. Brezina *Monatsh.* **97**, 1095 (1966).
[692] P. G. Manning, C. B. Monk *Trans. Faraday Soc.* **57**, 1996 (1961).
[693] H. B. Jonassen, L. Westerman *J. Am. Chem. Soc.* **79**, 4275 (1957).
[694] C. N. Reilley, J. Holloway *J. Am. Chem. Soc.* **80**, 2917 (1958).
[695] H. B. Jonassen, F. W. Frey, A. Schaafsma *J. Phys. Chem.* **62**, 1022 (1958).
[696] L. G. van Uitert, W. C. Fernelius, B. E. Doublas *U.S. At. Energy Comm. Rept.* NYD-626 (1951).
[697] T. Sekine, A. Koizumi, M. Saihairi *Bull. Chem. Soc. Japan* **39**, 2681 (1966).
[698] J. L. Walter, J. A. Ryan, T. J. Lane, E. F. Britten *J. Am. Chem. Soc.* **78**, 5560 (1956); **80**, 315 (1958).
[699] W. S. Fyfe *J. Chem. Soc.* 1032 (1955).
[700] O. S. Fedorova *Zh. Obshch. Khim.* **24**, 62 (1954.)
[701] E. I. Onstott, H. A. Laitinen *J. Am. Chem. Soc.* **72**, 4724 (1950).
[702] C. J. Nyman, E. P. Parry *Anal. Chem.* **30**, 1255 (1958).
[703] D. L. Leussing *J. Am. Chem. Soc.* **75**, 3904 (1953); **78**, 552 (1956); **80**, 4180 (1958).
[704] J. E. Powell, R. S. Kolat, G. S. Paul *Inorg. Chem.* **3**, 518 (1964).
[705] W. Stricks, I. M. Kolthoff, A. Heyndrick *J. Am. Chem. Soc.* **76**, 1515 (1954).
[706] N. C. Li, R. A. Manning *J. Am. Chem. Soc.* **77**, 5225 (1955).

[707] S. V. Larionov, V. M. Shulman, C. A. Podolskaya *Russ. J. Inorg. Chem.* **9,** 1264 (1964).
[708] R. Ramamam, S. Shanmuganathan *Current Sci. (India)* **37,** 39 (1968).
[709] H. W. Krause, F. W. Wilec, H. Mix, M. Langenbeck *Z. Phys. Chem. (Leipzig)* **225,** 342 (1964).
[710] D. J. Perkins *Biochem. J.* **55,** 649 (1953).
[711] A. Willi, G. Schwarzenbach *Helv. Chim. Acta* **34,** 528 (1951).
[712] R. Näsänen *Suomen Kemistilehti* **29** B, 91 (1956), **30** B, 61 (1957); **33,** 111 (1960); *Acta Chem. Scand.* **11,** 1308 (1957).
[713] R. N. Nanda, R. C. Das, R. K. Nanda *Indian J. Chem* **3,** 278 (1965).
714] R. C. Courtney, R. L. Gustafson, S. Chaberek, A. E. M.artell *J. Am. Chem. Soc.* **80,** 2121 (1958).
[715] L. Sommer *Collection Czech. Chem. Commun.* **28,** 2102 (1963).
[716] L. Sommer, T. Šepel, L. Kuřilová *Collection Czech. Chem. Commun.* **30,** 3834 (1965).
[717] R. A. Day, R. W. Stoughton *J. Am. Chem. Soc.* **72,** 5662 (1950).
[718] J. Bjerrum, S. Refu *Suomen Kemistilehti* **29** B, 68 (1956).
[719] M. Cadiot-Smith *J. Chim. Phys.* **60,** 957, 976, 991.(1963).
[720] G. Schwarzenbach *Helv. Chim. Acta* **33,** 974 (1950)
[721] R. L. Pecsok, R. A. Garber, L. D. Shields *Inorg. Chem.* **4,** 447 (1965).
[722] M. T. Beck, S. Görög *Magy. Kém. Folyóirat* **65,** 55 (1959).
[723] C. N. Reilley, R. W. Schmid *J. Elisha Mitchell Sci. Soc.* **73,** 279 (1957).
[724] T. A. Bohigian, A. E. Martell *Inorg. Chem.* **4,** 1264 (1965).
[725] L. Harju *Anal. Chim. Acta* **50,** 475 (1970).
[726] J. H. Grimes, A. J. Huggard, S. P. Wilford *J. Inorg. Nucl. Chem.* **25,** 1225 (1963), **15,** 849 (1958).
[727] H. Irving, J. J. R. F. de Silva *J. Chem. Soc.* 448, 458 (1963).
[728] B. Řehák, J. Körbl *Collection Czech. Chem. Commun.* **25,** 797 (1960).
[729] B. Buděšinský *Collection Czech. Chem. Commun.* **28,** 1858 (1963).
[730] M. Otomo *Bull. Chem. Soc. Japan* **37,** 504 (1964).
[731] B. Buděšinský *Z. Anal. Chem.* **188,** 266 (1962).
[732] B. Buděšinský, A. Bezděková *Z. Anal. Chem.* **196,** 172 (1963).
[733] V. G. Chukhlantsev *Zh. Neorgan Khim.* **1,** 1975 (1956).
[734] K. Masaki *Bull. Chem. Soc. Japan* **5,** 345 (1930).
[735] A. A. Noyes, C. G. Abbot *Z. Phys. Chem.* **16,** 125 (1895).
[736] A. M. Azzam, I. A. Shimi *Z. Anorg. Chem.* **321,** 284 (1963).
[737] L. Birckenbach, K. Huttner *Z. Anorg. Chem.* **190,** 1 (1930).
[738] J. M. Spencer, M. Le Pla *Z. Anorg. Chem.* **65,** 10 (1909).
[739] C. W. Davies, A. L. Jones *Trans. Faraday Soc.* **51,** 812 (1955).
[740] J. R. Howard, G. H. Nancollas *Trans. Faraday Soc.* **53,** 1449 (1957).
[741] A. Bellomo *Talanta* **17,** 1109 (1970).
[742] B. B. Owen, S. R. Brinkley *J. Am. Chem. Soc.* **60,** 2233 (1938).
[743] I. M. Kolthoff, J. J. Lingane *J. Phys. Chem.* **42,** 133 (1938).
[744] C. N. Muldrow, L. G. Hepler *J. Am. Chem. Soc.* **78,** 5989 (1956); *J. Phys. Chem.* **62,** 982 (1958).
[745] Y. Oka *J. Chem. Soc. Japan* **59,** 971 (1938).
[746] W. Böttger *Z. Phys. Chem.* **46,** 521 (1903).
[747] G. Schwarzenbach, M. Schellenberg *Helv. Chim. Acta* **48,** 28 (1965).
[748] I. Leden, R. Nilsson *Svensk Kem. Tidskr.* **66,** 126 (1954).
[749] R. W. Stoughton, M. H. Lietke *J. Phys. Chem.* **64,** 133 (1960).

[750] H. T. S. Britton R. A. Robinson *J. Chem. Soc.* 2328 (1930).
[751] H. T. S. Britton, W. L. German *J. Chem. Soc.* 1156 (1934).
[752] N. N. Mironov, A. I. Odnosertsev *Zh. Neorgan. Khim.* **2**, 2202 (1957).
[753] F. G. Sharovskii *Tr Komis po Analit. Khim. Akad. Nauk SSSR* **3**, 101 (1951).
[754] F. Jirsa, H. Jelinek *Z. Elektrochem.* **30**, 534 (1924).
[755] H. L. Johnston, H. L. Leland *J. Am. Chem. Soc.* **60**, 1439 (1938).
[756] K. K. Kelley, C. T. Anderson *Bur. Mines Bull.* 384 (1935).
[757] G. L. Beyer, W. Rieman *J. Am. Chem. Soc.* **65**, 971 (1943).
[758] F. Kohlrausch *Z. Phys. Chem. (Leipzig)* **64**, 129 (1908).
[759] C. W. Davies, P. A. H. Wyatt *Trans. Faraday Soc.* **45**, 770 (1949).
[760] L. E. Holt, J. A. Pierca, C. N. Kajdi *J. Colloid Sci.* **9**, 409 (1954).
[761] C. C. Templeton *J. Chem. Eng. Data* **5**, 514 (1960).
[762] T. O. Denney, C. B. Monk *Trans. Faraday Soc.* **47**, 992 (1951).
[763] N. M. Selivanova, V. A. Sneider *Nauk Dokl. Vysshei Shkoly Khim. i Khim. Tekhnol.* No. 2, 216 (1958).
[764] W. Feitknecht, P. Schindler *Pure Appl. Chem.* **6**, 130 (1963).
[765] L. B. Yeatts, W. L. Marshall *J. Phys. Chem.* **71**, 2641 (1967).
[766] Y. Kanko, S. Eyubi *Chem. Ztg.* **80**, 130 (1956).
[767] T. Rengemo, U. Brune, L. G. Sillén *Acta Chem. Scand.* **12**, 873 (1958).
[768] S. A. Greenberg, L. E. Copeland *J. Phys. Chem.* **64**, 1057 (1960).
[769] M. LeBlanc, O. Harnapp *Z. Phys. Chem.* **166** A, 321 (1933).
[770] I. V. Tananayev, M. A. Gluskova, G. B. Seifer *Zh. Neorgan Khim.* **1**, 66 (1956).
[771] J. Ste-Marie, A. E. Torma, A. O. Gübeli *Can. J. Chem.* **42**, 662 (1964).
[772] A. Hulanicki *Acta Chim. Acad. Sci. Hung.* **27**, 411 (1961).
[773] J. L. Weaver, W. C. Purdy *Anal. Chim. Acta* **20**, 376 (1959).
[774] T. W. Newton, G. M. Arcand *J. Am. Chem. Soc.* **75**, 2449 (1953).
[775] V. M. Tarayan, L. A. Eliazyan *Izvest. Akad. Nauk Arm. SSR, Ser. Khim. Nauk* **10**, 189 (1957).
[776] S. I. Sobol *Zh. Obshch. Khim.* **23**, 906 (1953).
[777] G. Bodländer, O. Storbeck *Z. Anorg. Chem.* **31**, 458 (1902).
[778] M. G. Vladimirova, I. A. Korovskii *Zh. Prikl Khim.* **23**, 580 (1950).
[779] I. D. Fridman, D. S. Sarbaev *Zh. Neorgan Khim.* **4**, 1849 (1959).
[780] A. M. Golub *Zh. Neorgan Chim.* **1**, 2517 (1956).
[781] S. Peterson, O. W. Cooper *Trans. Kentucky Acad. Sci.* **13**, 146 (1951).
[782] R. M. Keefer *J. Am. Chem. Soc.* **70**, 476 (1948).
[783] K. B. Yatsimirskii, V. P. Vasilev *Zh. Analit. Khim.* **11**, 536 (1956).
[784] L. H. N. Cooper *Proc. Roy. Soc.* **124** B, 299 (1937).
[785] P. I. Yakshova *Tr Voronyezhsk. Univ.* **42**, No. 2, 63 (1956).
[786] T. D. Turnquist, E. B. Sandell *Anal. Chim. Acta* **42**, 239 (1968).
[787] I. V. Pyastniskii, A. P. Kostyshina *Ukr. Khim. Zh.* **23**, 1957 (1957).
[788] P. O. Bethge, I. Jonevall-Westöö, L. G. Sillén *Acta Chem. Scand.* **2**, 828 (1948).
[789] A. E. Brodsky *Z. Elektrochem.* **35**, 833 (1929).
[790] J. T. Law *Thesis*, New Zealand Univ. (1946).
[791] T. De Vries, D. Cohen *J. Am. Chem. Soc.* **71**, 1114 (1949).
[792] H. Grossmann *Z. Anorg. Chem.* **43**, 356 (1905).
[793] G. Schwarzenbach, H. Widmer *Helv. Chim. Acta* **46**, 2613 (1963).
[794] J. Terpilowski, R. Staroścĭk *Chem. Anal. (Warsaw)* **7**, 629 (1962).
[795] R. Staroščik, H. Siaglo *Chem. Anal. (Warsaw)* **10**, 265 (1965).

[796] V. K. La Mer, F. H. Goldman *J. Am. Chem. Soc.* **51,** 2632 (1929).
[797] W. D. Kline *J. Am. Chem. Soc.* **51,** 2093 (1929).
[798] R. Näsänen *Z. Phys. Chem. (Leipzig)* **190,** 183 (1942).
[799] K. Bube *Z. Anal. Chem.* **49,** 525 (1910).
[800] H. Gämsjäger *Monatsh.* **98,** 1803 (1967).
[801] C. L. van Ende *Z. Anorg. Chem.* **26,** 129 (1901).
[802] F. H. MacDougall, E. J. Hoffman *J. Phys. Chem.* **40,** 317 (1936).
[803] H. Fromherz *Z. Phys. Chem. (Leipzig)* **153** A, 376 (1931).
[804] I. M. Kolthoff, R. W. Perlich, D. Weiblen *J. Phys. Chem.* **46,** 561 (1942).
[805] W. Geimann, R. Höltje *Z. Anorg. Chem.* **152,** 59 (1926).
[806] M. Jowett, H. I. Price *Trans. Faraday Soc.* **28,** 668 (1932).
[807] N. M. Selivanova, A. P. Kapustinskii, G. A. Zubova *Izvest. Akad. Nauk. SSSR, Otdel Khim. Nauk* 187 (1959).
[808] V. Bayerle *Rec. Trav. Chim.* **44,** 514 (1925).
[809] R. Akeret *Thesis*, Techn. Hochschule Zürich (1953).
[810] T. W. Davis *Ind. Eng. Chem. Anal. Ed.* **14,** 709 (1942).
[811] J. Wolfmann *Österr. Ungar Z. Zuckerind.* **25,** 286 (1896).
[812] N. S. Nikolaev, Yu. A. Lukýanichev *At. Energ. USSR* **12,** 334 (1962); *Zh. Neorgan Khim.* **8,** 1786 (1963).
[813] R. W. Soughton, A. J. Fry, J. E. Barney *J. Inorg. Nucl. Chem.* **19,** 286 (1961).
[814] A. I. Moskvin, L. N. Essen *Russ. J. Inorg. Chem.* **12,** 359 (1967).
[815] F. Ishikava, Y. Terui *Sci. Rept. Tohoku Univ.* **23,** 141 (1934).
[816] M. Randall, W. V. A. Vietti *J. Am. Chem. Soc.* **50,** 1526 (1928).
[817] S. Suzuki *J. Chem. Soc. Japan, Pure Chem. Section* **74,** 219 (1953).
[818] C. W. Davies, R. A. Robinson *Trans. Faraday Soc.* **33,** 633 (1937).
[819] A. Braibanti, I. Chierici *Gazz. Chim. Ital.* **88,** 793 (1958).
[820] M. Oosting *Rec. Trav. Chim.* **79,** 627 (1960).
[821] A. I. Moskvin, F. A. Zakharova *Russ. J. Inorg. Chem.* **4,** 975 (1959).
[822] A. O. Gubeli, J. Ste-Marie *Can. J. Chem.* **45,** 2101 (1967).
[823] W. H. Burgus *U.S. At. Energy Comm. Rept.* CN-2560 (1945).
[824] G. Jones, S. Baeckström *J. Am. Chem. Soc.* **56,** 1524 (1934).
[825] C. Bereczki—Biedermann, G. Biedermann, L. G. Sillén *Rept. Anal. Sec. IUPAC*, July (1953).
[826] H. L. Conley *U.S. At Energy Comm. Rept.* UCRL-9332 (1960).
[827] A. A. Noyes, T. J. Deahl *J. Am. Chem. Soc.* **59,** 1337 (1937).
[828] H. N. McCoy *J. Am. Chem. Soc.* **58,** 1577 (1936).
[829] W. M. Saltman, N. H. Nachtrieb *J. Electrochem. Soc.* **100,** 126 (1953).
[830] S. Hietanen, L. G. Sillén *Arkiv. Kemi* **10,** 103 (1956).
[831] A. L. Rotinan, I. I. Appenin *Zh. Obshch. Khim.* **10,** 1524 (1940).
[832] W. Kangro, F. Weingärtner *Z. Elektrochem.* **58,** 505 (1954).
[833] G. E. Coates *J. Chem. Soc.* 478 (1945).
[834] A. Walkley *Trans. Electrochem. Soc.* **93,** 316 (1948).
[835] K. J. Vetter, G. Manecke *Z. Phys. Chem. (Leipzig)* **195,** 270 (1950).
[836] A. W. Hutchison *J. Am. Chem. Soc.* **69,** 3051 (1947).
[837] G. Grube, H. L. Grube *Z. Elektrochem.* **44,** 771 (1938).
[838] A. Prokopcikas *Lietuvos TSR Mokslu Akad. Darbai Ser B* **2,** No. 29, 31 (1962).
[839] M. M. Haring, M. R. Hatfield, P. P. Zapponi *Trans. Electrochem. Soc.* **75,** 473 (1939).
[840] H. B. Wellman *J. Am. Chem. Soc.* **52,** 985 (1930).

iron, 237, 240
lead, 249
magnesium, 51, 80, 225, 231
nickel, 63
zinc, 151
conditional stability constants, 63
masking agent, 192
table of constants, 339
Ethylenediaminetetra-amminecopper (II), 21
Extraction,
constants, 261, 379–83
degree of, 264
error of, 266
methods, 155–65, 289
selectivity of, 264

Fluoride, complex with aluminium, 208
determination, 253
table of constants, 320
Formic acid, 208, 343
Fronaeus method, 121, 122

Gold, complex with chloride, 86
Gradient elution method, 286, 308
Gran plot, 205, 239, 246
Gravimetric analysis, 182
error, 184
examples, 187
tables of constants, 396

Half-wave potential, 149, 150, 151,
shift of, 149, 248
Hammett acidity function, 101
Height equivalent to one theoretical plate, 278, 302
Hexa-amminenickel(II), 21
Histidine, 315, 346
Hydrogen ion concentration, 37
calculation of, in acid–base titrations, 194
examples, 208
role in complex formation, 59
Hydrogen peroxide, 208, 324
Hydroquinone, 246
Hydroxo-complexes with
aluminium, 63, 76, 208
calcium, 264
iron, 50, 60, 77, 187

lead, 264
mercury, 63
nickel, 49
thorium, 33
table of constants, 323
α-Hydroxyisobutyric acid complex with erbium, 65
8-Hydroxyquinoline, 263
complex with aluminium, 184, 267
mole-fraction distribution, 38
table of constants, 347

Ilkovič equation, 148
Indicator, acid–base, 200
complexometric, 224
error, 200, 211, 218, 225, 229, 241
metallochromic, 224, 225
redox, 238, 239, 240
Ion-exchange separation, 270, 277
anion-exchange equilibria, 281, 291
calculations, 282
efficiency of separation, 283
resins, 285
separation of metal ions, 284, 285
cation-exchange equilibria, 270, 290
efficiency of separation, 274, 276
plate theory, 277
rate of migration, 273
use of complexing agents, 275
determination of stability constants, 165
examples, 297
gradient elution, 286, 308
liquid, 289
stepwise change in eluent, 286
table of constants, 383
Ionic product of water, temperature dependence, 80
Ionic strength, 23, 46
example, 46
Iridium, complexes with bromide, 47
Iron, complexes with acetylacetone, 46
2,2'-bipyridyl, 255
cyanide, 17-8, 70, 245, 321
EDTA, 19, 237, 240
hydroxide, 50, 60, 187
thiocyanate, 219, 266
redox couple, 41, 42, 83
separation from manganese, 297
titration with dichromate, 233

Isomerism, 20-1

Job's method, 137-40

Kinetics of reaction of complexes, 74, 75

Lactate, 46, 250, 256, 349
Lead, complex with acetate, 83
   EDTA, 249
   hydroxide, 269
   oxalate, 191
Least squares method, 44, 121, 173
Leden successive extrapolation method, 114-5, 172
Lewis acid-base theory, 18, 70
Ligand concentration, free, 28, 47, 59, 116, 118
   total, 26, 32, 59
Ligand-field theory, 20, 68
Ligand number, see Average ligand number
Ligands, classification, 17, 19, 70
   multidentate, 19
   protonation, 89
   tables of constants, 317
   unidentate, 19
Linearization of titration curves, 205, 239, 246
Liquid-liquid extraction, 258-70, 289-96
   calculations, 261
   degree of extraction, 265
   error, 266
   examples, 266
   extraction of chelates, 260
   selectivity, 264, 268
Liquid ion-exchangers, 258, 289, 310

Magnesium, 251
   complex with EDTA, 80, 225, 231
      hydroxide, 251
   determination, 251
   separation from calcium, 276
Malonic acid, 210, 350
Manganese, separation from iron, 297
Mannitol, 206
Masking, 184, 191, 228, 248, 253, 268

Mercury, complex with ammonia, 177
   bromide, 268
   chloride, 307
   EDTA, 61
   hydroxide, 63, 177
   thiocyanate, 254
Metal ions, chromatographic separation, 273, 284
   classification, 69
   concentration, 26, 32
   mole-fraction of free, 29
Methyl isobutyl ketone, 164
Methyl Orange, 200
Mixed constants, 24
Mixed-ligand complexes, 20, 61, 174, 177
Multidentate-ligand complexes, 19
Mole-fraction, 29, 37
   distribution, 30, 31, 38, 42
   example, 47
   free metal ion, 29
   non-protonated species, 37
   oxidized and reduced form, 41
   protonated species, 37
Mononuclear wall, 33
Murexide, 230, 352

Nernst equation, 40, 115, 186, 226
Neutral Red, 213
Newton's approximation, 43, 44, 48
   example, 48
Nickel complex with acetylacetone, 211
     dimethylglyoxime, 184, 189
     EDTA, 63
     ethylenediamine, 128
     hydroxide, 49
     sulphosalicylate, 170
     thiocyanate, 49
   perchlorate, 18
   sulphate, 18
β-Nitroso-α-naphthol, 102

Octoammine-μ-amido-μ-nitro-dicobalt (II), 19
Oxalate, complex with calcium, 188, 191
   copper, 308
   lead, 188
   magnesium, 308
   zinc, 308
   table of constants, 354

Oxalic acid 209, 288, 308
Oxine, see 8-Hydroxyquinoline

PAN, [1-(2-pyridylazo)-2-naphthol], 226, 256, 355
Paper chromatography, ion-exchange, 296–7, 311
Pauling's valence bond theory, 67
pe definition, 40
   standard, 40
pH, see Hydrogen ion concentration
1,10-Phenanthroline, 73, 238, 356
   cadmium complex, 153
Phenolphthalein, 200, 210
Phosphoric acid, 36
Phthalic acid, 306
π-Bonding, 69
Platinum, complex with chloride, 20
Polarography, 247–52
   determination of stability constants, 147
   examples, 249
Potential, cell, 78
   conditional standard, 237, 240
   redox, 40
Potentiometric methods, determination of protonation constants, 89–101, 105–10, 112, 135–6
   stability constants, 112–36
Precipitation, quantitative, 184
Precipitation titrations, 213–9
Precipitation formation constant, 182, 213, 369–74
   conditional, 183
Precipitation reactions, 183, 185, 213
Protonation constant, 34, 36, 59, 318–68
   conditional, 206
   determination of, 89–111
   potentiometric method, 89–101
      apparatus, 100
      average proton number, 97
      polyacidic bases, 93, 95
      practical instructions, 98
      Schwarzenbach's graphical method, 95
      uncharged bases, 95
   spectrophotometric methods, 101
      practical advice, 104
Pyridine, 303, 359

Pyridine-2-aldoxime, cobalt complexes, 144
   table of constants, 359
Pyrophosphate, 298, 324

Rare earths, 249, 254, 256
Redox equilibria, 39, 41, 42, 375–9
   conditional, 235
Redox potential, 40
Redox titrations, 232–47
   table of constants, 375
Relaxation method, 114, 131, 135
Retention factor, 296, 311

Samarium, 249
Schwarzenbach's graphical method, 95
Selectivity, of extraction, 264
   of membrane electrodes, 218–9
Separation factor, 275
Side-reactions, 52, 53, see also Conditional equilibrium constants
Silver, electro-deposition, 193
   titration with chloride, 213
Solubility product, 182, 213
   calculation, 187, 188
Solvent, effect on equilibrium constant, 86
   non-aqueous, 86
Spectrochemical series, 68
Spectrophotometry, 252–7
   complex displacement, 253
   determination of protonation constants, 101–5, 111
      practical advice, 104
   determination of stability constants, 137–47
   examples, 255
Spiro and Hume's method, 177
Stability constants, 25
   calculation, curve-fitting methods, 123–5, 141, 145, 158, 160
   from [L] and $A_{M(L)}$ data, 114–5, 150, 167, 175
   from [L] and $\bar{n}$ data, 119–25, 128–35, 168
   from [L] and Φ data, 125
   conditional, 51, 53, 61, 62–64, 221
   data collection, 45
   determination, extraction methods, 155

414

Index

    examples, 164
    free-ligand concentration, 116
        examples, 124
    ion-exchange methods, 165–70
        example, 170
    mixed-ligand complexes, 174, 177, 180–1
    mononuclear complexes, 111
    pH measurements, 116–36
    polarographic methods, 147–55
        practical aspects, 151
    polynuclear complexes, 181
    potentiometric methods, 112–36
    practical instructions, 98, 104, 151, 162
    spectrophotometric methods, 137, 139–41, 143–7
        Bjerrum's method, 141–3
        continuous variation, 137
        examples, 143
        mole-ratio method, 140–1
        other methods, 143
    examples, 46, 63, 73, 127, 143, 164, 170
    overall, 23, 25, 26, 31, 46, 317
    successive, 75

Stability of complexes, 69
Sulphosalicylate ion, 108, 363
    complexes with beryllium, 253
    copper, 127, 138, 143, 176, 299
    nickel, 170

Tartaric acid, 80, 363
Temperature, effect on equilibrium constant, 79
Tetra-ammine zinc(II), 20
Tetrachloroaurate(III), 86
Tetrachloroplatinate(II), 20
Tetracyanozincate(II), 21, 85
Thallium, determination by polarography, 249
    redox couple, 50
    titration with ascorbic acid, 236, 241
Thermodynamic relationships, 22, 77
Thorium, complex with hydroxide, 33
Thiocyanate, 325
    complex with iron, 266

    mercury, 254
Tiron (pyrocatechol disulphonic acid) 293, 295, 310, 365
Titration, acid–base, 194–213
    curve, 93, 94
    determination of complex-forming ions, 207
    end-point, 200
    equivalence point, 199, 200, 204
    error, 200, 202–5
    Gran plot, 205
    indicators, 200
    weak acids, 206
complexometric, 219–32
    analysis of a mixture of metals 227
    curve, 221
    end-point detection, 224–7
    equivalence point, 220
    error, 221
    examples, 229
    indicators, 224
precipitation, 319–9
    curve, 214
    equivalence point, 214
    error, 214, 215
    examples, 218
redox, 232–47
    curve, 233
    end-point detection, 238
    equivalence point, 233, 234
    error, 234, 235, 236
    examples, 240
    indicators, 238
p-Tolidine, 303
Triethylenetetraminehexa-acetic acid (TTHA) 72, 367
    complex with calcium, 181
Trichloroacetic acid, 283, 304, 366
Trichlorotriammine cobalt(III), 20

Unidentate ligand, 19

Vanadium, 164
van't Hoff equation, 79
Variamine Blue, 240

Watters' method, 176